Chapman & Hall/CRC Biostatistics Series

Fundamental Concepts for New Clinical Trialists

Scott Evans

Harvard University
Boston, Massachusetts, USA

Naitee Ting

Boehringer-Ingelheim
Ridgefield, Connecticut, USA

CRC Press
Taylor & Francis (
Boca Raton Londc

CRC Press is an imprint of the
Taylor & Francis Group, an in
A CHAPMAN & HA

Chapman & Hall/CRC Biostatistics Series

Editor-in-Chief

Shein-Chung Chow, Ph.D., Professor, Department of Biostatistics and Bioinformatics, Duke University School of Medicine, Durham, North Carolina

Series Editors

Byron Jones, Biometrical Fellow, Statistical Methodology, Integrated Information Sciences, Novartis Pharma AG, Basel, Switzerland

Jen-pei Liu, Professor, Division of Biometry, Department of Agronomy, National Taiwan University, Taipei, Taiwan

Karl E. Peace, Georgia Cancer Coalition, Distinguished Cancer Scholar, Senior Research Scientist and Professor of Biostatistics, Jiann-Ping Hsu College of Public Health, Georgia Southern University, Statesboro, Georgia

Bruce W. Turnbull, Professor, School of Operations Research and Industrial Engineering, Cornell University, Ithaca, New York

Published Titles

Adaptive Design Methods in Clinical Trials, Second Edition
Shein-Chung Chow and Mark Chang

Adaptive Designs for Sequential Treatment Allocation
Alessandro Baldi Antognini and Alessandra Giovagnoli

Adaptive Design Theory and Implementation Using SAS and R, Second Edition
Mark Chang

Advanced Bayesian Methods for Medical Test Accuracy
Lyle D. Broemeling

Advances in Clinical Trial Biostatistics
Nancy L. Geller

Applied Meta-Analysis with R
Ding-Geng (Din) Chen and Karl E. Peace

Basic Statistics and Pharmaceutical Statistical Applications, Second Edition
James E. De Muth

Bayesian Adaptive Methods for Clinical Trials
Scott M. Berry, Bradley P. Carlin, J. Jack Lee, and Peter Muller

Bayesian Analysis Made Simple: An Excel GUI for WinBUGS
Phil Woodward

Bayesian Methods for Measures of Agreement
Lyle D. Broemeling

Bayesian Methods for Repeated Measures
Lyle D. Broemeling

Bayesian Methods in Epidemiology
Lyle D. Broemeling

Bayesian Methods in Health Economics
Gianluca Baio

Bayesian Missing Data Problems: EM, Data Augmentation and Noniterative Computation
Ming T. Tan, Guo-Liang Tian, and Kai Wang Ng

Published Titles

Bayesian Modeling in Bioinformatics
Dipak K. Dey, Samiran Ghosh,
and Bani K. Mallick

Benefit-Risk Assessment in Pharmaceutical Research and Development
Andreas Sashegyi, James Felli,
and Rebecca Noel

Biosimilars: Design and Analysis of Follow-on Biologics
Shein-Chung Chow

Biostatistics: A Computing Approach
Stewart J. Anderson

Causal Analysis in Biomedicine and Epidemiology: Based on Minimal Sufficient Causation
Mikel Aickin

Clinical and Statistical Considerations in Personalized Medicine
Claudio Carini, Sandeep Menon,
and Mark Chang

Clinical Trial Data Analysis using R
Ding-Geng (Din) Chen and
Karl E. Peace

Clinical Trial Methodology
Karl E. Peace and
Ding-Geng (Din) Chen

Computational Methods in Biomedical Research
Ravindra Khattree and
Dayanand N. Naik

Computational Pharmacokinetics
Anders Källén

Confidence Intervals for Proportions and Related Measures of Effect Size
Robert G. Newcombe

Controversial Statistical Issues in Clinical Trials
Shein-Chung Chow

Data Analysis with Competing Risks and Intermediate States
Ronald B. Geskus

Data and Safety Monitoring Committees in Clinical Trials
Jay Herson

Design and Analysis of Animal Studies in Pharmaceutical Development
Shein-Chung Chow and
Jen-pei Liu

Design and Analysis of Bioavailability and Bioequivalence Studies, Third Edition
Shein-Chung Chow and
Jen-pei Liu

Design and Analysis of Bridging Studies
Jen-pei Liu, Shein-Chung Chow,
and Chin-Fu Hsiao

Design and Analysis of Clinical Trials for Predictive Medicine
Shigeyuki Matsui, Marc Buyse,
and Richard Simon

Design and Analysis of Clinical Trials with Time-to-Event Endpoints
Karl E. Peace

Design and Analysis of Non-Inferiority Trials
Mark D. Rothmann, Brian L. Wiens,
and Ivan S. F. Chan

Difference Equations with Public Health Applications
Lemuel A. Moyé and
Asha Seth Kapadia

DNA Methylation Microarrays: Experimental Design and Statistical Analysis
Sun-Chong Wang and
Arturas Petronis

Published Titles

DNA Microarrays and Related Genomics Techniques: Design, Analysis, and Interpretation of Experiments
David B. Allison, Grier P. Page,
T. Mark Beasley, and Jode W. Edwards

Dose Finding by the Continual Reassessment Method
Ying Kuen Cheung

Dynamical Biostatistical Models
Daniel Commenges and
Hélène Jacqmin-Gadda

Elementary Bayesian Biostatistics
Lemuel A. Moyé

Empirical Likelihood Method in Survival Analysis
Mai Zhou

Exposure–Response Modeling: Methods and Practical Implementation
Jixian Wang

Frailty Models in Survival Analysis
Andreas Wienke

Fundamental Concepts for New Clinical Trialists
Scott Evans and Naitee Ting

Generalized Linear Models: A Bayesian Perspective
Dipak K. Dey, Sujit K. Ghosh,
and Bani K. Mallick

Handbook of Regression and Modeling: Applications for the Clinical and Pharmaceutical Industries
Daryl S. Paulson

Inference Principles for Biostatisticians
Ian C. Marschner

Interval-Censored Time-to-Event Data: Methods and Applications
Ding-Geng (Din) Chen, Jianguo Sun,
and Karl E. Peace

Introductory Adaptive Trial Designs: A Practical Guide with R
Mark Chang

Joint Models for Longitudinal and Time-to-Event Data: With Applications in R
Dimitris Rizopoulos

Measures of Interobserver Agreement and Reliability, Second Edition
Mohamed M. Shoukri

Medical Biostatistics, Third Edition
A. Indrayan

Meta-Analysis in Medicine and Health Policy
Dalene Stangl and
Donald A. Berry

Mixed Effects Models for the Population Approach: Models, Tasks, Methods and Tools
Marc Lavielle

Modeling to Inform Infectious Disease Control
Niels G. Becker

Modern Adaptive Randomized Clinical Trials: Statistical and Practical Aspects
Oleksandr Sverdlov

Monte Carlo Simulation for the Pharmaceutical Industry: Concepts, Algorithms, and Case Studies
Mark Chang

Multiple Testing Problems in Pharmaceutical Statistics
Alex Dmitrienko, Ajit C. Tamhane,
and Frank Bretz

Published Titles

Noninferiority Testing in Clinical Trials: Issues and Challenges
Tie-Hua Ng

Optimal Design for Nonlinear Response Models
Valerii V. Fedorov and Sergei L. Leonov

Patient-Reported Outcomes: Measurement, Implementation and Interpretation
Joseph C. Cappelleri, Kelly H. Zou, Andrew G. Bushmakin,
Jose Ma. J. Alvir, Demissie Alemayehu, and Tara Symonds

Quantitative Evaluation of Safety in Drug Development: Design, Analysis and Reporting
Qi Jiang and H. Amy Xia

Quantitative Methods for Traditional Chinese Medicine Development
Shein-Chung Chow

Randomized Clinical Trials of Nonpharmacological Treatments
Isabelle Boutron, Philippe Ravaud, and David Moher

Randomized Phase II Cancer Clinical Trials
Sin-Ho Jung

Sample Size Calculations for Clustered and Longitudinal Outcomes in Clinical Research
Chul Ahn, Moonseong Heo, and Song Zhang

Sample Size Calculations in Clinical Research, Second Edition
Shein-Chung Chow, Jun Shao, and Hansheng Wang

Statistical Analysis of Human Growth and Development
Yin Bun Cheung

Statistical Design and Analysis of Clinical Trials: Principles and Methods
Weichung Joe Shih and Joseph Aisner

Statistical Design and Analysis of Stability Studies
Shein-Chung Chow

Statistical Evaluation of Diagnostic Performance: Topics in ROC Analysis
Kelly H. Zou, Aiyi Liu, Andriy Bandos, Lucila Ohno-Machado, and Howard Rockette

Statistical Methods for Clinical Trials
Mark X. Norleans

Statistical Methods for Drug Safety
Robert D. Gibbons and Anup K. Amatya

Statistical Methods for Immunogenicity Assessment
Harry Yang, Jianchun Zhang, Binbing Yu, and Wei Zhao

Statistical Methods in Drug Combination Studies
Wei Zhao and Harry Yang

Statistics in Drug Research: Methodologies and Recent Developments
Shein-Chung Chow and Jun Shao

Statistics in the Pharmaceutical Industry, Third Edition
Ralph Buncher and Jia-Yeong Tsay

Survival Analysis in Medicine and Genetics
Jialiang Li and Shuangge Ma

Theory of Drug Development
Eric B. Holmgren

Translational Medicine: Strategies and Statistical Methods
Dennis Cosmatos and Shein-Chung Chow

CRC Press
Taylor & Francis Group
6000 Broken Sound Parkway NW, Suite 300
Boca Raton, FL 33487-2742

First issued in paperback 2020

© 2016 by Taylor & Francis Group, LLC
CRC Press is an imprint of Taylor & Francis Group, an Informa business

No claim to original U.S. Government works

ISBN-13: 978-1-4200-9087-1 (hbk)
ISBN-13: 978-0-367-78339-6 (pbk)

Visit the Taylor & Francis Web site at
http://www.taylorandfrancis.com

and the CRC Press Web site at
http://www.crcpress.com

Contents

Preface

Although imperfect, clinical trials remain the gold standard research study for evaluating the effects of a medical intervention. The approval and appropriate use of drugs, devices, biologics, and other interventions are primarily based on findings established from clinical trials.

There are many books that describe particular aspects of clinical trials or cover specific statistical methodologies. This book is not designed to present detailed statistical methodology. Instead, the aim of this book is (1) to describe the fundamental scientific concepts of the design, data monitoring, analyses, and reporting of clinical trials with the hope that the understanding of these concepts will provide a foundation for addressing the challenges encountered in future trials and (2) to discuss the practical aspects of trials that are not typically part of statistical methodology courses, providing guidance for avoiding potential pitfalls and maximizing a statistician's contribution to a clinical trial. Although parts of the book target statisticians (Chapter 4), the fundamental scientific issues and the practical aspects of clinical trials that are described are of interest and considerable value to nonstatisticians. Thus, the book is well-suited for a broader audience as might be met in a general course on clinical trials targeting clinicians and other clinical trialists. The book focuses on important concepts and promotes "thinking clinical trials" while limiting complex statistical notation. Dr. Evans uses this book as part of his "Principles of Clinical Trials" course at the Harvard School of Public Health.

This book evolved in part from a workshop sponsored by the Pharmaceutical Education and Research Institute (PERI). This three-day workshop was designed to train new clinical trial statisticians after they had received academic training in graduate schools. The authors, Dr. Scott Evans and Dr. Naitee Ting, were codirectors of the workshop. We offered unique but complementary experiences and perspectives. Dr. Evans is from academia, teaches a course in clinical trials in the Department of Biostatistics at Harvard University, and is the principal investigator for a Statistical and Data Management Center for an NIH-funded clinical trials network. Dr. Evans also serves on an FDA (Food and Drug Administration) Advisory Committee and on several data monitoring committees for industry and NIH-sponsored clinical trials. Dr. Ting has more than 20 years of experience in the pharmaceutical industry and has also taught courses in clinical trials at the Department of Statistics at the University of Connecticut and the University of Rhode Island. The book was prepared in great part based upon our career experiences. Thus, the book covers real-world topics and offers practical perspectives.

Introduction of Each Chapter

The book has two sections. Section I provides background information, including an introduction to clinical trials (Chapter 1), a description of the product development process (Chapter 2), a description of the regulatory processes (Chapter 3), and a discussion of some of the valuable attributes that statisticians can develop to enhance their contribution to a clinical trial (Chapter 4). Section II discusses scientific issues, following the temporal flow of the clinical trial: trial design (Chapters 5 and 6), data monitoring (Chapter 7), analyses of efficacy, safety, and benefit:risk (Chapters 8 and 9), and the reporting/publication of clinical trials (Chapter 10).

Chapter 1 provides an introduction to the clinical trial. The clinical trial is defined and described, and a contrast with epidemiological studies (advantages and disadvantages) is provided. The types of interventions that can be evaluated in clinical trials are discussed. Trial phases are described, and the fundamental elements of the protocol document are outlined. The modern issue of clinical trial registration is discussed, including the rationale, process, and primary registries. Finally, ethical issues are discussed, including historical ethical failures, landmark documents designed to protect trial participants, informed consent, and sound statistical ethics.

Chapter 2 discusses the medicinal product development process. An appreciation of this process is critical when designing and implementing clinical trials that are part of a development program. The chapter begins by introducing a product label and then discusses nonclinical and clinical development. Trial phases are discussed within the context of the development process.

The medicinal product development industries (e.g., pharmaceutical, biologic, and device industries) are highly regulated. In the United States, the FDA is the regulatory review body. Before a new product is approved and made available for general patient use, regulators review the clinical data and clinical study reports. Regulators then make a decision to reject or to approve the intervention. An understanding of this process is helpful when designing and analyzing trials as part of a development program. Chapter 3 briefly describes the regulatory review process.

Studying statistics is one thing, but learning to be a statistician is another. In most of the biostatistics/statistics graduate programs, there is a paucity of practical training to prepare statisticians to serve as important contributory members within clinical trial research teams. Chapter 4 targets statisticians, discussing the nontechnical attributes of effective clinical trial statisticians. This chapter also offers suggestions for becoming a better statistician by maximizing a statistician's contribution to the clinical trial effort.

Most problems in clinical trials are a result of poor planning. Fancy statistical methods cannot rescue design flaws. Thus, careful planning with clear foresight is crucial. Issues in trial conduct and analyses should be anticipated during trial design and proactively addressed. Chapters 5 and 6 are devoted to clinical trial design.

There are many issues that must be carefully considered when designing clinical trials. Fundamental issues include clearly defining the research question, minimizing variation, randomization and stratification, blinding, selection of a control group, selection of the target population, selection of endpoints, sample size, and planning for interim analyses. These issues are discussed in detail in Chapter 5.

The selection and formulation of a clinical trial design require logic and creativity. Many designs can be considered when planning a clinical trial. Common clinical trial designs include single-arm trials, placebo-controlled trials, parallel trials, crossover trials, factorial trials, noninferiority trials, and designs for validating a diagnostic device. The choice of the design depends on the specific research questions of interest, characteristics of the disease and intervention, the endpoints, the availability of a control group, and on practical issues such as the availability of funding. Chapter 6 discusses common clinical designs; highlights their strengths, limitations, and assumptions; and provides guidance regarding when these designs may be considered in practice.

Interim data monitoring is an important part of clinical trials. Interim data monitoring helps to ensure the safety of trial participants and provides an opportunity to improve trial efficiency, and thus is an area where statisticians can make significant contributions. Methods for interim data monitoring and the use of data monitoring committees (DMCs) are discussed in Chapter 7.

The analyses of efficacy of a clinical trial are fraught with many challenges including the definition and selection of analysis sets, subgroup analyses, multiplicity, missing data, competing risks, and adherence issues. We discuss these issues and preparation of the statistical analysis plan (SAP) in Chapter 8.

When evaluating interventions, researchers must go beyond measuring the beneficial effects. It is critical to thoroughly evaluate the potentially harmful effects and the effects on quality-of-life (QoL). The analyses of safety and QoL are discussed in Chapter 9. Furthermore, it is important to provide a synthesized overall evaluation that simultaneously considers all the benefits and harms to provide an overall picture of the value and utility of the intervention. This benefit:risk evaluation lies at the root of clinical trials but has not been well-formalized. Chapter 9 includes a unique discussion of benefit:risk evaluation based upon short courses presented on invitation by Dr. Evans to the FDA.

Clinical trial results are communicated to the medical community primarily in the form of publications in the medical literature. Appropriate reporting of clinical trial results is crucial for safe and effective use of an intervention, and thus, for ensuring optimal patient care. Chapter 10 summarizes guidelines and provides suggestions for the proper reporting of clinical trials in the medical literature, providing a unique reference that is not typically provided in other textbooks.

Authors

Dr. Scott Evans teaches clinical trials at Harvard University where he is the director of the Statistical and Data Management Center for the Antibacterial Resistance Leadership Group, an NIH-funded clinical trials network. He serves on a U.S. FDA Advisory Committee and several data monitoring committees for industry and NIH-sponsored clinical trials. He has been a recipient of the Mosteller Statistician of the Year award and is a fellow of the American Statistical Association. Dr. Evans is a visiting professor at the Department of Medical Statistics at Osaka University in Japan and serves as the executive editor for *CHANCE*, and the editor-in-chief of *Statistical Communications in Infectious Diseases*.

Dr. Naitee Ting has nearly 30 years of experience in the pharmaceutical industry and currently works at Boehringer Ingelheim. He has taught courses on clinical trials at Columbia University, the University of Connecticut, and the University of Rhode Island. He is a fellow of the American Statistical Association.

Section I

Background

1

Clinical Trials

1.1 Introduction

Less than half of all medical interventions that are delivered today are supported by evidence (Institutes of Medicine [IOM], 2010). The consequence of this is suboptimal patient care and an inefficient use of valuable resources. This frequent misguided and uninformed use of interventions illustrates the need for reliable clinical research studies to be designed, conducted, analyzed, and reported in a transparent and timely manner.

Many clinical research studies are observational in which investigators observe individuals and measure their health outcomes. Examples include case–control studies, cohort studies, and case series. Although potentially informative, these studies are subject to many biases including selection bias in the study and biases induced by selective intervention assignment.

Another type of clinical research study is the clinical trial (Figure 1.1). In 2014, the NIH revised its definition of a clinical trial to "a research study in which one or more human subjects are prospectively assigned to one or more interventions (which may include placebo or other control) to evaluate the effects of those interventions on health-related biomedical or behavioral outcomes." More broadly, a clinical trial is generally considered to be a prospective biomedical or health-related research study that is conducted in humans, follows a predefined protocol, and is designed to answer specific questions regarding the health outcome effects of new interventions, or new ways of using known interventions. Interventions can be therapeutic, preventive, or diagnostic in nature.

Interventions can come in many forms including drugs, biologics, devices, dietary supplements, surgical or radiologic procedures, and behavioral or process-of-care interventions. A recent study (Krall, 2009) summarized the frequencies of trials and trial participants being conducted in the United States by the type of interventions being utilized (Table 1.1). These interventions are evaluated to treat, prevent, or diagnose many diseases. Nearly 50% of the active clinical trials evaluate interventions for cancer while 10% evaluate interventions for cardiovascular diseases. However, there are more participants in cardiovascular disease trials (~320,000) compared to 130,000 for cancer trial participants.

How research is classified

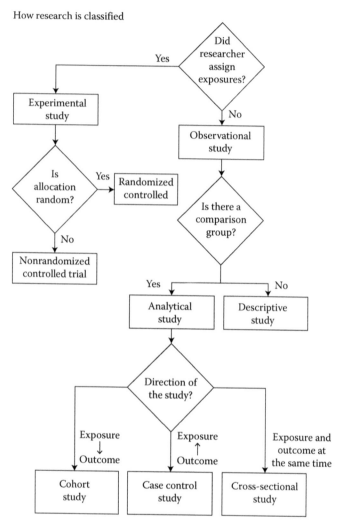

FIGURE 1.1
Classification tree of clinical research.

Randomized clinical trials are distinct from observational studies. In the former, there is control particularly with respect to intervention assignment and methods for evaluation. The control over the intervention assignment allows the clinical trial to avoid many of the potential biases associated with observational studies and thus have higher relative validity. Standardized evaluation improves efficiency. For this reason, a clinical trial is the gold standard clinical study evaluating the safety and effectiveness of an intervention. In other words, carefully conducted clinical trials are the most reliable studies to identify interventions that work in people.

TABLE 1.1

Summary of 10,974 Ongoing Clinical Trials and 2.8 Million
Study Participants

Intervention	% of Trials	% of Participants
Drugs	59	50
Biologics	9	7
Devices	7	7
Behavioral	10	16
Dietary	2	2
Genetic	<1	1
Procedural	7	8
Radiation	1	1
Other	5	8

Source: Adapted from Krall, R., 2009, US Clinical Research, *IOM Drug Forum Clinical Trials Workshop*, Washington, DC.

The disadvantages of clinical trials are that they are costly and resource-intensive, complex to design and conduct, and sometimes are not feasible due to ethical or other practical constraints. Data summarized as part of the FDA Critical Path Initiative (CPI) suggested that the cost of bringing a new medicine to market was $0.8–1.7 billion. These costs often force sponsors to concentrate on interventions with a high expected financial return. Thus, there are fewer trials for rare diseases, critical third world diseases, prevention indications, individualized therapy, and public health concerns (e.g., counterterrorism). A more recent IOM report summarized the results from another study that suggested that drug development costs for an approved compound were $15.2 million in phase I, $41.7 million in phase II, and $115.2 million in phase III. The report further stated that a large global clinical trial involving 14,000 participants and 300 research sites can reach total cost of $300–600 million to implement, conduct, and monitor to completion, with site payments being approximately half of the total costs and monitoring nearly a third. The report also highlighted the extensive development time for many clinical trials. For example, the occluded artery trial (OAT; Hochman et al., 2006) took 3 years from the first planning meeting to trial initiation. It has also been estimated that only 30% of approved drugs will generate sufficient revenue to cover the costs of research and development.

A second disadvantage of clinical trials is that trial participants may not be representative of the targeted population. Trial participants are usually volunteers, must be referred and located near the setting in which the trial is being conducted. In some cases, they may be healthier than the typical population as they are often selected based on expected success but could be less healthy in other cases if alternatives are unavailable. People with a lower socioeconomic status may be referred to trials less frequently (e.g., trials for new cancer drugs). One survey suggested that 2% of the U.S. population is always involved in clinical trials (and 4% of clinicians) but 57% would like to be involved. Thus, the generalizability of trial results must be evaluated.

The objective of a clinical trial is to isolate the effects of an intervention (or differences between interventions) and estimate these effects with acceptable precision. A treatment effect is efficiently isolated by controlling potential bias and confounding and minimizing extraneous variation (i.e., variation due to factors other than treatment response). Several key features of clinical trials are used to pursue this goal including the use of prospective evaluation, randomization (possibly with stratification), adherence to the intent-to-treat (ITT) principle, use of a control group, blinding, and standardized definitions, measurement, and evaluation.

Clinical trials are typically sponsored by industry (e.g., private pharmaceutical, biotechnology, or device companies) or government agencies (e.g., the National Institutes of Health [NIH]). The NIH is a part of the United States Department of Health and Human Services (DHHS) and is the nation's medical research agency. The NIH is the largest source of funding for medical in the world with more than 83% of its $31 billion dollar budget going to researchers at universities and research organizations. Thanks in large part to NIH researchers, life expectancy in 2009 increased to 78 years compared to 47 years in 1900. The NIH has 27 institutes and centers (Figure 1.2) each with a specific research agenda often focusing on particular disease areas.

Industry-sponsored trials are development-driven and are conducted to pursue regulatory approval to market a new intervention or an approved intervention for a new indication. In contrast, government-sponsored trials are designed to answer important medical questions that may not be related to regulatory approval. For example, a government-sponsored trial may be conducted to compare two or more approved interventions in head-to-head fashion to see which one produces the best health outcomes. In these pressing economic times, biomedical and health research and development spending from all sources declined by more than 4 billion dollars (3%) in fiscal year 2011 compared to 2010.

Rapid scientific advancements (e.g., the mapping of the human genome) and the increased availability of technologies provide unprecedented potential for the translation of basic discoveries into medical interventions. Despite this, the success of innovative interventions has slowed in recent years as evidenced by a decreasing trend in the number of regulatory applications for new drugs, biologics, and devices (Critical Path, 2004). A new medical compound entering phase I, often representing the accumulation of a decade of preclinical evaluation, is estimated to have only 8% chance of reaching the market (down from a historical rate of 14%). During the period from 1996 to 1999, the FDA approved 157 new drugs but only 74 drug approvals were made in 2006–2009 and 21 in 2010. There was better news in 2011 and 2012 with 30 and 39 drug/biologic approvals, respectively. A recent BioMedTracker study evaluating approximately 4000 drug/biologic interventions from 2004 to 2010 concluded that, of interventions that were evaluated in

National Institutes of Health

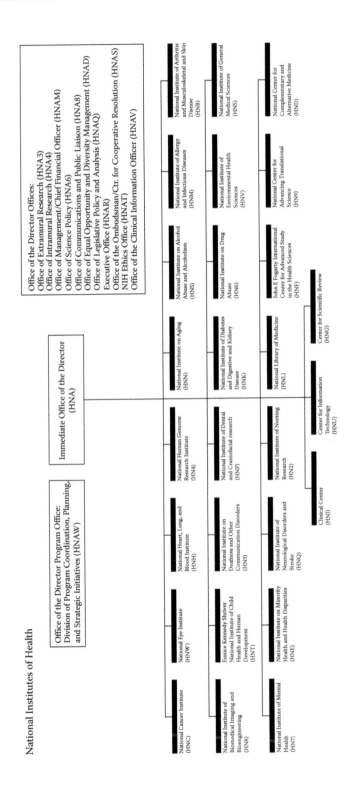

FIGURE 1.2
The NIH organizational chart.

phase I trials, biologics had a better FDA approval rate (15%) than drugs (7%). Infectious disease (e.g., HIV, HCV) interventions had the highest approval rate (12%) followed by endocrine system drugs featuring diabetes treatments (10.4%), and autoimmune diseases, for example, rheumatoid arthritis (9.4%), despite lower rates for oncology drugs being a very active area for development. Pharmaceutical industry efforts will also be greatly affected by patent expirations (i.e., at least 21 medicines had expired patents in 2012).

To address the need to reverse the slowing trends for innovative discovery and development, cross-sector collaborations among industry, government, academia, and other sectors, are increasing. For example, the NIH has created a new National Center for Advancing Translational Sciences (NCATS) that will play a key role in convening cross-sector collaborations to advance development.

Some clinical trials are conducted by clinical trial networks (CTNs). CTNs are organizations that have resources (i.e., funding to support an infrastructure of disease-area expertise, trial participants, clinical sites, clinical coordination, biostatistics, data management, and standardized protocols) to facilitate trials. CTNs are frequently funded by NIH to conduct a series of trials in a specific area of medical need. Examples of CTNs include the AIDS Clinical Trials Group (ACTG) funded by the National Institute of Allergy and Infectious Diseases and the Eastern Cooperative Oncology Group (ECOG) funded by the National Cancer Institute. These CTNs have provided many contributions and greatly advanced the field in their respective disease areas. In many cases, CTNs strive to answer research questions that industry does not address. The advantages of a CTN include operational efficiency, development and maintenance of expertise, the promotion of cutting-edge research, and the fostering of collaboration of disease-area experts.

1.2 Phases

The frequently used phase denotations provide a framework for classifying clinical trials based on their associated objectives and size. Drug development is generally considered to have four phases. Phase definitions are not completely well-defined, but various definitions share the same basic elements. The phase definitions by NIH via clinicaltrials.gov and the World Health Organization (WHO) are summarized in Table 1.2.

Each phase seeks to answer a different research question. It can be difficult to classify some trials as they have key elements that are typically associated with more than one phase. Thus, some trials are classified as "multi-phase," for example, phase II/III.

Phase I trials are the first trials in humans. Clinicaltrials.gov defines phase I trials as "initial studies to determine the metabolism and pharmacologic actions of drugs in humans, the side effects associated with increasing doses, and

TABLE 1.2

Phase Definitions by Clinicaltrials.gov and WHO

Phase	Definition
I	*Clinicaltrials.gov*
	Initial studies to determine the metabolism and pharmacologic actions of drugs in humans, the side effects associated with increasing doses, and gaining an early evidence of effectiveness and may include health participants.
	WHO
	Clinical trials to test a new biomedical intervention in a small group of people (e.g., 20–80) for the first time to evaluate safety (e.g., to determine a safe dosage range and identify side effects.
II	*Clinicaltrials.gov*
	Controlled clinical studies conducted to evaluate the effectiveness of an intervention for a particular indication or indications in patients with the disease or condition under study and to determine the common short-term side effects and risks.
	WHO
	Clinical trials to study the biomedical or behavioral intervention in a larger group of people (several hundred) to determine efficacy and to further evaluate its efficacy.
III	*Clinicaltrials.gov*
	Expanded controlled and uncontrolled trials after preliminary evidence suggesting effectiveness of the drug has been obtained and are intended to gather additional information to evaluate the overall benefit–risk relationship of the drug and provide adequate basis for physician labeling.
	WHO
	Studies to investigate the efficacy of the biomedical or behavioral intervention in large groups of human subjects (from several hundred to several thousand) by comparing the intervention to other standards or experimental interventions as well as to monitor adverse effects, and to collect information that will allow the intervention to be used safely.
IV	*Clinicaltrials.gov*
	Postmarketing studies to delineate additional information including the drug's risks, benefits, and optimal use.
	WHO
	Studies conducted after the intervention has been marketed; these studies are designed to monitor the effectiveness of the approved intervention in the general population and to collect information about any adverse effects associated with widespread use.

to gain early evidence of effectiveness and may include health participants."
While the WHO defines phase I trials as "clinical trials to test a new biomedical
intervention in a small group of people (e.g., 20–80) for the first time to evaluate safety (e.g., to determine a safe dosage range and to identify side effects."
In general, phase I trials assess the tolerance of a drug; evaluate toxicity, often

with a goal to find the maximum tolerated dose (MTD). Phase I trials are generally conducted on healthy people and thus the evaluation of efficacy is not a primary objective. Phase I trials frequently evaluate bioavailability through the evaluation of pharmacokinetics (PK), the effect of the body on the intervention. PK evaluation consists of the study of absorption, distribution, metabolism, and elimination (ADME). Phase I trials may also evaluate pharmacodynamics (PD), the effect of the drug on the body (e.g., respiration, heart rate). Bioequivalence studies are phase I studies to evaluate if a test formulation (e.g., a generic) displays similar PK characteristics to a reference formulation.

Phase II trials are conducted after phase I studies in intervention development. Clinicaltrials.gov defines phase II trials as "controlled clinical studies conducted to evaluate the effectiveness of an intervention for a particular indication or indications in patients with the disease or condition under study and to determine the common short-term side effects and risks." While WHO defines phase II trials as "clinical trials to study the biomedical or behavioral intervention in a larger group of people (several hundred) to determine efficacy and to further evaluate its efficacy." Phase II trials are frequently conducted to help recommend a range of doses to be studied in phase III. Phase II studies may be conducted on a homogeneous population and may employ surrogate endpoints.

Phase III trials are conducted after phase II trials. Clinicaltrials.gov defines phase III trials as "expanded controlled and uncontrolled trials after preliminary evidence suggesting effectiveness of the drug has been obtained, and are intended to gather additional information to evaluate the overall benefit–risk relationship of the drug and provide adequate basis for physician labeling." While WHO defines phase III trials as "studies to investigate the efficacy of the biomedical or behavioral intervention in large groups of human subjects (from several hundred to several thousand) by comparing the intervention to other standards or experimental interventions as well as to monitor adverse effects, and to collect information that will allow the intervention to be used safely." Phase III trials use randomization to assign an intervention, are often blinded, multicenter, and use the intended target population. Sponsors often submit trial results to regulators after phase III trials to request approval.

Phase IV trials are large postapproval trials. Clinicaltrials.gov defines phase IV trials as "postmarketing studies to delineate additional information including the drug's risks, benefits, and optimal use." While WHO defines phase IV trials as "studies conducted after the intervention has been marketed... these studies are designed to monitor the effectiveness of the approved intervention in the general population and to collect information about any adverse effects associated with widespread use." Phase IV studies investigate long-term effects through a longer duration of follow-up than earlier-phase trials.

In 2009, Krall reported the phases of trials being conducted in the United States (Table 1.3).

How each phase relates to a product development program is discussed in Chapter 2.

TABLE 1.3

Summary of the Phases of 10,974 Ongoing Clinical Trials and 2.8 Million Study Participants

Intervention	% of Trials	% of Participants
Phase 0	1	<1
Phase I	15	4
Phase I/II	8	2
Phase II	27	8
Phase II/III	3	3
Phase III	12	41
Phase IV	9	10
Other	25	32

Source: Adapted from Krall, R., 2009, US Clinical Research, *IOM Drug Forum Clinical Trials Workshop*, Washington, DC.

1.3 Protocol

A protocol is a scientific document describing the study plan on which a clinical trial is based. The plan is carefully designed to safeguard the health of the participants as well as answer specific research questions. The protocol has several purposes including assisting the project team in thinking through the research, ensuring that both participant and trial management are considered during the planning stage, providing a sounding board for external comments including regulators and ethics committees, orienting appropriate staff for the preparation of case report forms and data collection procedures, guiding the treatment of study participants, and providing a document that can be used as a foundation or starting point for future studies.

A protocol describes the background, the justification and rationale for conducting the trial, and elements of trial design (e.g., what types of people may participate in the trial; the schedule of tests, procedures, medications, and dosages; the length of the study, and how to care, treat, and monitor the participants). A statistical section of the protocol describes important design issues, defines the trial endpoints, outlines how the trial was sized, discusses data monitoring methods, and outlines the plan for data analyses when the trial is completed.

A typical clinical trial protocol often begins by defining a team roster, sites participating in the trial and a glossary of relevant terms. The Standard Protocol Items Recommendations for Intervention Trials (SPIRIT) statement (www.spirit-statement.org) provides recommendations for a minimum set of scientific, ethical, and administrative elements that should be addressed in a clinical trial protocol. The recommendations are outlined in a 33-item checklist (Table 1.4) and a figure (Figure 1.3) representing an example template of recommended content for the schedule of enrollment, interventions, and assessments.

TABLE 1.4

SPIRIT Checklist of Recommended Items to Address in a Clinical
Trial Protocol and Related Documents

Administrative Information

- Title
- Trial registration
- Protocol version
- Funding
- Roles and responsibilities

Introduction

- Background and rationale
- Objectives
- Trial design
- Study setting
- Eligibility criteria
- Interventions
- Outcomes
- Participant timeline
- Sample size
- Recruitment

Methods: Assignment of Interventions (for Controlled Trials)

- Allocation: sequence generation; concealment mechanism; implementation
- Blinding (masking)

Methods: Data Collection, Management, and Analysis

- Data collection methods
- Data management
- Statistical methods

Methods: Monitoring

- Data monitoring
- Harms
- Auditing

Ethics and Dissemination

- Research ethics approval
- Protocol amendments
- Consent or assent
- Confidentiality
- Declaration of interests
- Access to data
- Ancillary and posttrial care
- Dissemination policy

Appendices

- Informed consent materials
- Biological specimens

Note: Complete explanations can be found at www.spirit-statement.org.

| | STUDY PERIOD | | | | | | | |
| | Enrollment | Allocation | Postallocation | | | | | Close-out |
TIMEPOINT**	$-t_1$	0	t_1	t_2	t_3	t_4	etc.	t_x
ENROLLMENT:								
Eligibility screen	×							
Informed consent	×							
[List other procedures]	×							
Allocation		×						
INTERVENTIONS:								
[Intervention A]			←——————→					
[Intervention B]			×		×			
[List other study groups]			←——————————→					
ASSESSMENTS:								
[List baseline variables]	×	×						
[List outcome variables]				×		×	etc.	×
[List other data variables]			×	×	×	×		×

**List specific time points in this row.

FIGURE 1.3
SPIRIT statement example template of recommended content for the schedule of enrollment, interventions, and assessments.

Development of a protocol is a joint effort with input from people from multiple disciplines (e.g., clinicians, statisticians, nurses, data managers, drug supply personnel, clinical research associates, and project managers). It is also an iterative process with several rounds of reviews and revisions. Development of the protocol is guided by the primary clinical research question of interest, expressed as the primary objective. The clinician and the statistician work together and transform the primary objective into a statistical hypothesis. A vision for the final analyses can then be developed, and the trial can be appropriately sized and powered.

A protocol team also defines secondary objectives and associated endpoints. The clinicians describe the clinical procedures and necessary monitoring of participant safety. The statisticians develop plans for data monitoring and analyses in response to the challenges of the protocol. This may include developing a plan to handle multiplicity concerns from subgroup analyses or a plan for missing data imputation. The data managers work with the statisticians, programmers, and clinicians to design the case report forms (CRF) for data collection that will be necessary for monitoring and analyses.

After a protocol is finalized, typically, there is an "investigator meeting" where the trial sponsor explains and discusses the protocol with prospective investigators. The clinical trial starts shortly after the investigator meeting. Occasionally, a protocol needs to be amended (termed a *protocol amendment*), for example, to revise the participant entry criteria due to enrollment issues or safety concerns, or to reflect newly discovered information regarding the interventions or disease conditions under study.

1.4 Clinical Trial Registration

Medical practice should be based on the totality of the evidence, often consisting of several studies. Selective reporting (e.g., only "positive" trials) distorts the evidence as perceived by the medical community, resulting in suboptimal decision making and patient care. Indeed, negative evidence is frequently underreported. For example, Tam et al. (2011) reported that a substantial number of phase III clinical trials of systemic cancer treatment remain unpublished after 6.5 years or more with 71% of these unpublished trials being negative trials. Many other publications are delayed by 5 years or more.

Poor practices regarding the dissemination of trial results also break an implicit contract with trial participants, sponsors, and institutional review boards (IRBs), by jeopardizing the scientific integrity, quality, and public health value of the research. Researchers have an ethical and scientific obligation to conduct and report research in an honest and transparent manner. Thus, it is important to acknowledge all clinical trials that are conducted including "negative" ones.

To address this issue, in November 1997, the United States Congress enacted the Food and Drug Administration Modernization Act (FDAMA) requiring NIH to establish a database of clinical trials. The NIH established www.clinicaltrials.gov in February 2000. Milestones in the history of clinicaltrials.gov are given in Table 1.5. The idea is that trials should be registered as a proof of their existence and transparency of their original design.

Criteria for the trials affected are listed in Table 1.6. Responsible parties (i.e., the study sponsor or responsible principal investigator [PI]) must register the trial within 21 days of the enrollment of the first participant. Registration information must include primary and secondary outcomes. Submissions are reviewed, and revisions may be requested prior to publication. Results are required to be reported within 12 months of the date of the final data collection for the prespecified primary outcome measure. The required results to be reported are listed in Table 1.6. Phase I trials and noninterventional trials should be registered, but results do not need to be submitted.

Other clinical trial registries have since been created in various countries. Criteria for a clinical trials registry are displayed in Table 1.6. Trial

TABLE 1.5

Clinicaltrials.gov Milestones

- November 1997: United States Congress enacts the Food and Drug Administration Modernization Act (FDAMA) requiring NIH to establish a database of clinical trials.
- February 2000: NIH establishes www.clinicaltrials.gov.
- June 2004: The state of New York sues GlaxoSmithKline for failing to report trial results that demonstrated harmful effects of an antidepressant.
- September 2004: The ICMJE requires prospective registration of trials as a precondition for publication.
- May 2006: The World Health Organization (WHO) international clinical trials registry platform is launched (www.who.int/ictrp).
- September 2007: United States Congress enacts the Food and Drug Administration Amendments Act (FDAAA) which expanded the legal requirements for trial registration and mandated the reporting of trial results with penalties for noncompliance.
- December 2007: The scope of registration is clarified to include all phase II–IV drug and device trials.
- September 2008: A requirement is made to report basic results.
- September 2009: A requirement is made to report adverse events.

Source: Adapted from Dickerson, K., and Rennie, D., 2012, *JAMA*, 307(17):1861–1864.

registration consists of the publication of international-agreed information regarding the design, conduct, and administration of clinical trials. Trials are provided a unique ID. The required data to be entered are listed in Table 1.5.

A particularly important development occurred in September 2004 when the International Committee of Medical Journal Editors (ICMJE) required prospective registration of trials (prior to participant enrollment) as a precondition for publication. During the years 2001–2004 less than 2000 trials were registered in clinicaltrials.gov each year. However, beginning in 2005 after the ICMJE requirement, more than 10,000 trials have been registered each year. In May 2005, there were 13,153 trials registered in clinicaltrials.gov and in April 2012 there were 124,196 trials registered. Califf et al. (2012) reported that the number of registered interventional clinical trials increased from 28,881 (October 2004 to September 2007) to 40,970 (October 2007 to September 2010) and the number of missing data elements has generally declined, but 52% of trials were registered after the first participant had enrolled.

Registration of all clinical trials is important and should be viewed as a scientific and ethical responsibility. Trial registration helps to improve the transparency and consistency in reporting and results in higher quality clinical trials by avoiding problems created by publication bias and selective reporting. This helps to ensure that medical decision-making is well-informed and guided by all available evidence. Registration may also result in more effective collaboration between researchers, improve participant recruitment, and make it easier to identify research gaps, facilitate systematic reviews, and to avoid duplicative trial efforts. Registration is also consistent with the Declaration of Helsinki which states "every clinical trial must be registered in a publically accessible database before recruitment of the first subject."

TABLE 1.6

Clinical Trial Registry Information: Registry Criteria, Required Registration Data, Criteria for Trials with Legal Registration Requirements, Required Results Data

Registry Criteria

- Accessible to the public at no charge
- Managed by a not-for-profit
- Open to all prospective registrants
- Electronically searchable
- Has a mechanism to ensure the validity of registration data

Required Data to Be Entered upon Registration

- The intervention
- Comparison studied
- Hypotheses
- Primary and secondary outcomes
- Eligibility criteria
- Targeted sample size
- Key trial dates (registration date, start date, date of last follow-up, date of closure to data entry, date trial data is considered complete)
- Funding sources
- Contact information

Criteria for Trials with Legal Requirements for Trial Registration and Mandated Results Reporting into www.clinicaltrials.gov as per FDAAA

- Phase II–IV intervention trials
- Trials involving drugs, biologics, and devices regulated by the FDA
- Trials with at least one U.S. site or are conducted under an IND or IDE
- Trials initiated or ongoing as of September 27, 2007

Required Results to Be Reported (within 12 Months of the Date of the Final Data Collection for the Prespecified Primary Outcome Measure)

- Administrative information (e.g., contact information)
- Scientific information (four modules)
  - Participant flow
  - Baseline characteristics
  - Outcome measures and statistical analyses
  - Adverse events

1.5 Ethical Issues

Clinical trials are human studies for which the outcomes are unknown. This uncertainty applies to the effectiveness and safety of the interventions as well as how interventions compare with one another. The interventions being studied may be ineffective or unsafe. Trial participants may receive little or no benefit and could potentially be harmed. Hopefully, the knowledge gained from clinical trials will benefit *future* patients.

The uncertainty of the outcome is indeed the very reason that trials are conducted, and provides a basis for randomized trials. Randomized trials are only ethical in a state of equipoise (i.e., when there is insufficient scientific evidence that one intervention is clearly superior to another). It would not be ethical to randomize participants to an intervention that is known to be inferior or continue to follow participants on interventions that are discovered to be inferior without alerting the participants to this information. Equipoise must remain in order for a trial to continue although opinions regarding the level of required uncertainty often vary, with some believing that without reliable evidence from well-conducted randomized clinical trials, uncertainty remains.

In clinical trials, a conflict can exist between what is best for the study and what is best for the trial participant. Thus, extreme care must be taken to ensure that ethical concerns take priority over scientific interests and that the health and rights of trial participants are protected. Clinical trials should be designed to minimize risks and include sufficient monitoring to ensure the fair and safe treatment and evaluation of trial participants. This goal is complicated by the cultural variation in ethics that exist, for example, country variation in ethics in multinational studies.

1.5.1 Historical Ethical Failures

There have been several cases of the unethical exploitation of people in research studies. A few of the well-documented ones are briefly described in Table 1.7.

1.5.2 Landmark Documents

As a consequence of these ethical tragedies, many documents and guidelines have been developed to help guide and ensure the ethical treatment of human research participants. Hippocrates was believed to have created the first guideline for medical practice, termed the Hippocratic Oath, whereby clinicians took an oath to "do no harm." However, more recent important documents include the Declaration of Helsinki, the Nuremberg Code, and the Belmont Report. Key elements of these documents are summarized in Table 1.8.

In 1964, the World Medical Association (WMA) established a code of ethics for clinical research, known as the Declaration of Helsinki. The key principles are that the research must: conform to general scientific principles, be formulated in a written protocol, be conducted by qualified individuals, and include written informed consent of research participants.

In 1974, the judges of the Nuremberg trials outlined 10 principles (listed in Table 1.8) for the ethical conduct for human experimentation, known as the *Nuremberg Code.*

In 1978, the Belmont Report was published outlining ethical guidelines and the definition and nature of informed consent. The report outlined three principles for the protection of human subjects: autonomy, beneficence, and justice (described in Table 1.7). These ethical principles provide the basis for federal regulations.

TABLE 1.7

Examples of the Unethical Exploitation of People in Research Studies

Nuremberg Trials

During World War II, there were no standards for human experimentation. In Nazi Germany, dangerous and torturous clinical experiments were conducted on Jewish people in concentration camps. In 1946–1947, 20 people were tried as a part of the Nuremberg Trials, with 16 being convicted and seven of which were executed, including four physicians.

The Tuskegee Study

Between 1932 and 1972, 600 impoverished African Americans (399 with syphilis and 201 without syphilis) from Tuskegee, Alabama, were enrolled in an observational research study. By 1947, penicillin had become the standard treatment for syphilis. However, the study participants were not provided with penicillin nor were they informed of its availability. This study illustrates that even observational studies can be unethical.

Jewish Chronic Disease Hospital

In 1963, at the Jewish Chronic Disease Hospital in Brooklyn, New York, cancer cells were deliberately injected into elderly patients to evaluate immunologic response. Consent was said to have been obtained, however, no record of consent was documented.

Willowbrook State Hospital

Children with mental retardation at the Willowbrook State Hospital in New York were exploited in hepatitis studies from 1956 to 1971. Some children were deliberately infected with viral hepatitis (e.g., fed stool samples from infected individuals) to study the natural history of hepatitis disease. Researchers attempted to defend these actions by saying that they had parental consent and that the children had a high likelihood of contracting the disease anyway.

Unit 731

Unit 731 was a covert biological and chemical warfare research and development unit of the Imperial Japanese Army that undertook lethal human experimentation during the Second Sino-Japanese War (1937–1945) and World War II. Estimates of the number of deaths range as high as 200,000 military personnel and civilians in China.

In 1981, the DHHS codified the Policy for the Protection of Human Subjects (45 CFR 46). Subpart A ("the common rule") of these regulations provides the basic foundation for IRBs. Additional subparts provide protections to vulnerable populations such as pregnant women, fetuses, and neonates (subpart B), prisoners (subpart C), and children (subpart D) involved in human subjects research.

1.5.3 Institutional Review Boards

In 1974, the U.S. Congress passed the National Research Act, which established the National Commission for the Protection of Human Subjects in BioMedical and Behavioral Research. The act also required the establishment of IRBs to review and approve clinical trials before they could be conducted.

An IRB is a group of individuals that has been formally designated to approve, monitor, and review biomedical and behavioral research involving

TABLE 1.8

Important Ethics Documents

Declaration of Helsinki (1964)

Established a code of ethics for clinical research. The key principles are that the research must
- Conform to general scientific principles
- Be formulated in a written protocol
- Be conducted by qualified individuals and
- Include written informed consent of research participants

Nuremberg Code (1974)

Outlined 10 principles for the ethical conduct for human experimentation:
1. Voluntary consent
2. No reasonable alternative to human experimentation
3. Based on biological knowledge and animal experimentation
4. Avoid unnecessary suffering and injury
5. No expectation of death or disability
6. Risk is consistent with humanitarian importance
7. Protection against a remote chance of death or injury
8. Conducted by qualified scientists
9. Participants have the option to discontinue and
10. Experiment is terminated if injury is likely

Belmont Report (1978)

Outlined ethical guidelines and the definition and nature of informed consent. The report outlined three principles for the protection of human subjects:
- *Autonomy*: The respect for persons including the right of self-governance and the protection of persons who are unable to make informed decisions for themselves.
- *Beneficence* (and nonmalfeasance): The patients' right to receive the benefit and the researchers' obligation to avoid harm.
- *Justice*: The fair distribution of benefits and burdens.

human subjects. The IRB consists of at least five individuals from varying backgrounds (e.g., clinicians, other scientists, nonscientists, community representatives) who are free from conflicts of interest, and include at least one man and one woman. IRBs are regulated by the Office for Human Research Protections (OHRP) within the U.S. Department of Health and Human Services (HHS). The mission of the IRB is to protect the rights and welfare of study participants. The IRB reviews research protocols and related materials (e.g., informed consent documents and investigator brochures) for scientific, ethical, and regulatory acceptability, providing an oversight function for research conducted on human subjects. The IRB has the authority to approve, to require modifications, or to disapprove the research. Approvability is based upon evaluation of whether: risks are minimized, risks are reasonable relative to potential benefits and the importance of the knowledge to be gained, the selection of participants is equitable, informed consent and appropriate documentation is obtained for all participants, data monitoring is adequate to ensure patient safety, and privacy and confidentiality are appropriately protected (Table 1.9). Most research institutions have their own IRBs.

TABLE 1.9

IRB Approvability Criteria

- Risks are minimized
- Risks are reasonable relative to potential benefits and the importance of the knowledge to be gained
- The selection of participants is equitable
- Informed consent and appropriate documentation are obtained for all participants
- Data monitoring is adequate to ensure patient safety
- Privacy and confidentiality are appropriately protected

However, there have been recent efforts to establish central IRBs that would provide oversight for multiple institutions for a specific project.

1.5.4 Informed Consent

Researchers are required to obtain a signed and dated Informed Consent Form (ICF) for each participant before they can be enrolled in a clinical study. There are eight basic elements (and six additional elements) to an ICF Table 1.10.

An ICF can be waived in emergency medicine trials (e.g., stroke, epilepsy, and traumatic brain injury) as participants would not be able to provide consent. Often in these cases, community leaders provide consent to conduct such studies in specific areas or hospitals. An ICF is prepared for parents

TABLE 1.10

Basic Elements of an Informed Consent Form

---

1. A statement that the study involves research, an explanation of the purposes of the research, and the expected duration of the subject's participation, a description of the procedures to be followed, and identification of any procedures which are experimental.
2. A description of any reasonably foreseeable risks or discomforts to the subject.
3. A description of any benefits to the subject or to others that may reasonably be expected from the research.
4. A disclosure of appropriate alternative procedures or courses of treatment, if any, that might be advantageous to the subject.
5. A statement describing the extent, if any, to which confidentiality of records identifying the subject will be maintained.
6. A research involving more than minimal risk, an explanation as to whether any compensation, and an explanation as to whether any medical treatments are available if injury occurs and, if so, what they consist of, or where further information may be obtained.
7. An explanation of whom to contact for answers to pertinent questions about the research and research subjects' rights, and whom to contact in the event of a research-related injury to the subject and
8. A statement that participation is voluntary, refusal to participate will involve no penalty or loss of benefits to which the subject is otherwise entitled, and the subject may discontinue participation at any time without penalty or loss of benefits to which the subject is otherwise entitled.

---

or legal guardians in pediatric trials or other vulnerable populations (e.g., elderly) when people are unable to understand the study.

1.5.5 Modern Cases of Negligence

Despite the development of guidelines and oversight of human research, there have been recent cases of ethical negligence. We describe a few in Table 1.11.

1.5.6 Statistical Ethics

Statisticians are also responsible for maintaining ethical standards including the integrity of the trial design, data monitoring, transparent analyses, objective interpretation, and honest reporting. This responsibility can be challenging

TABLE 1.11

Examples of the Ethical Negligence

The Jesse Gelsinger Case

In 1999, Jesse Gelsinger was the first person that died in a clinical trial for gene therapy. He was 18 years old. Gelsinger suffered from ornithine transcarbamylase deficiency, a genetic disease of the liver. Gelsinger joined a clinical trial run by the University of Pennsylvania that aimed at developing a treatment for infants born with severe disease. On September 13, 1999, Gelsinger was injected with an adenoviral vector carrying a corrected gene to test the safety of the procedure. He died 4 days later, apparently suffering a massive immune response triggered by the use of the viral vector used to transport the gene into his cells, leading to multiple organ failure and brain death. An FDA investigation concluded that the researchers involved in the trial, broke several rules of conduct including (1) the inclusion of Gelsinger as a substitute for another volunteer who dropped out, despite having high ammonia levels that should have led to his exclusion from the trial, (2) failure to report that two patients had experienced serious side effects from the gene therapy, and (3) failure to disclose, in the informed-consent documentation, the deaths of monkeys who were given a similar treatment.

Administrative Failings at Duke University

In 1999, the U.S. government temporarily shut down federally funded research on humans at Duke University Medical Center, one of the nation's largest and most prestigious medical research facilities, after federal investigators determined that the university could not ensure the safety of participants. Problems cited included an oversight committee's failure to keep track of human studies after they began and a failure to document that special, federally mandated protections for children were in place.

The Ellen Roche Case

Ellen Roche was a 24-year-old technician from the Johns Hopkins Asthma and Allergy Centre who volunteered in 2001 to take part in a study designed to provoke a mild asthma attack in order to study the reflex that protects the lungs of healthy people against asthma attacks. After inhaling the hexamethonium, a medication used for treating high blood pressure in the 1950s and 1960s, Roche became ill and eventually died. The OHRP investigated the case and accused the Hopkins IRB of failing to take proper precautions. It also found that prior to approving the study, the researchers, and the IRB failed to uncover published literature about the toxic effects of inhaling hexamethonium, the study was not reviewed at a duly convened IRB meeting, and volunteers were not warned of the risks.

TABLE 1.12

The American Statistical Association's Ethical Guidelines for Statistical Practice

1. *Professionalism:* Competence, judgment, diligence, self-respect, and worthiness of the respect of other people.
2. *Responsibilities to funders, clients, and employers:* Responsibility for assuring that statistical work is suitable to the needs of those who are paying for it, that funders understand the capabilities and limitations of statistics in addressing their problem, and that the funder's confidential information is protected.
3. *Responsibilities in publications and testimony:* Report sufficient information to give readers a clear understanding of the intent of the work, how and by whom it was performed, and any limitations on its validity.
4. *Responsibilities to research subjects:* Responsibility for protecting the interests of human and animal subjects of research during data collection and analysis, interpretation, and publication of the resulting findings.
5. *Responsibilities to research team colleagues:* Mutual responsibilities of professionals participating in multidisciplinary research teams.
6. *Responsibilities to other statisticians or statistical practitioners:* Contribute to the strength of the profession overall by sharing nonproprietary data and methods, participating in peer review, and respecting differing professional opinions.
7. *Responsibilities regarding allegations of misconduct:* Addresses the process of investigating potential ethical violations and treating those involved with both justice and respect.
8. *Responsibilities of employers, including organizations, individuals, attorneys, or other clients employing statistical practitioners:* Encourages employers to recognize the highly interdependent nature of statistical ethics and statistical validity. Statistical practitioners should not be pressurized to produce a desirable result.

particularly when there are incentives for particular outcomes that could jeopardize objectivity and honesty. General guidelines for statisticians consist of maintaining professional knowledge, having regard for human rights, presenting findings objectively and honestly, avoiding deceptive or unsupported statements, disclosing conflicts of interest, being motivated by furthering knowledge and understanding, professional courtesy, open-mindedness, documentation of methods, and supporting fellow researchers.

The American Statistical Association has developed ethical guidelines for statistical research. The guidelines (http://www.amstat.org/about/ethical guidelines.cfm) consist of the following eight topic areas outlined in Table 1.12.

References

Califf RM, Zarin DA, Kramer JM, Sherman RE, Aberle LH, Tasneem A. 2012. Characteristics of clinical trials registered in ClinicalTrials.gov, 2007–2010. *JAMA* 307(17):1838–1847.

Challenge and Opportunity on the Critical Path to New Medicinal Products. March 2004. U.S. Department of Health and Human Services. Food and Drug Administration.

Dickerson K, Rennie D. 2012. The evolution of trail registries and their use to assess the clinical trial enterprise. *JAMA* 307(17):1861–1864.

Hochman JS, Lamas GA, Buller CE, Dzavik V, Reynolds HR, Abramsky SJ, Forman S et al. 2006. Coronary intervention for persistent occlusion after myocardial infarction. *NEJM* 355(23):2395–2407.

IOM (Institutes of Medicine). 2010. *Transforming Clinical Research in the United States: Challenges and Opportunities: Workshop Summary.* Washington, DC: The National Academies Press.

Krall R. 2009. US Clinical Research. *IOM Drug Forum Clinical Trials Workshop.* Washington, DC.

Tam VC, Tannock IF, Massey C, Rauw J, Kryzyzanowska MK. 2011. Compendium of unpublished phase III trials in oncology: Characteristics and impact on clinical practice. *J Clin Oncol* 29(23):3133–3139.

2

Product Development Process

A new pharmaceutical product can be broadly classified into one of the two categories—a synthetic chemical compound (i.e., drug), or a biologic. A drug is usually a new chemical entity synthesized by scientists, often working in a drug company (the sponsor). Examples of drugs include Lipitor® [Atorvastatin Calcium] for the treatment of high cholesterol and Prozac® [Fluoxetine] for the treatment of depression. A biologic can be a protein, a part of a protein, DNA, or a different form that is either extracted from tissues of a live body, or cultured (e.g., by a type of bacteria), that is discovered and developed by the sponsor (usually a biotechnology company). An example of a biologic is Epogen® [Epoetin Alfa Injection] to help increase red blood cells.

In the United States, approval by the Food and Drug Administration (FDA) is a key step in allowing the new pharmaceutical product to become available to physicians and the patient population. For new drugs, the FDA evaluation is based on the new drug application (NDA) while, for new biologics, evaluation is based on the biological license application (BLA). An NDA or a BLA is a large, comprehensive document that includes a description of the manufacturing process, as well as all the results from the nonclinical experiments and clinical trials. This document is developed by the sponsor and submitted to the FDA for approval.

Although many of the scientific concerns are different for the development of a drug and a biologic, the clinical development processes for them are very similar. Both must pass the investigational new drug (IND) review process before they can be tested in humans, and after the review, both usually need to go through phase I, phase II, and phase III clinical development. The development processes for drugs and biologics also require interactions with regulatory agencies such as the FDA (in the United States) including end-of-phase II meetings, and pre-NDA meetings (for drugs) or pre-BLA meetings (for biologics). These FDA–industry interactions are described in Chapter 3 and interactions with other regulatory agencies (e.g., Europe) are also covered in Chapter 3. The discussion in this chapter focuses on drug development since most of the pharmaceutical products are drugs. Development concerns specific to biologics will be mentioned when necessary.

This chapter is divided into three sections: the drug label (Section 2.1), nonclinical development (Section 2.2), and clinical development (Section 2.3).

2.1 The Drug Label

The drug label must be finalized after a new drug completes the premarketing development process but before it receives regulatory approval and is ready for general patient population use. A drug label contains important efficacy and safety information to assist physicians in prescribing the drug. Information contained on the drug label informs patients regarding appropriate drug use and potential adverse effects.

A drug label contains all the clinical, nonclinical, and manufacturing information about the drug. It provides detailed drug characteristics and describes how the drug should be dosed. A complete label usually includes sections with the following titles: description, clinical pharmacology, indications and usage, contraindications, warnings, precautions, adverse reactions, overdosage, dosage and administration, and methods of supply. The sponsor and the regulatory agency (e.g., FDA) agree upon the content of the label.

A draft of the drug label is frequently constructed early in the development stages and is used to guide the drug development process. The draft drug label specifies the anticipated and desired characteristics of the new drug. The quantitative information in a drug label is presented using descriptive and inferential statistics.

As an example, consider the drug label for Lipitor [Atorvastatin Calcium]. High LDL (low-density lipoprotein) cholesterol (i.e., the bad cholesterol) affects approximately 17% of Americans aged 20 and older, contributing to atherosclerotic cardiovascular disease (the single leading cause of death and disability in the developed world) and stroke. Lipitor has been shown to lower LDL cholesterol and triglycerides, and raise HCL cholesterol (i.e., the good cholesterol); thus reducing the risk of atherosclerotic cardiovascular disease and stroke. Selected sections of the Lipitor label (Lipitor Prescribing Information, 2005) are presented in Appendix I.

Statistical summaries are utilized in many places in the Lipitor label. For example, after describing the chemical structure of Lipitor, the label summarizes the clinical pharmacology including detailed pharmacokinetics and pharmacodynamics (PK/PD). These data are based on phase I PK/PD clinical trials. Design and analyses of PK/PD studies are discussed in Chapters 6 and 8, respectively.

Clinical data must support every claim on a drug label. In the case of Lipitor, the label claims "prevention of cardiovascular disease, hypercholesterolemia and mixed dyslipidemia, hypertriglyceridemia, dysbetalipoproteinemia (*Fredrickson* type III), homozygous familial hypercholesterolemia, and heterozygous familial hypercholesterolemia in pediatric patients." Data supporting this claim are summarized in Section A.3 in the Appendix.

The Lipitor label also contains a summary of adverse events that were observed in the clinical studies. The percent of participants experiencing various types of adverse events are summarized for placebo and each of the approved doses. We will discuss drug safety and benefit:risk considerations in Chapter 10.

2.2 Nonclinical Development

The drug development process (Ting, 2006) can be broadly classified into two major stages: (1) nonclinical development and (2) clinical development. We discuss nonclinical (also known as preclinical) development in this section and clinical development in the following Section 2.3.

Nonclinical development includes all testing performed outside the human body. The goal of nonclinical development is to identify promising drugs with desirable pharmacologic activity to take into clinical development. In nonclinical development, the focus is primarily on cells, tissues, organs, or animal bodies. Experiments are frequently performed in laboratories or pilot plants.

Throughout the drug development process, two important scientific questions are constantly being addressed: (1) Does the drug work? and (2) Is the drug safe? Beginning in the laboratory where the drug is first discovered, the drug endures many tests to assess its efficacy and safety. Only drugs that appear promising progress to the next stage of development. In the United States, after a promising drug passes all the nonclinical tests, an IND is filed with the FDA. Only after the IND is approved, can clinical development (i.e., clinical trials) proceed.

Three fundamental areas of nonclinical development are covered in this section: pharmacology (Section 2.2.1), toxicology (Section 2.2.2), and formulation (Section 2.2.3).

2.2.1 Pharmacology

Pharmacology is the study of the selective biological activity of a drug on a living matter. A drug has biological activities when, in appropriate doses, it causes a cellular response. It is selective when the response occurs in some cells and not in others. A drug has to demonstrate these activities before it can be further developed. In the early stage of drug testing, it is important to differentiate an "active" candidate (i.e., a drug with appropriate biological activity that may be desirable to develop further) from an "inactive" candidate. Screening procedures are used to identify active candidates. The sensitivity (i.e., the conditional probability that the screening instrument will identify a truly active candidate as active), specificity (i.e., the conditional probability that the screening instrument will identify a truly inactive candidate as inactive), positive predictive value (i.e., the conditional probability that a positive screening result is identifying a truly active candidate), and negative predictive value (i.e., the conditional probability that a negative screening result is identifying a truly inactive candidate) are used to help make decision; whether to progress with development of the drug candidates. Sensitivity and specificity are generally inversely related. However, if the screening instrument is very good then both sensitivity and specificity values can be high (i.e., close to 1).

Quantification of pharmacological activities may be viewed as a measure of the drug potency or strength. Drug potency can be measured using a bioassay, an experiment for estimating the potency of a drug by means of the reaction that follows its application to living matters (Finney, 1978).

One important relationship that must be studied is the dose–response relationship. In dose–response experiments, several doses (or various concentration levels) of the drug are selected, and the responses are measured for each corresponding dose. After response data are collected, statistical methods are applied to estimate the relationship between the dose and the response. If the pharmacological response is low and does not change with increases in dose, indicating that the test drug is not very potent, then the sponsor is less likely to proceed with future development. However, if the drug candidate is active, then the drug should demonstrate a clear increase in pharmacologic activity with increasing doses. Information regarding the dose response relationship can then be used to help guide the selection of doses used in clinical trials.

2.2.2 Toxicology/Drug Safety

Drug safety is a paramount concern throughout all stages of development. In the preclinical stage, drug safety must be studied for a few different species of animals (e.g., rabbits, pigs, mice, or other rodents). Animal studies are designed to examine the toxic or adverse drug effects when animals are treated with varying doses and during varying lengths of time to evaluate cumulative effects of the drug. If the results of animal studies indicate potentially serious side effects, then the drug development is often terminated or suspended pending further investigations.

A single-dose study is an experiment in which each animal is dosed only once with the drug. When more than one dose is given, the study is denoted as a multiple-dose study. Depending on the duration of exposure to the drug, animal toxicity studies are classified as acute studies, subchronic studies, chronic studies, and reproductive studies (Selwyn, 1988). Often the first few safety studies are acute studies (i.e., the animal is given one or more doses of the drug). Acute studies are typically approximately 2 week in duration (can be single-dose or multiple-dose studies). A single-dose study is an acute study. Single-dose acute studies in animals are primarily used to identify the doses to be considered in subchronic and chronic studies. Only drugs that appear to be safe in the single-dose studies progress into multiple-dose studies. Repeat dose studies of 30–90 days duration are called subchronic studies while chronic studies are usually designed with more than 90 days of duration. These studies are conducted in rodents and at least one non-rodent species. Some chronic studies are viewed as carcinogenicity studies since the rodent studies consider tumor incidence as an important endpoint. Reproductive studies are carried out to assess the drug's effect on fertility and conception, and are used to study the drug effect on the fetus and developing offspring.

Data collected from toxicology studies help to estimate the relationship between dose and toxicity. The information is often used to identify a "no observed adverse event level" (NOAEL) for the drug and provides insight into the types of adverse events that might be anticipated when the drug is used in humans. Data from the animal toxicity studies are then used to help design phase I and phase II clinical trials.

2.2.3 Drug Formulation Development

One of the primary differences between the development of a drug and a biologic is the way they are formulated and manufactured. If the therapeutic under development is a biologic, then the formulation is typically a solution that contains a high concentration of a biologic that is injected into the subject.

If the potential pharmaceutical product is a drug, then the formulation can be tablets, capsules, a solution, patches, a suspension, or another form. A drug is a mixture of the synthesized chemical compound (active ingredients) and other inactive ingredients designed to improve the absorption of the active ingredients. How the mixture is prepared depends on the results of a series of experiments. Often these experiments are performed under physical constraints (e.g., the amount of the supply of the raw materials, the capacity of the container, the size and shape of the tablets). In the early stage of drug development, the drug formulation must be flexible so that various doses can be tested in animals and humans. Often in the nonclinical development stage or in early phase of clinical trials, the drug is supplied in a powder form or as a solution to allow flexible dosing. However, after the drug progresses into late phase I or early phase II clinical trials, then fixed dosage forms such as tablets, capsules, or other formulations are desirable.

The dose strength depends on both nonclinical and clinical information. Frequently, a "drug formulation group" works in close collaboration with laboratory scientists, toxicologists, and clinical pharmacologists to determine the possible dose strengths for the drug. In many cases, the originally proposed dose strengths needs to be revised based on results of phase II studies. These formulations are developed for clinical trial usage and may be different from the commercial formulation. When the drug is approved for the market, then the commercial formulation should be readily available for distribution.

2.3 Clinical Development

The results of nonclinical studies for the drug will be summarized as part of an IND submission. An IND is a document that contains all the information known about the drug that was learned in nonclinical development. A typical IND includes the name and description of the drug (such as the chemical structure and other ingredients); how the drug is processed; preclinical safety information, marketing information; and future plans for investigating the drug in the United States and in foreign countries. It also contains a description of the clinical development plan (CDP, refer to Section 2.3.5). If there are no concerns from the FDA after 30 days from the date of the IND filing, then clinical development can proceed. The drug may then be referred to as the "test drug" or the "study drug."

Clinical development is distinguished from nonclinical development by the use of humans as the experimental unit. Clinical trials are designed to collect data from normal healthy volunteers and subjects with the target disease, in order to help understand how the drug acts upon the human body, how the human body acts on the drug, and how the drug helps patients with the target disease. Clinical development can be divided into phases I, II, III, and IV.

2.3.1 Phase I Clinical Trials

Phase I clinical trials are designed to study the short-term effects (e.g., PK, what does a human body do to the drug; and PD and what does a drug do to the human body). These studies also investigate the initial dose range for the drug and help determine how often the drug should be used (e.g., once/day vs. twice/day). Participants in phase I studies are often healthy (i.e., non-diseased) volunteers.

In phase I PK studies, the goal is to understand the PK properties and to estimate PK parameters (e.g., area under the curve [AUC], maximum concentration [Cmax], the time to maximum concentration [Tmax], described in the next paragraph) of the drug. In many cases, phase I trials are designed to study the bioavailability of a drug (i.e., the rate and extent to which the drug ingredient is absorbed and becomes available at the site of drug action, Chow and Liu, 1999) or the bioequivalence among different formulations of the same drug.

A bioavailability or a bioequivalence study is carried out by measuring the drug concentration levels in the blood or serum over time from participants. These measurements are summarized into one value per subject per treatment period. Data are collected at discrete time points. Figure 2.1 presents a drug concentration–time curve (for one subject after orally taking a single-dose of study drug). Typical variables used for analysis of PK activities include the AUC, Cmax, minimum concentration (Cmin), and the time

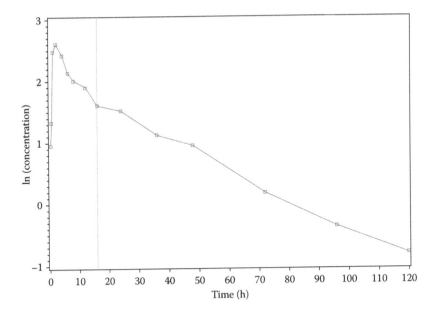

FIGURE 2.1
Example of a time–concentration curve.

to maximum concentration (Tmax), each computed from the drug concentration levels as shown in Figure 2.1.

2.3.2 Phase II/III Clinical Trials

Phase II/III trials are usually controlled studies utilizing parallel treatment groups designed to study the efficacy and safety of a test drug. Unlike phase I studies, subjects recruited into phase II/III studies have the targeted disease for which the drug is being developed. Both efficacy and safety measures are evaluated. For example, in a trial for the evaluation of hypertension (high blood pressure), the efficacy endpoints are blood pressure measurements. For an anti-infective trial, the endpoint may be "cure" or the "time to cure."

Phase II trials are often designed to compare one or more doses of a test drug against placebo in a well-defined population. The objective of phase II studies is generally estimation oriented (e.g., to estimate the appropriate doses to be used for phase III trials). Phase II trials are usually short term (e.g., several weeks) and designed with a small or moderate sample size. Subjects who are recruited for phase II trials are often homogeneous and somewhat restricted (i.e., they tend to have a disease severity that is not too severe and not too mild, are free from other diseases, and are not taking other medications).

One important type of phase II study is the dose–response study. Drug efficacy is often anticipated to increase as dose increases. A phase II dose–response study addresses the following fundamental questions (Ruberg, 1995):

1. Is there evidence of a drug effect?
2. How doses exhibit a response different from the control response?
3. What is the nature of the dose–response relationship?
4. What is the optimal dose?

Typical dose–response studies are designed with fixed doses and parallel treatment groups. For example, in a four-treatment group trial designed to study dose–response relationships, three test doses (low, medium, and high) are compared to placebo. Analyses of phase II dose–response studies are discussed in Chapter 8.

The development process for interventions to treat life-threatening diseases such as cancer are, in general, different from interventions to treat chronic diseases. For example, consider chemotherapy interventions for cancer that can be very toxic. It may not be ethical to enroll healthy participants for phase I cancer trials. Typically phase I oncology trials recruit cancer patients and escalate doses of study drug to help estimate maximal tolerable dose (MTD). Phase II oncology trials are designed as proof-of-concept (PoC) studies.

Phase III trials are generally randomized, long term (can last up to a few years), large scale (several hundreds of participants), conducted in a heterogeneous population but in subjects with the disease in question, blinded when possible, and controlled (comparative against a known active control or placebo). Phase III trials tend to be confirmatory trials designed to verify findings established from earlier studies.

Although dose recommendations are primarily made during phase II, there may be safety concerns with exposure for a longer period. For example, cumulative drug exposure over time may cause adverse events. Thus, the dose selection may be further refined in phase III trials because the drug exposure is long term and within a large patient population. Thus, it is good practice to consider incorporating more than one dose in phase III trials. It may be prudent to have a dose higher than the target dose so that in case the target dose is not efficacious; a higher dose can be considered as an alternative. It may also be prudent to have a dose that is lower than the target dose, so that if the target dose is not safe, then a lower dose can be considered as an alternative.

When a clinical trial is completed, all of the collected data are stored in a database. Statistical analyses are performed on data extracted from the database, and a study report is prepared for each completed clinical trial. The preparation of the tables, listings, and graphs for the report is a joint effort between statisticians, data managers, and programmers. Statisticians, clinicians, and technical/medical writers then develop the clinical interpretations from analyses. Study reports from individual clinical trials eventually become part of the NDA.

It is important to note, however, that the development is not always a linear process. For example, a phase I trial for a capsule may be in progress while a phase II trial is being conducted for the same intervention with injection as the method of delivery. Lessons learned from the clinical studies can be circulated back to the discovery and preclinical experts. Clinical data can inform reformulation of the intervention.

2.3.3 New Drug Application

If the drug is shown to be safe and efficacious through phase I, II, and III clinical trials, then the sponsor files an NDA (or a BLA for a biologic) to the FDA. An NDA is a document summarizing the results obtained from nonclinical experiments and clinical trials. An NDA typically contains sections on the proposed drug label, the pharmacological class, foreign marketing history, chemistry, manufacturing and controls, nonclinical pharmacology and toxicology, human PK and bioavailability, microbiology, clinical trial data, results of the statistical analyses, and the benefit:risk relationship. If the sponsor intends to market the new drug in other countries, then documents must be prepared for submission to those corresponding countries. For example, a new drug submission (NDS) must be filed with Canadian regulatory agency and a marketing authorization application (MAA) must be filed with the European regulatory agencies (EMEA—European Medicines Agency). Recently, an effort to harmonize the submissions has led to a common technical document (CTD). The CTD can be submitted to multiple countries and regulatory agencies.

The NDA frequently includes individual clinical study reports and combined study results. These results may be summarized using meta-analyses or pooled data analyses across studies. Such analyses are performed on efficacy data to produce a summary of clinical efficacy (SCE, also known as the integrated analysis of efficacy [IAE]) and on safety data to produce a summary of clinical safety (SCS, also known as the integrated analysis of safety [IAS]).

Increasingly, electronic submissions are filed as a part of the NDA. Electronic submissions usually include individual subject data, programs to process the data, and software/hardware to assist the regulatory agency in reviewing the data.

Often, an NDA is filed while phase III studies are ongoing. Sponsors carefully select the "data cut-off date" (a date by which all the data in the database is "frozen" and stored). This occurs so that the data summaries that are presented as a part of the NDA can be replicated. These data (often stored in an "NDA database") may have to be retrieved and reanalyzed after filing in response to queries from the regulatory agency.

The drug can only be made available for general public use in the United States if the NDA (or BLA) is approved by the FDA.

2.3.4 Accelerated Approval and Unique Clinical Development Methods

In some cases, drugs for serious or life-threatening diseases can be approved before large-scale phase III studies are completed to meet public need (e.g., due to a lack of alternative treatments). The FDA has implemented a mechanism (Accelerated Approval Products Submission of Promotional Materials—Subpart H, 1999) to speed approval of promising therapies for serious and life-threatening diseases. Under the accelerated approval regulations established in 1991, the drug effectiveness can be assessed in a preliminary manner using laboratory endpoints as a surrogate for clinical endpoints. Substantial evidence from trials using the surrogate endpoint is necessary, and then the clinical endpoints are examined after (often provisional) marketing approval.

For example, the FDA has generally considered an effect on survival or relief of patient symptoms as evidence of clinical benefit in oncology. However, objective tumor response rates and time-to-progression have been viewed as surrogate endpoints that are reasonably likely to predict clinical benefit. Accelerated approval of oncologic drugs has been based upon the demonstration of tumor size reduction in patients with refractory disease or in patients whose disease had no effective therapy. Postapproval confirmatory studies of products approved on the basis of tumor shrinkage are often required to further define the utility of the new drug for the approved indication. For accelerated approval of drugs that ameliorate treatment-associated toxicities, postapproval studies may be required to examine the effect of the therapy on survival and to demonstrate that the surrogate measures correspond to clinical benefit. In 1996, the FDA approved Gemzar® (gemcitabine for injection) to treat pancreatic cancer. At the time, data supported the claim that Gemzar® improved disease-related symptoms such as pain, improved the performance of daily activities, and resulted in beneficial weight change. Later research demonstrated that Gemzar prolonged survival of people with pancreatic cancer.

Accelerated approval has been used regularly in the evaluation of anti-HIV drugs. Two large randomized studies are usually required for standard evaluations. The surrogate endpoint for accelerated approval is HIV RNA viral load assessed at 24 weeks. Regular approval is granted based on the effect on viral load at 48 weeks. The same study thus provides support for accelerated and regular approval. It should be noted that the use of an analogous laboratory surrogate to support accelerated approval as described for HIV has not yet occurred in the setting of oncology drug approvals.

Fast Track is a formal mechanism to interact with the FDA that is described in the Food and Drug Administration Modernization Act of 1997 (FDAMA). The benefits of Fast Track include the scheduling of meetings to seek FDA input into development plans, the option of submitting an NDA in sections rather than all components simultaneously, and the option of requesting the evaluation of studies using surrogate endpoints (i.e., accelerated approval).

The Fast Track designation is intended for the combination of a product and a claim that addresses an unmet medical need.

Under the FDAMA, reviews for NDAs are designated as either *standard* or *priority*. A *standard* designation implies that the FDA must complete the review of the application and take action (approve on not approve) in less than 10 months after the filing date. A *priority* review designation sets the target date for the FDA action at 6 months after the filing date.

Expanded access programs are designed to ensure that people with serious and life-threatening diseases that lack satisfactory therapies can have access to promising therapeutic agents even before the completion of controlled clinical trials.

An *orphan drug* refers to a product that treats a rare disease affecting fewer than 200,000 Americans. For example, drugs developed to treat Huntington's disease, myoclonus, amyotrophic lateral sclerosis (ALS—Lou Gehrig's disease), Tourette's syndrome, and muscular dystrophy are considered as disease targets for the development of orphan drugs. Adequate drugs for many such diseases have not been developed, in part because development of these drugs is not profitable. The intent of the Orphan Drug Act (signed into law on January 4, 1983) is to stimulate the research, development, and approval of products that treat rare diseases. This mission is accomplished by offering incentives for orphan drug development including granting 7 years of marketing exclusivity after approval of its orphan drug product, providing tax incentives for orphan drug clinical research, offering grant funding, and offering waivers of drug approval application fees and annual product fees. Since the Orphan Drug Act passed, over 100 orphan drugs and biological products have been brought to market. Examples include Colazal® (Balsalazide) for the treatment of pediatric patients with ulcerative colitis and CroFab® (Crotalidae) for the treatment of envenomations inflicted by North America crotalid snakes.

2.3.5 Clinical Development Plan

In the early stages of drug development, a CDP is drafted, often before IND submission. A draft drug label is prepared and used to guide the clinical development process. The draft drug label specifies the anticipated and desired characteristics of the drug, thus describing the target profile of the drug.

The CDP includes a description of the phase I, II, and III clinical trials that are to be conducted to support the drug label, along with the associated timeline for these trials. The CDP describes the strategy for identifying the recommend dosage. For example, if the drug will be used with a single fixed dose, then the CDP will describe the clinical studies that will identify the dose. Similarly, the CDP describes studies to determine the dosing frequency. For example, patients with chronic diseases may take multiple medications per day, but prefer and tend to comply with a once-a-day (QD)

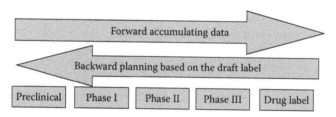

FIGURE 2.2
Clinical development plan—Thinking process.

dosing. If the preliminary data for the test drug indicate that it will have to be used three times a day, then the CDP may include studies to reformulate the test drug so that it can be used QD, before it progresses into later phase development.

As a drug progresses through the clinical development process, the CDP will be updated to reflect the most current information about the drug. The sponsor will also assess whether the draft drug label should be revised. If a new draft of the drug label is prepared, then the CDP should be revised so that studies can be designed to support the new drug label.

The CDP is experienced prospectively but planned retrospectively (Figure 2.2). The scientific process is prospective. As data accumulate, more knowledge is obtained about the drug, and later phase studies can be designed. However, the planning is based on the draft drug label and a target profile for the drug. Depending on the desired claims that the sponsor wishes to make about the drug, the sponsor must design phase III studies to support those claims. In order to design the phase III studies, specific data must be obtained from corresponding phase II or phase I studies.

2.3.6 Postmarketing Development

The development process does not cease when an NDA is submitted or approved. Often the approved drug is continually studied to gather additional safety and efficacy data. Studies performed after the drug is approved are typically called postmarketing studies or phase IV studies. Many of the phase IV studies are required by the FDA to examine safety issues associated with long-term use of the drug. Other phase IV studies are used to establish new uses of the drug. For example, the drug may be studied in a different population or for a different indication.

One of the major objectives in postmarketing development is to establish a more complete safety profile for the new drug. Large-scale drug safety surveillance studies are very common in phase IV. Subjects recruited into phases I, II, and III are "selected" in the sense that subjects have to be within a certain age range, have a specified disease severity and have other entry criteria restrictions. However, after drug approval, all patients with the disease can be exposed to the drug. Using postmarketing studies, drug safety data can be obtained from the broader general patient population in this setting in which the drug is used in practice.

Since participants who are recruited into phases I–III trials may not be representative of the real-world patient population given eligibility constraints, volunteer status, and proximity to enrolling sites, phase IV postmarketing trials may be designed to enroll a broader patient population including participants from special populations (e.g., pregnant women, diabetics, elderly) that would have been ineligible for earlier phase trials. In some situations, a registry may be created to follow-up patients to study the "real world experience" as to how the drug is actually used after regulatory approval. However, interpretation of registry data can be challenging given nonrandomized assignment, selective reporting, and nonstandardized follow-up and evaluation.

Another objective of a phase IV study is to demonstrate an improvement in patients' quality of life (QoL) and establish its economic value. Studies designed to achieve these objectives include QoL studies and pharmacoeconomic studies. Studies of this nature are often referred to as "outcomes research" studies. Results obtained from outcomes research studies can be used by the sponsor to promote the new drug if the drug displays a superior QoL profile relative to alternative treatments.

Pharmacoeconomic studies are designed to study the direct and indirect costs of treating a disease. In these studies, costs of various alternative therapies are compared. Costs may include the price of the medication, expenses for monitoring the patient (physician costs, costs of lab tests, etc.), costs for treating side effects caused by a treatment, and hospital charges. Sponsors comparatively evaluate costs of the treatment options to identify the most economically efficient treatment.

A drug that has been developed for a specific disease may also be useful for treating other diseases. However, the sponsor may not have sufficient resources to develop the drug simultaneously for multiple indications. Thus, the sponsor may develop the drug for a specific disease, obtain regulatory approval, and then develop the drug to treat other diseases. Postmarketing studies are commonly used to assess the efficacy and safety of the drug for treating new indications (symptoms or diseases). This is also known as *expanding the label* (or *line extension*).

Approval by the FDA for a new indication can be based on a supplemental NDA (sNDA). An sNDA is typically submitted after an initial NDA has been approved for its primary indication. Approval for the primary indication implies that the drug is efficacious for the approved indication, and importantly, that the drug is relatively safe for patient use. Since the safety has been established, the focus of the sNDA is generally on efficacy for the new indication. sNDAs are typically used to support new indications, for establishing a new formulations, new regimens, or new routes of application.

Regulatory agencies and sponsors continually monitor all approved drugs. For each drug, an initial safety database is constructed based on its initial

NDA. This database will include all the adverse events data, laboratory data, and other safety data collected from the phases I, II, and III clinical studies. After approval, additional safety data will be collected from ongoing or new clinical trials. All safety data are combined and monitored to construct a comprehensive safety profile of the drug. In the evaluation of an sNDA for a new indication, the regulatory agencies will assess the overall safety profile of the drug. However, the evaluation of drug efficacy for the sNDA is based only on data from the sNDA studies.

Occasionally in postmarketing studies, efficacy is observed at a lower dose than the dosage that is recommended on the drug label. Lower doses tend to provide a better safety profile than higher doses. A drug label could then be revised to include the lower dose as one of the recommended doses. Conversely, the recommended dose may work for some patients, but is not high enough for other patients. In these cases, a dosing increase may be necessary, and the sponsor may negotiate with the regulatory agency regarding modification of the drug label to allow a higher dose to be prescribed.

References

Accelerated Approval Products Submission of Promotional Materials—Subpart H. 1999. http://www.fda.gov/cder/guidance/2197dft.pdf.

Chow SC, Liu JP. 1999. *Design and Analysis of Bioavailability and Bioequvalence Studies.* New York: Marcel Dekker, Inc.

Finney DJ. 1978. *Statistical Methods in Biological Assay*, 3rd edn. London: Charles Griffin.

Lipitor Prescribing Information. 2005. http://www.lipitor.com/content/LipitorPI.pdf.

Ruberg SJ. 1995. Dose–response studies I. Some design considerations. *J Biopharm Stat* 5(1):1–14.

Selwyn MR. 1988. Preclinical safety assessment. In: *Biopharmaceutical Statistics for Drug Development*, Peace KE (ed.). New York: Marcel Dekker, Inc.

Ting N (ed.). 2006. *Introduction and New Drug Development Process, Dose Finding in Drug Development.* New York: Springer.

3

Regulatory Review Organizations

Pharmaceutical Research and Development (R&D) is a highly regulated industry. In order to allow for any new drug, biologic, or device to be approved for public use, the regulatory agency must go through an intensive scientific review and regulatory inspection. In the United States, the Food and Drug Administration (FDA) is the regulatory agency. Nearly 25 cents of every dollar spent by Americans are on products that are regulated by the FDA. Other regulatory agencies throughout the world include the Therapeutic Products Directorate (TPD) in Canada, the European Medicines Agency (EMA) in the European Union, and the Pharmaceutical and Medical Devices Agency (PMDA) in Japan. Among these agencies, the FDA was the first to require evidence of efficacy obtained from clinical trials. Thus, it is of interest to review a brief history of FDA.

3.1 Food and Drug Administration

In 1906, the Food and Drugs Act was passed in order to protect against misbranding and adulteration of foods and drugs. In 1912, the Sherley Amendment was established to prohibit labeling medicines with false and fraudulent claims. In 1930, the FDA was formed to ensure that (1) food is safe and wholesome; (2) drugs, biological products, and medical devices are safe and effective; (3) cosmetics are unadulterated; (4) the use of radiological products does not result in unnecessary exposure to radiation; and (5) all these products are honestly and informatively labeled (Fairweather, 1994 and Chow and Liu, 2004). In 1962, the Kefauver-Harris Drug Amendment was passed. This amendment not only strengthened the safety requirements for new drugs but also established an efficacy requirement for new drugs for the first time. Today, the mission of the FDA is to protect the public by assuring the safety, efficacy, and security of drugs, biologics, and devices. The FDA is also responsible for advancing public health by helping to speed innovations that make medicines more effective, safer, more affordable, as well as help the public to get accurate, science-based information to use medical interventions appropriately and improve their health (www.fda.gov).

The FDA currently has seven centers (Figure 3.1): the Center for Drug Evaluation and Research (CDER), the Center for Biologics Evaluation and

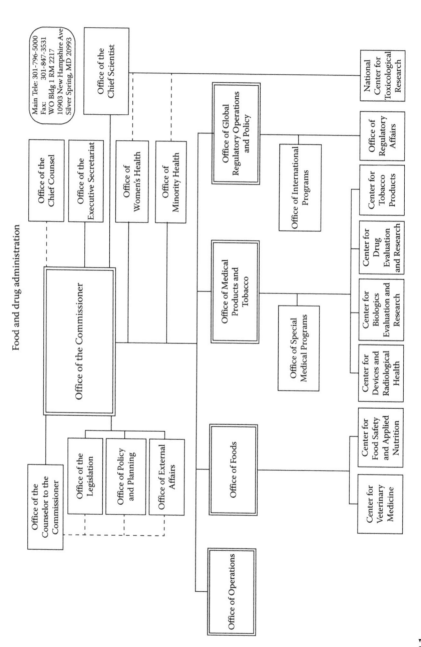

FIGURE 3.1
FDA organizational chart.

Research (CBER), the Center for Devices and Radiological Health (CDRH), the Center for Veterinary Medicine, the Center for Tobacco Products, the Center for Safety and Applied Nutrition, and the Center for Toxicological Research. The CDER, CDRH, and CBER are the centers that are most relevant to clinical trials as they review data for drug, device, and biological interventions, respectively.

3.1.1 Drugs

Chapter 2 outlined the development process for drugs in the United States and the role of the FDA in this process. When phase III trials are completed, an NDA is assembled and submitted to the FDA (CDER) for review. An NDA is a large document that includes many key components. Historically it took the FDA a very long time to review an NDA. In 1992, the U.S. Congress passed the Prescription Drug User Fee Act (PDUFA, http://www.fda.gov/forindustry/userfees/prescriptiondruguserfee/default.htm) to help decrease review time. PDUFA authorizes the FDA to utilize the "user fee" (paid by the pharmaceutical industry when submitting NDA) to hire additional staff to help speed up the review process. From 1992 to 1997, PDUFA helped FDA to reduce the average time required for review of a new drug application from 30 to 15 months. In 1997, the U.S. Congress passed the FDA Act (FDAMA, www.fda.gov/RegulatoryInformation/Legislation/FederalFoodDrugandCosmeticActFDCAct/Significant AmendmentstotheFDCAct/FDAMA/FullTextofFDAMAlaw/default. htm) to enhance the FDA's mission and its operations to deal with challenges in the twenty-first century. In 2004, the FDA launched a Critical Path Initiative (http://www.fda.gov/ScienceResearch/SpecialTopics/Critical PathInitiative/default.htm), a document that outlines the scientific reasons for the decrease in the number of innovative medical products submitted for approval, a puzzling trend in light of advances in biomedicine and disappointing from a public health perspective. Important events in the FDA history are summarized in Table 3.1.

3.1.2 Biologics

Biologics include blood and blood products, vaccines, allergenic products, and therapeutic products such as cytokines, monoclonal antibodies, and cellular or gene therapies. The Code of Federal Regulations (CFR) defines a biologic as "any virus, therapeutic serum, toxin, antitoxin, or analogous product applicable to the prevention, treatment, and cure of diseases or injuries of man." Biologics are derived from biological sources and present special safety concerns related to the immune system and inflammatory response. The biotechnology industry is generally regarded as the most likely source for innovative products.

The regulatory process begins with an investigational new drug (IND) filed with CBER after preclinical testing suggests activity in animals. Phase I–III

TABLE 3.1

Important Events in Recent FDA History (Relevant to Product Development)

Year	Event(s)
1906	The original *Food and Drugs Act* is passed by Congress on June 30 and signed by President Theodore Roosevelt. It prohibits interstate commerce in misbranded and adulterated foods, drinks, and drugs.
1912	Congress enacts the *Sherley Amendment* to overcome the ruling in *U.S. v. Johnson*. It prohibits labeling medicines with false therapeutic claims intended to defraud the purchaser, a standard difficult to prove.
1930	The name of the Food, Drug, and Insecticide Administration is shortened to *Food and Drug Administration (FDA)* under an agricultural appropriations act.
1962	*Kefauver-Harris Drug Amendments* passed to ensure drug efficacy and greater drug safety. For the first time, drug manufacturers are required to prove to FDA the effectiveness of their products before marketing them. The new law also exempts from the Delaney proviso animal drugs and animal feed additives shown to induce cancer but which leave no detectable levels of residue in the human food supply.
1976	*Medical Device Amendments* passed to ensure safety and effectiveness of medical devices, including diagnostic products. The amendments require manufacturers to register with FDA and follow quality control procedures. Some products must have premarket approval by FDA; others must meet performance standards before marketing.
1983	*Orphan Drug Act* passed, enabling FDA to promote research and marketing of drugs needed for treating rare diseases.
1984	*Drug Price Competition and Patent Term Restoration Act* expedites the availability of less costly generic drugs by permitting FDA to approve applications to market generic versions of brand-name drugs without repeating the research done to prove them safe and effective. At the same time, the brand-name companies can apply for up to 5 years additional patent protection for the new medicines they developed to make up for time lost while their products were going through FDA's approval process.
1988	*Food and Drug Administration Act* of 1988 officially establishes FDA as an agency of the Department of Health and Human Services with a Commissioner of Food and Drugs appointed by the President with the advice and consent of the Senate, and broadly spells out the responsibilities of the Secretary and the Commissioner of research, enforcement, education, and information.
1991	Regulations published to *Accelerate the Review of Drugs* for life-threatening diseases.
1992	*Prescription Drug User Fee Act* requires drug and biologics manufacturers to pay fees for product applications and supplements, and other services. The act also requires FDA to use these funds to hire more reviewers to assess applications.
1993	Revising a policy from 1977 that excluded women of childbearing potential from early drug studies, FDA issues guidelines calling for improved assessments of *medication responses as a function of gender*. Companies are encouraged to include patients of both sexes in their investigations of drugs and to analyze any gender-specific phenomena.
1997	*Food and Drug Administration Modernization Act* reauthorizes the Prescription Drug User Fee Act of 1992 and mandates the most wide-ranging reforms in agency practices since 1938. Provisions include measures to accelerate review of devices, regulate advertising of unapproved uses of approved drugs and devices, and regulate health claims for foods.

(Continued)

TABLE 3.1 (*Continued*)

Important Events in Recent FDA History (Relevant to Product Development)

Year	Event(s)
1998	FDA promulgates the *Pediatric Rule*, a regulation that requires manufacturers of selected new and extant drug and biological products to conduct studies to assess their safety and efficacy in children.
1999	*ClinicalTrials.gov* is founded to provide the public with updated information on enrollment in federally and privately supported clinical research, thereby expanding patient access to studies of promising therapies.
2002	Under the *Medical Device User Fee and Modernization Act*, fees are assessed sponsors of medical device applications for evaluation, provisions are established for device establishment inspections by accredited third parties, and new requirements emerge for reprocessed single-use devices.
2005	Formation of the *Drug Safety Board* is announced, consisting of FDA staff and representatives from the National Institutes of Health and the Veterans Administration. The Board advises the Director, Center for Drug Evaluation and Research, FDA, on drug safety issues and work with the agency in communicating safety information to health professionals and patients.

trials are conducted similar to drug development. A biological license application (BLA) is then submitted to market the product. The BLA must demonstrate that the biologic meets standards of safety, purity, potency, and consistency of manufacture in addition to demonstrating efficacy.

The concept of a generic in biologics is called a *biosimilar* and is an especially challenging area due to the inherent variability in biologic products. The manufacturing process is a fundamental characteristic of the biologic, and thus biologics that are manufactured differently are considered to be distinct. The process of developing a biosimilar is not well developed but is a rapidly evolving area.

3.1.3 Devices

A simple definition of a medical device is that it is a medical item that is not a drug or biologic. There is a very broad range of products that fall under the definition of a medical device and, as a result, there are many more devices than drugs. Devices can be therapeutic, diagnostic, or something else. The mechanism of action of a device is usually physical. Examples include wheelchairs; breast implants; tongue depressors; thermometers; apnea monitors; blood pressure machines; lasers for eye surgery; hearing aids; artificial hips and knees; stents; latex gloves; condoms; *in vitro* diagnostics devices that test samples of blood, sputum, urine for pregnancy, human papilloma virus, strep or influenza; and diagnostic imaging such as computed tomography (CT) scans, positron emission tomography (PET), and magnetic resonance imaging (MRI). There are also products that are device–drug or device–biologic combinations such as drug-eluding coronary stents. In excess of

8000 new medical devices are introduced to the United States market each year with approximately 4% of Americans having at least one implanted medical device.

Regulators classify devices based on their potential risks and design complexity. Each regulatory body has their classification system. The FDA defines three classes of medical devices: class I (general controls), class II (special controls), and class III (premarket approval). As the potential risks increase, the classification level increases indicating a higher level of regulatory control and evaluation. A summary of the FDA device classification and examples are provided in Table 3.2.

The development of devices is not easily categorized into phases. There is also usually no preclinical analog for devices. A single "confirmatory device study" is often sufficient for devices. "Condition-of-approval" or other postmarketing studies may be required.

The IND analog for devices is the investigational device exemption (IDE) for "significant risk devices" (i.e., a device that presents a potential serious risk to the health, safety, and welfare of a subject, and is (1) an implant, (2) used in supporting or sustaining life, or (3) of substantial importance in

TABLE 3.2

FDA Device Classification and Examples

Class	Definition	Examples
I	Devices for which *general controls* are sufficient to provide reasonable assurance of the safety and effectiveness. Class I devices typically do not require FDA premarket review prior to being marketed. *General controls* include prohibition against adulterated or misbranded devices, good manufacturing practices, registration of manufacturing facilities, listing of device types, recordkeeping, and so forth.	Adhesive bandages, stethoscope, exam light, crutches, patient scale, tongue depressors, arm slings, and examination gloves.
II	Devices that cannot be classified as class I because general controls are insufficient to provide a reasonable assurance of safety and effectiveness, and there is sufficient information to establish *special controls* to provide such assurance. Class II devices typically require premarket notification to FDA prior to being marketed. *Special controls* include performance standards, postmarket surveillance, patient registries, development and dissemination of guidance documents, and so on.	X-ray systems, surgical drapes, pumps, echocardiograph, ventilator, syringes, powered wheelchairs, surgical sutures, CT machines, and hemodialysis system.
III	Devices that cannot be classified as class II because general and special controls are insufficient to provide a reasonable assurance of safety and effectiveness and the devices are (1) life-sustaining or supporting, (2) of substantial importance in preventing impairment of human health, or (3) present potential or unreasonable risk of illness or injury. Class III devices require premarket approval prior to being marketed.	Replacement heart valves, implantable pacemakers, HIV diagnostic tests, contraceptive intrauterine devices, extended wear soft contact lenses.

diagnosing, curing, mitigating, treating disease, or preventing impairment of human health). Many diagnostic devices can be exempt from an IDE provided that the testing is noninvasive, does not pose significant risk, does not induce energy in the subject, and is not used as a diagnostic without confirmation by another established diagnostic.

The NDA analog for medical devices is the premarket approval application (PMA) that is required for class III devices. The FDA directive for PMAs for medical devices is to "rely upon valid scientific evidence to determine whether there is a reasonable assurance that the device is safe and effective." The source of this evidence may be well-controlled studies but can also be from other types of studies.

A premarket notification (also called a 510(k)) is also available for non-class III devices. This provision allows marketing of a device that is "substantially equivalent" to an existing device. To establish substantial equivalence, the experimental device does not need to be as effective as the predicate device, if the clinical data demonstrated that any reduction in effectiveness was offset by an improvement in patient safety/risk. The 510 process has come under recent scrutiny (e.g., a report by the Government Accountability Office) because some devices (e.g., hip replacement devices, external defibrillators) have received approval with no or limited trial data. In many cases, safety signals have been seen postapproval resulting in several device recalls.

3.1.4 FDA–Industry Interactions

Intervention development starts with nonclinical, and then moves to clinical phases. During the development process, there are important meetings between the FDA and the sponsor. For illustration, we describe the FDA–industry interactions in drug/biologic development.

There are three important meetings between the sponsor and the FDA. The first potential meeting is the pre-IND meeting occurring prior to the IND submission. The next key meeting is the end-of-phase II meeting occurring after phase II clinical data are available but prior to launching phase III trials. The third meeting is pre-NDA/BLA meeting occurring after phase III trials are completed but prior to NDA/BLA submission. If the phase III data support the safety and efficacy of the intervention, then the sponsor submits the NDA or BLA to the FDA for review. Of course, the FDA–industry interactions prior to the NDA/BLA submission are not limited to these three meetings. Other meetings can also be scheduled on an as-needed basis. The format of FDA–industry interactions can be email, fax, phone conversation, telecoms, or face-to-face meetings. In order to help facilitate a face-to-face meeting, the sponsor submits a briefing document to the FDA a few weeks before the meeting and sometimes the sponsor also submits their presentation slides prior to the meeting. We discuss each of the three primary meetings in more detail.

3.1.4.1 Pre-IND Meeting

The pre-IND meeting is not necessary to every IND. Prior to IND submission, the sponsor may work with the FDA to discuss the product concept. Discussions of development strategy may include whether the animal findings can translate to humans, how the preclinical findings can help in clinical development, and what clinical studies will be necessary to bring this product forward. During the pre-IND meeting, statisticians from both the sponsor and the FDA may discuss any potential statistical issues relating to the development (e.g., clinical endpoints, study objectives).

An IND submission is based on both scientific and strategic considerations. The sponsor needs to have a proper understanding of the preclinical data from *in vivo, in vitro* studies, and is willing to take the potential risks of clinical development as the intervention may not deliver the expected efficacy or it may carry certain safety concerns. Meanwhile, the sponsor should also consider the timing of development, other interventions under development, resource availability, and alternative treatments. Before submission, the sponsor evaluates whether it is "reasonably safe to proceed." After the IND is submitted, clinical protocols are prepared and reviewed by institutional review boards (IRBs). Once protocols are approved by IRBs, studies can be initiated, and subjects can be recruited into trials. Figure 3.2 outlines the FDA's IND review process.

3.1.4.2 End of Phase II Meeting

As described in Chapter 2, a clinical development plan (CDP) should be drafted around the time of IND submission. Early-phase (phase I and II) clinical trials are designed according to this CDP. After the clinical data are

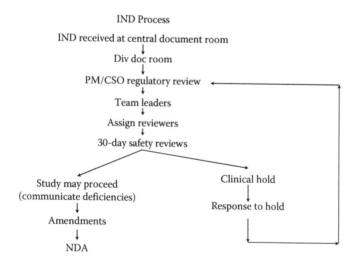

FIGURE 3.2
IND process.

available from phase I and II trials, the sponsor, and the FDA prepare for the end of phase II meeting. End of phase II meetings are common to new chemical (or biological) entities, and sometimes are set up for new uses of marketed interventions. Typically, an end of phase II meeting is a face-to-face meeting. This meeting is important because this is the time when the FDA reviews both preclinical and clinical findings and provides advice to the sponsor regarding phase III development. In the development of some interventions, the time between submission of the IND and end of phase II can be very long (e.g., several years). During this period, the sponsor or the FDA can request additional meetings. Often the FDA–industry interactions are initiative based on requests from the sponsor, seeking guidance from the FDA on the development program. The end of phase II meeting is also critical because

1. The sponsor will be making a major commitment to invest on large, expensive phase III trials.
2. Both the FDA and the sponsor must evaluate whether the phase II data provide a high level of confidence that continued development will eventually result in a successful intervention.
3. Phase III design features will be outlined (e.g., the primary endpoint definition, the selected dose(s), and the noninferiority margin).
4. Agreement must be secured regarding whether the amount of intervention exposure (e.g., number of patients, length of study, and doses) are sufficient to evaluate safety.
5. Safety concerns are outlined in a plan to capture appropriate data to evaluate these concerns.

Typically, after the end of phase II meeting, the sponsor may engage in phase III development, may run additional phase I/II studies to collect more data, or may decide to stop further development.

3.1.4.3 Pre-NDA/BLA Meeting

When phase III results are available, the sponsor requests a pre-NDA meeting for a new drug or a pre-BLA meeting for a biologics. This meeting typically takes place about 2–12 months prior to the actual submission and is a face-to-face meeting. In preparing this meeting, the sponsor submits a briefing document summarizing the efficacy and safety findings from phase III studies, and this document may also include phase I and phase II study results. In this meeting, the sponsor and the FDA discuss whether the evidence of efficacy can be established, the need for risk management and any plans to address potential issues. Other agenda items in this meeting may also include statistical issues such as missing data, multiple comparison adjustment, or issues relating to data standards for electronic submission of analysis data sets. The NDA or BLA submission occurs after the pre-NDA/BLA meeting. Figure 3.3 outlines the NDA/BLA review timeline.

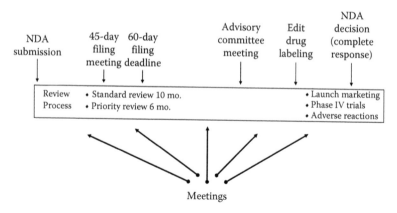

FIGURE 3.3
NDA/BLA review timeline.

3.1.4.4 Advisory Committee Meetings

The Federal Advisory Committee Act (FACA, Public Law 92-463) provides the legal foundation for federal agencies to call for advisory committees (ACs). An AC meeting may be called to inform FDA decisions regarding a particular challenging submission under review or if there is an important clinical issue being discussed (e.g., what an appropriate endpoint is for trials within a particular disease area). An AC offers independent advice on scientific issues to help inform decision making for the FDA. Most AC meetings are organized to discuss a challenging submission under review. Most AC meetings are open to public and provide a forum for public input.

The FDA currently employs many ACs delineated by FDA Center and disease area. There are 17 from CDER, 5 for CBER, and 1 medical devices AC that has 18 panels (http://www.fda.gov/advisorycommittees/default.htm). AC members are appointed by the FDA commissioner and serve for fixed terms. ACs typically consist of scientific or technical experts (clinicians and statisticians), a nonvoting consumer representative, and a nonvoting industry representative with particular experience in the relevant disease area. It is important to AC members to be objective and free from conflicts of interest so that objective recommendations can be made. AC members should be knowledgeable in the disease area of interest, the types of trials being discussed, and should not be shy about discussing important and challenging issues in public.

An AC meeting is typically a 1-day meeting and involves several parties including the sponsor, the FDA, the AC, and sometimes the public. An example of an agenda for the FDA AC meeting is provided in Table 3.3. AC members are provided with briefing reports from the sponsor and the FDA prior to the AC meeting. They are also supplied with the list of targeted discussion questions. The meeting usually begins with administrative remarks and

TABLE 3.3

Example FDA Advisory Committee Meeting Agenda

DEPARTMENT OF HEALTH & HUMAN SERVICES   Public Health Service

AGENDA

April 25, 2012
Circulatory System Devices Panel
HeartWare Ventricular Assist System (VAS)
Gaithersburg Holiday Inn

Panel Chairperson   **Designated Federal Officer**
Richard Page, M.D.   Jamie Waterhouse

08:00	am	Call to Order Conflict of Interest and Deputization of the Voting Members Panel Introductions
08:15	am	Sponsor Presentation
09:30	am	Sponsor Q&A
10:00	am	Break
10:15	am	FDA Presentation
11:30	am	FDA Q & A
12:00	pm	Lunch
01:00	pm	Open Public Hearing*
02:00	pm	Panel Deliberations and FDA Questions
03:30	pm	Break
03:45	pm	Panel Deliberations and FDA Questions
05:00	pm	FDA and Sponsor Summations
05:30	pm	Panel Vote
06:00	pm	Adjournment

* **Open Public Hearing** – Interested persons may present data, information, or views, orally or in writing, on the issue pending before the panel. Scheduled speakers who have requested time to address the panel will speak at this time. After they have spoken, the Chair may ask them to remain if the panel wishes to question them. Then the Chair will recognize unscheduled speakers as time allows. Only the panel may question speakers during the open public hearing.

then the sponsor makes a presentation including the relevant background and a summary of their analyses of the data from the trials. The FDA then presents their analyses of the data highlighting any discrepancies between their analyses and that of the sponsor. In the afternoon, there is frequently an open hearing at which the public (typically patients, interested clinicians, and patient interest groups) can make comments. Later in the afternoon the AC discusses the following:

- Study design issues (e.g., the appropriateness of endpoints, entry criteria, and dosing).
- Study conduct issues (e.g., the appropriateness of patient monitoring, the prevalence of loss-to-follow-up, and adherence to the intervention).
- Analysis issues (e.g., multiplicity concerns, subgroup analyses, handling of missing data, appropriateness of pooling data across studies, clinical relevance of effects, and analyzes populations).
- Post-approval studies (e.g., are they required and if so how should they be designed? i.e., entry criteria).
- Labeling issues (e.g., limitations to generalizing to certain subgroups, safety warnings, and appropriate patient monitoring).
- Appeals to FDA decisions.

Issues specific to the case will also be discussed. Examples include (1) when the primary analyses for one of the trials did not meet statistical significance but was close (e.g., $p = 0.06$), (2) when statistical significance was reached but the magnitude of the observed effect was very modest, (3) when significance was observed in a subgroup of participants but not overall and there was concern for multiplicity and type I error inflation, (4) when efficacy was demonstrated but there was still a concern for safety because of observed toxicity or insufficient exposure data, or (5) when a single-arm trial was conducted without a control group hindering interpretation of the data without a reference comparison. AC members ask questions to the sponsor, FDA, and other presenters at the AC meeting to help better understand the trials, the data, and the presented analyses. AC members need to understand how the intervention works and what the treatment alternatives are so that they can respond to FDA questions.

The AC is asked targeted questions during the discussion. Typical questions include the following:

- Do the data demonstrate the efficacy of the intervention?
- Do the data demonstrate the safety of the intervention?
- Do the expected benefits of the intervention outweigh the expected risks?

At the end of many meetings, AC members are asked to vote on approvability (i.e., approvable vs. approvable with conditions vs. not approvable) after considering all the data (from trials as well as testimonials). If the vote for approvable with conditions is made, then the conditions are discussed. AC meetings are recorded. Videos and slides from the AC meeting are posted on the FDA AC website (http://www.fda.gov/advisorycommittees/default.htm).

Typically, a sponsor invests a large amount of manpower to prepare for an AC meeting. The preparation includes presentation slides production and a briefing document. The sponsor often has a mock AC meeting to help them prepare for critical review and questions from the AC members. The mock panelists are typically people with AC experience, often invited from academia or elsewhere.

3.2 European Medicines Agency

The regulatory body in Europe is the EMA. The committee within EMA that is responsible for reviewing submissions is the Committee for Medicinal Products for Human Use (CHMP). Currently, the CHMP includes 29 member states. Submission of a new product is known as the marketing authorization application (MAA). The CHMP review follows either a mutual recognition procedure or a centralized procedure.

In the community or centralized procedure, the CHMP is responsible for conducting the initial assessment of medicinal products for which a community-wide marketing authorization is sought. The CHMP is also responsible for several postauthorization and maintenance activities, including the assessment of any modifications or extensions to the existing marketing authorization. In the mutual recognition and decentralized procedures, the CHMP arbitrates in cases where there is a disagreement between member states concerning the marketing authorization of a particular medicinal product. The CHMP also acts in referral cases, initiated when there are concerns relating to the protection of public health or where other community interests are at stake.

Assessments conducted by the CHMP are based on purely scientific criteria and determine whether or not the products concerned meet the necessary quality, safety, and efficacy requirements (in accordance with EU legislation, particularly Directive 2001/83/EC). These processes ensure that medicinal products have a positive benefit–risk balance in favor of patients/users of these products once they reach the marketplace.

Subsequent monitoring of the safety of authorized products is conducted through the EU's network of national medicines agencies, in close cooperation with healthcare professionals and the pharmaceutical companies themselves. The CHMP plays an important role in this EU-wide pharmaco-vigilance activity by closely monitoring reports of potential safety concerns (adverse drug reaction reports or ADRs) and, when necessary, making recommendations to the European Commission regarding changes to a product's marketing authorization or the product's suspension/withdrawal from the market.

In cases, when there is an urgent requirement to modify the authorization of a medicinal product due to safety concerns, the CHMP can issue an "urgent safety restriction" (USR) to inform healthcare professionals about changes as to how or in what circumstances the medication may be used.

The CHMP publishes a European Public Assessment Report (EPAR) for every centrally authorized product that is granted a marketing authorization, setting out the scientific grounds for the committee's opinion in favor of granting the authorization, plus a "summary of product characteristics" (SPC), labeling and packaging requirements for the product, and details of the procedural steps taken during the assessment process. EPARs are published on the EMA's website and are available in all official languages of the European Union.

Scientific assessment work conducted by the CHMP is subject to an internal peer-review system to safeguard the accuracy and validity of opinions reached by the committee. The EMA's integrated quality-management system ensures effective planning, operation, and control of the CHMP's processes and records.

Other important activities of the CHMP and its working parties include the following:

- The provision of assistance to companies researching and developing new medicines
- The preparation of scientific and regulatory guidelines for the pharmaceuticals industry
- Cooperation with international partners on the harmonization of regulatory requirements for medicines

Recently, most of the reviews are performed under the centralized procedure for each MAA submission, a rapporteur (one of the member states) and a corapporteur (another member state) is first identified by the CHMP, and then the applicant (the sponsor) submits the MAA to the rapporteur and co-rapporteur. The review process follows a well-defined schedule. A timetable is prepared by the EMEA in consultation with rapporteur and corapporteur and is proposed to the CHMP for adoption. The EMA ensures that the opinion of the CHMP is given within 210 days (less any clock stops for the applicant to provide answers to questions from the CHMP) in accordance with the following standard timetable, which can be shortened in exceptional cases. Some of the key days are specified in Table 3.4.

TABLE 3.4

EMA Action Days

Day	Action
1[a]	Start of the procedure
80	Receipt of the Assessment Report(s) or critique from rapporteur and corapporteur(s) by CHMP members (which includes the peer reviewers) and EMEA. EMEA sends Rapporteur and Corapporteur Assessment Report/critique to the applicant making it clear that it only sets out their preliminary conclusions and that it is sent for information only and does not yet represent the position of the CHMP
100	Rapporteur, corapporteur, other CHMP members and EMEA receive comments from members of the CHMP (including peer reviewers)
115	Receipt of the draft list of questions (including the CHMP recommendation and scientific discussions), from rapporteur and corapporteur, as discussed by the peer reviewers, by CHMP members and EMEA
120	CHMP adopts the list of questions (LoQ) as well as the overall conclusions and review of the scientific data to be sent to the applicant by the EMEA
	Clock stop. At the latest by day 120, adoption by CHMP of request for GMP/GLP/GCP inspection, if necessary (inspection procedure starts)
121[a]	Submission of the responses, including revised SPC, labeling and package leaflet texts in English
	Restart of the clock

After receipt of the responses, the CHMP adopts a timetable for the evaluation of the responses. In general, the following timetable applies:

150	Joint Response Assessment Report from rapporteur and corapporteur received by CHMP members and the EMEA. EMEA sends this Joint Assessment Report to the applicant making clear that it is sent for information only and does not yet represent the position of the CHMP. Where applicable inspection to be carried out. EMEA/QRD subgroup meeting for the review of English product information with participation of the applicant (optional) around day 165
170	Deadline for comments from CHMP members to rapporteur and corapporteur, EMEA, and other CHMP members
180	CHMP discussion and decision on the need for adoption of a list of "outstanding issues" and/or an oral explanation by the applicant. If an oral explanation is needed, the clock is stopped to allow the applicant to prepare the oral explanation
	Submission of final inspection report to the EMEA, rapporteur and corapporteur by the inspection team (at the latest by day 180)
181	Restart of the clock and oral explanation (if needed)
181–210	Final draft of English SPC, labeling and package leaflet sent by the applicant to the rapporteur and corapporteur, EMEA, and other CHMP members
By 210	Adoption of CHMP Opinion + CHMP Assessment Report.
	Adoption of a timetable for the provision of product information translations

(Continued)

TABLE 3.4 (*Continued*)

EMA Action Days

Day	Action
After the adoption of a CHMP opinion, the preparation of the Annexes to the Commission Decision is carried out in accordance with the following timetable:	
215 at the latest	Applicant provides to the EMEA the product information and annex A in the 20 languages (all EU languages including Norwegian) and the "QRD Form 1" by e-mail[b] or on CD-ROM for review by member states
229	Member states send linguistic comments on the product information by e-mail with a copy to the EMEA together with QRD Form 1
232 at the latest	Applicant provides EMEA with final translations of SPC, Annex II, labeling and package leaflet in the 20 languages (+ "QRD Form 2")
237	Transmission of opinion and annexes in all EU languages to applicant, commission, and members of the Standing Committee, and Norway and Iceland
246	Applicant provides EMEA with one paper final full color "worst-case" mock-up of outer and inner packaging for each pharmaceutical form
239–261	Draft Commission Decision
	Standing Committee Consultation
By 277	Finalization of EPAR in consultation with rapporteur, corapporteur, CHMP and applicant (the latter for confidentiality aspects)
277	Final Commission Decision

[a] Target dates for the submission of the responses are published on the EMEA website CHMP meeting.

[b] By e-mail: qrd@emea.europa.eu.

3.3 Guidances

Regulators publish guidance documents covering various aspects of development. For clinical development, there are guidelines for drugs or biologics developed to treat certain diseases, good clinical practice (GCP), the preparation of clinical study reports, drug safety, bioequivalence, and on statistics. The FDA posts their guidance documents at www.fda.gov/regulatoryinformation/guidances/default.htm.

In Europe, the CHMP publishes a number of guidelines for clinical efficacy and safety. Similar to the FDA guidance, these documents cover a wide range of clinical issues including clinical PK, blood products, various human body systems (e.g., cardiovascular system, nervous system), and many general guidelines. Many of the general guidelines are related to statistics and presented in the form of concise "points to consider" (PtC) documents, for example, "points to consider on missing data." The guidance documents are very educational and reviewing them is well worthwhile (http://www.emea.europa.eu/htms/human/humanguidelines/efficacy.htm).

In the early 1990s, there were three major regions involved in the clinical development of new drugs, biologics or medical devices—North America, Europe, and Asia. In Asia, the primary country that engaged in clinical trials and product regulations was Japan. The three regions had different requirements for product approvals. A sponsor would have to prepare required region-specific documents depending on where approval was being sought. Differences in the approval status of a product (approved in one region but not another) created confusion and inconsistency spurring a global initiative called the International Conference on Harmonization (ICH). The ICH was formulated in the earlier part of 1990 and was a cooperative effort of all three regions put forth. Participants of the ICH include regulatory and industry experts, as well as academia in all the three regions. The ICH effort has resulted in many guidelines that can now be considered as an international standard for product development.

ICH guidances can be broadly divided into these three parts: quality (manufacturing); safety (nonclinical); and efficacy (clinical). Based on this

TABLE 3.5

ICH E Guidelines

- E1A The Extent of Population Exposure to Assess Clinical Safety: For Drugs Intended for Long-term Treatment of Nonlife-Threatening Conditions
- E2A Clinical Safety Data Management: Definitions and Standards for Expedited Reporting
- E2B International Conference on Harmonization; Guidance on Data Elements for Transmission of Individual Case Safety Reports
- E2B(R) Clinical Safety Data Management: Data Elements for Transmission of Individual Case Safety Reports
- E2C Clinical Safety Data Management: Periodic Safety Update Reports for Marketed Drugs
- E2C Addendum to ICH E2C Clinical Safety Data Management: Periodic Safety Update Reports for Marketed Drugs
- E2D Postapproval Safety Data Management: Definitions and Standards for Expedited Reporting
- E2E Pharmacovigilance Planning
- E3 Structure and Content of Clinical Study Reports
- E4 Dose-Response Information to Support Drug Registration
- E5 Ethnic Factors in the Acceptability of Foreign Clinical Data
- E6 Good Clinical Practice: Consolidated Guideline
- E7 Studies in Support of Special Populations: Geriatrics
- E8 General Considerations for Clinical Trials
- E9 Statistical Principles for Clinical Trials
- E10 Choice of Control Group and Related Issues in Clinical Trials
- E11 Clinical Investigation of Medicinal Products in the Pediatric Population
- E12A Principles for Clinical Evaluation of New Antihypertensive Drugs
- E14 Clinical Evaluation of QT/QTc Interval Prolongation and Proarrhythmic Potential for Nonantiarrhythmic Drugs
- E15 Pharmacogenomics Definitions and Sample Coding

classification, there are four types of ICH guidelines: Q, S, E, and M (multi-disciplinary). We focus on the ICH E (efficacy, relating to clinical) guidelines. It is important to note, however, that ICH E guidance is not limited to efficacy (e.g., ICH E1 and E2 guidelines are about clinical safety). It is critical that clinical statisticians become familiar with ICH E guidance covering clinical issues (Table 3.5). ICH E9 covers statistical issues.

The analog to ICH for devices is the global harmonization task force (www.ghtf.org) which was created to improve the uniformity between regional medical device regulatory systems.

References

Chow SC, Liu JP. 2004. *Design and Analysis of Clinical Trials*. Wiley, Hoboken, NJ.
Fairweather WR. 1994. Statisticians, the FDA and a time of transition. *Presented at Pharmaceutical Manufacturers Associated Education and Research Institute Training Course in Non-clinical statistics*. Georgetown University Conference Center, February 6–8, Washington DC.
Federal Advisory Committee Act (FACA, Public Law 92-463).

4

Clinical Trial Statisticians

> The quiet statisticians have changed our world—not by discovering new facts or technical developments, but by changing the ways we reason, experiment, and form our opinions about it.
>
> **Ian Hacking**
> *Contemporary Philosopher*

Much of the academic training for statisticians focuses on data analysis methods and associated statistical theory. However, clinical trial statisticians perform many functions other than data analyses. A statistician's time is spent designing clinical trials, running simulations, researching, data monitoring, data cleaning, interacting with other team members regarding important trial issues, participating in project meetings (or teleconferences), preparing for analyses, interpreting results, writing reports, and working on development programs. The academic training of statisticians does not prepare them for these diverse activities. However, there are several attributes and qualities that statisticians can develop to help prepare them for the breadth of these responsibilities and optimize their contribution to clinical trial research (Chuang-Stein 1996, 1999, 2005; Johnson 2007; Phillips 1999; Zelen 2006). Learning statistics is one thing, but learning to be a statistician is another.

In this chapter, we outline the broad and multifaceted roles of clinical trial statisticians, discuss attributes and habits they should develop to optimize their effectiveness, and offer suggestions that can help them make important contributions to their research teams. The details of specific roles will be discussed in later chapters.

4.1 Roles of the Clinical Trial Statistician

A typical image of the statistician is that of a data analyst, performing sample size calculations, and calculating p-values and confidence intervals. They are often viewed by other researchers as technicians, methodologists, mathematicians, and programmers. However, clinical trial statisticians are *collaborative strategists* and experts in decision-making under uncertainty. The roles of statisticians have evolved into strategic and leadership positions, as the value of critical statistical thought has been realized by the research community.

The roles of the clinical trial statistician have expanded to that of a scientist, researcher, teacher, student, consultant, and communicator. Highlights of the roles of the clinical trial statistician are provided in Table 4.1.

As well-rounded scientists who have intricate knowledge of the quantitative and conceptual issues, statisticians can make important contributions to all facets of a clinical trial. In fact, it is critical that statisticians be involved in many aspects of a clinical trial in order to optimize the scientific

TABLE 4.1

Highlights of the Roles of a Clinical Trial Statistician

- Design
    - Clarifying the research question
    - Reducing extraneous variation
    - Creating a randomization schedule possibly with stratification
    - Assisting with selection of an appropriate control group
    - Assisting with selection of an appropriate target population and entry criteria
    - Assisting with appropriate blinding measures
    - Assisting with selection of appropriate endpoints
    - Calculating sample size (with selection of type I and II error rates, estimating variability, and defining effect sizes that are "clinically relevant")
    - Evaluating power sensitivity
    - Anticipating addressing issues with adherence or loss-to-follow-up
    - Clarifying assumptions and limitations of the design
    - Assisting with case report form (CRF) development
    - Assisting with writing of the protocol
    - Assisting with trial registration
- Data monitoring
    - Developing a data monitoring plan
    - Monitoring data issues
    - Helping to ensure the safe and fair treatment of trial participants
    - Assisting with DMC organization and the development of a charter
    - Preparing reports for DMCs
- Analyses
    - Preparing a statistical analysis plan with careful attention to particularly challenging issues such as subgroup analyses, multiplicity, and missing data
    - Assisting with data cleaning
    - Writing a final report
    - Helping interpret the results of statistical output
- Dissemination of study results
    - Assisting with the development of manuscripts
    - Assisting with reporting results into a clinical trial registry
    - Assisting with the preparation of materials for regulatory submission
- General
    - Assisting with development strategy and project prioritization
    - Being an active and cooperative collaborator and team player

contributions of the trial and to avoid common pitfalls in a clinical trial development. Statisticians will soon realize that they must educate their research colleagues regarding the contributions that they can make throughout the life of a clinical trial. H.G. Wells once wrote, "Statistical thinking will one day be as necessary for efficient citizenship as the ability to read and write." While the world may be gradually moving toward this lofty vision, it is clear that statistical thinking is critical for efficient research.

The value of statisticians in clinical trial methodology research is showcased when statisticians create efficient designs, avoid potential pitfalls with intelligent foresight, and apply creative data monitoring techniques that allow for faster identification of efficacious or futile products while maintaining trial integrity. The FDA's increased rigor of statistical reviews has also highlighted the need for proper statistical input into clinical trials.

Statisticians can contribute immensely to the design of a clinical trial. Since the design of experiments and clinical trials is often taught in statistics and biostatistics departments, it only seems natural that statisticians play a lead role in trial designs. Unfortunately, some organizations do not capitalize on the skill of qualified statisticians to help direct the design of clinical trials; viewing statisticians' contribution as occurring only at the end of the trial when statistical analyses are needed. This often results in inefficient designs and trials that are unable to answer the research question of interest. Statisticians can help ask important questions such as, "what is the appropriate control group?" "what are the assumptions of a specific design?" or "what is the appropriate endpoint?" They can help construct objectives to be precise, clear, and phrased such that an associated hypothesis can be tested, or a quantity (e.g., the treatment effect) can be estimated. Optimally, statisticians should take a lead in developing the design of a trial, ensuring that the design is the most efficient to achieve the study objectives.

Careful attention should be paid to potential sources of bias and confounding, as the optimal place to control for these problems is in study design. Variation in responses should be minimized as much as possible in order to identify treatment effects if they exist. The design may have associated assumptions that serve as a guide for the analyses to be performed, and thus, affects the interpretation of trial results and associated limitations. Statisticians understand these issues and, therefore, need to play a role in educating research colleagues with respect to such assumptions and limitations. Possible outcomes, interpretations, and limitations can be conveyed prior to initiation of a clinical trial so that the interpretation of trial results is transparent, and unpleasant surprises can be avoided.

Every well-designed clinical trial has a clear protocol and a statistical analysis plan (SAP). Statisticians are the primary authors of SAPs and contribute to writing sections of the clinical trial protocol. Statisticians typically develop randomization methods, help decide if stratification is beneficial and feasible, if blinding can be achieved, contribute to the selection of an appropriate control group, decide on the choice of a target population, select endpoints and objectives, and construct sample size estimates for appropriately powering a study. Inexperienced statisticians frequently err by only reviewing the statistical section of the protocol. As integral members of the research team, statisticians should read and review the entire protocol and associated documentation for a clinical trial. Statisticians may not understand all the details of the clinical sections, but they should understand the basic concepts of each section.

Statisticians play a lead role in data monitoring (not to be confused with clinical monitoring at sites) of an ongoing clinical trial. Statisticians should develop a monitoring plan (MP) that outlines how the study will be monitored (e.g., what reports will be produced during the conduct of a trial, who will prepare and receive them, the contents of the report, and when the reports will be prepared). Statisticians should discuss the potential safety concerns with the project clinician and medical officer to determine how frequently interim analyses are required and what data will be important to review. Statisticians lead the preparation of the interim analysis reports for Data Monitoring Committees (DMCs) and present interim results at DMC meetings.

Statisticians are most recognized for conducting data analyses. Notably, the preparation for analyses begins in the trial design stage, and details regarding analyses are documented in a SAP developed by the protocol statisticians. The work culminates with the statistician helping to write a final study report and presenting the results to collaborating team members.

After an analysis is conducted, statisticians should play a central role in the overall interpretation and reporting of the data. Since the statistician has expertise and knowledge of the assumptions, limitations, and interpretation of the statistical methods that have been applied, it is important that statisticians take a lead role in the interpretation of the data and help other researchers interpret the data and avoid overinterpretation of the data by optimistic researchers.

4.2 Important Attributes and Suggestions for Development

Academic training often prepares clinical trial statisticians with appropriate applied and theoretical skills in statistics, often with associated statistical programming techniques. Nevertheless, there are several additional qualities and attributes that are important for statisticians to develop that are not typically the focus of statistics courses. These "intangibles" are essential in order for statisticians to be able to accomplish their multifaceted roles effectively. In this section, we discuss these qualities and offer suggestions to help statisticians accomplish their diverse roles, develop the attributes necessary for accomplishing these roles and enhance their contribution to clinical trial

TABLE 4.2

Suggestions for Statisticians to Maximize Their Contributions

- Improve communication skills (writing, listening, speaking, and presenting)
- Keep learning (statistics and medicine)
- Know the medical literature
- Think first (before researching) and keep thinking
- Educate colleagues regarding fundamental statistical concepts
- Identify options and their pros and cons
- Be proactive
- Become Detective Sherlock Holmes
- Avoid being isolated
- Ask lots of questions; question the question
- Voice scientific opinions
- Protect scientific integrity
- Use your references and resources
- Identify mentors
- Learn from your mistakes
- Do not rush your answers
- Be open-minded and compassionate; practice humility and professionalism
- Finish the job
- Participate in professional societies, attend professional meetings, and take short courses

research. A summary of suggestions for statisticians to maximize their contributions is provided in Table 4.2.

4.2.1 Improve Communication Skills (Writing, Listening, Speaking, and Presenting)

Effective communication between statistical and nonstatistical colleagues is essential for effective clinical trial design, conduct, and interpretation. Different types of people whom a clinical trials statistician might come into contact with include physicians, nurses, other biostatisticians, data managers, medical writers, pharmacologists, preclinical scientists, regulatory personnel, technicians, ethicists, information technology staff, patients and their families, risk management experts, translators, international relations staff, manufacturing, quality control engineers, marketing specialists, and public relations staff.

Complex statistical analyses need to be conveyed in a manner that is easily interpretable for nonstatisticians. It is important not to speak in a technical statistical jargon when communicating with nonstatisticians. Statisticians must also gain the confidence in their convictions and not change their mind because a senior physician disagrees with a statistical concept. Conversely, when clinical colleagues are explaining medical concepts, statisticians may need to ask the clinician for simpler explanations.

4.2.2 Keep Learning (Statistics and Medicine)

Working statisticians should not stop learning only because they are out of school. Statisticians should continually view themselves as students. Clinical trials are an evolving scientific field, and statisticians need to remain current with the advances in statistical methods, tools, and software.

4.2.3 Know the Medical Literature

Early in my statistical career, I looked up to some of the leading clinical trial statisticians in the world as examples to emulate. One thing that impressed me about these statisticians was how well they knew the medical literature. This gave them insights into the challenges in clinical trials in particular disease areas and also enhanced their understanding of the medical issues that helped to shape strategies for addressing the challenges. Statisticians should also "learn the medicine" by improving their knowledge of the disease area and associated interventions. This learning will help statisticians communicate, design better studies, provide more effective data monitoring, and provide more informative analyses.

4.2.4 Think First (before Researching) and Keep Thinking

When approaching a research problem, it is not uncommon to research what others have done in similar situations (e.g., search the statistical or medical literature). This is a very reasonable approach as other people may have learned a few lessons regarding how to approach an issue. However, before reviewing what others have done, think about how you might approach the problem. You might be able to avoid pitfalls that other researchers have fallen into with a little novel thinking of your own. Researching what others have done can be conducted after your critical thinking about the problem.

Statisticians often automate processes during programming and report writing for efficiency and to reduce errors. However, statistics is more than merely implementing preplanned analyses. It is imperative that statisticians do not stop thinking about the research questions. Statisticians are one of the creative and critical thinkers of a clinical trial team. Statisticians should take a lead in offering innovative ideas. Statisticians should think ahead (e.g., envision the final analyses of a trial during the design stage), prepare for the unexpected (e.g., subject dropout, poor adherence), and develop a plan for these complications (e.g., appropriate adjustment of a sample size or a plan for sensitivity analyses).

4.2.5 Educate Colleagues regarding Fundamental Statistical Concepts

Clinicians and other researchers rely on statisticians for education regarding statistical issues that affect the design or conduct of a trial. Statisticians should be aware that statistical principles may be an important part of a decision, and this may not be readily apparent to medical colleagues.

4.2.6 Identify Options and Their Pros and Cons

Unfortunately, statistics is often viewed by other researchers as a "necessary evil" or an obstacle in clinical trial research and statisticians are often quoted as saying "you can't do that!" This occurs frequently when statisticians are trying to protect trial integrity. For example, when clinical colleagues desire to exclude nonadherent patients, statisticians remind them of the principle of intention-to-treat (ITT). However, effective collaboration requires dialogue and understanding. Thus, when issues arise, it is important to not only tactfully highlight the problematic issue, but also offer potential options to address the issue, with the associated pros, cons, potential ramifications, and limitations of these options.

4.2.7 Be Proactive

Proactive efforts by statisticians by getting involved early in the design stages of the trial and staying involved in all facets of a trial is critically important. Such involvement helps statisticians understand the trial development process, the roles of other team members, and issues that statisticians need to be aware of during data monitoring and analyses.

4.2.8 Become Detective Sherlock Holmes

The purpose of a scientific study is to find the truth. It is important to have a hunger for the truth and the energy to pursue it by following leads and chasing down clues.

4.2.9 Avoid Being Isolated

Interactions with clinical trial research colleagues increase opportunities for learning, making contributions, and creating team unity. Conversations with research colleagues including clinicians, data managers, project managers, medical writers, clinical research associates, and others, help these colleagues understand the roles of a statistician, and statisticians the roles of others. Moreover, statisticians should strive to understand the roles of others, just as others should understand the roles of the statistician.

4.2.10 Ask Lots of Questions; Question the Question

Questions facilitate understanding, learning, and communication. For example, statisticians are often asked to calculate a sample size. However, many issues should be discussed before a sample size is calculated including clarification of the research question, the primary endpoint, the control group, the appropriate selection of a "minimum clinically relevant difference," estimates of variability identified using relevant data from prior studies, the plausibility of assumptions, and limitations of the anticipated analyses.

Furthermore, it is important to "question the question." It is not uncommon for members of a project team to believe that they understand a common research question when in fact individuals have varying perspectives regarding what the research question truly is. Making sure that the entire project team has a unified and deep understanding of what the objectives of the trial are and what they are not is extremely important. Discussing the research question and carefully clarifying each team member's understanding elucidates which research questions are truly important, and keeps a project team focused on a common goal.

4.2.11 Voice Scientific Opinions

Statisticians have a unique understanding of concepts and fundamental principles that research colleagues may not fully understand. Statisticians should not be afraid to offer ideas that are contrary to popular opinion if those ideas will improve or reexamine the scientific validity of the trial. For example, clinical colleagues may wish to discontinue nonadherent patients from a trial and exclude them from analyses. Statisticians should remind and educate their research colleagues about the ITT principle and the importance of continuing to follow these patients.

4.2.12 Protect Scientific Integrity

Statisticians should closely monitor and protect the integrity of the clinical trial. Statisticians should strive to remain objective (e.g., discussing limitations and potential biases so that data are not overinterpreted) despite pressures for a "positive" trial. The American Statistical Association (ASA) has drafted guidelines for good statistical ethics (see Chapter 1). Statistician should adhere to these principles to practice honest reporting and analyses (e.g., checking assumptions, adherence to multiple testing principles). Decisions should optimally be based on scientific fundamentals but with the realization of the practical limitations. Remember that the research conducted directly influences how patients are treated. Thus, statisticians have an obligation to maintain the highest level of scientific integrity. Scientific integrity benefits everyone including researchers, sponsors, and, of course, the patients.

4.2.13 Use Your References and Resources

Statisticians should use several references and resources to aid in decision making. These references are not limited to textbooks, journals, and websites.

They also include regulatory guidelines and standard operating procedures, as well as mentors and colleagues.

4.2.14 Identify Mentors

Statisticians should identify mentors that can serve as teachers, role models, and an important resource for asking questions. These mentors may be internal to the statistician's place of employment or external (e.g., in academia, industry, or government). Such mentorship is often identified by participating in professional activities such as conferences or short courses.

4.2.15 Learn from Your Mistakes

Mistakes are a part of research and the learning process. Statisticians should take responsibility and attempt to learn from these mistakes. This helps to prevent future errors. Statisticians can ask colleagues to review important material prior to general distribution to help minimize the frequency of such errors.

4.2.16 Do Not Rush with Answers

Statisticians should take the appropriate time to prepare sound answers to research questions. It is acceptable to not know the immediate answers to questions. Research takes time for critical thought and statisticians should take the necessary time to construct an appropriate solution to a problem. The solution begins with a clear understanding of the research question. Often, after a detailed discussion to define the problem, it is beneficial for statisticians to repeat their understanding of the problem, "Let me make sure that I understand ..." to ensure comprehension of the issues. Once comprehension has been established, it is perfectly acceptable to say, "I do not know the answer yet. Please let me do some research and think about that issue. I will then get back to you as soon as I can." Do not panic and remember the scientific foundation (i.e., the scientific method and how cause and effect are established). Finding a solution may require extensive research or development of novel methods. Once the statistician has a draft solution, he or she may want to review it with a colleague before distribution to team members.

4.2.17 Be Open-Minded and Compassionate; Practice Humility and Professionalism

Statisticians are a part of a team working towards a common goal. In fact, statistics has been called "the unselfish profession" with statisticians often helping other scientists with difficult analytical challenges. Statisticians should practice humility and approach discussions with an open mind to avoid isolating other team members. Carefully listen to the ideas of research colleagues with appropriate tact as they educate you with their expertise. Make yourself available for questions and welcome the opportunity for future discussion. Communication of relevant information is important but how information is communicated is also important for being a good research colleague.

4.2.18 Finish the Job

Think about the trials results and how to interpret them. The construction of tables, figures, data listings, test statistics, and p-values are not the end, but the *beginning* of the challenge. There is an enormous amount of data collected in clinical trials, and clinical study reports are very dense with information. Statisticians need to parse this information and turn it into knowledge within the context of the research question. Statisticians should take a lead in interpretation of the trial results, as they have the best understanding of the statistical methods, associated assumptions, and resulting limitations. Focus on how to process and use the information derived from trial analyses in addition to imparting statistical analysis results.

4.2.19 Participate in Professional Societies, Attend Professional Meetings, and Take Short Courses

Statisticians should participate in professional societies. Several excellent professional societies of interest to statisticians are available. These include the ASA, the International Biometric Society (IBS) including regional groups such as the Eastern North American Region, the Western North American Region, the British and Irish Region, and the Japan Region, the Society for Clinical Trials (SCT), the Drug Information Association (DIA), the International Society for Clinical Biostatistics (ISCB), the International Chinese Statistical Association (ICSA), the International Society for Biopharmaceutical Statistics (ISBS), the American Association for the Advancement of Science (AAAS), and societies dedicated to particular disease areas (e.g., the International AIDS Society for HIV/AIDS) or therapies (e.g., the DIA). Also, the ASA has local chapters that organize statistical activities within a geographic region and sections (e.g., Biopharmaceutical, Biometrics) that organize activities focusing on a specific application or methodology area. Statisticians may wish to take short courses offered from these or other organizations and attend the annual meetings of these organizations. Statisticians should also attend medical meetings in their disease-areas of interest (e.g., the Conference on Retroviruses and Opportunistic Infections for statisticians that work in HIV/AIDS). These conferences provide the opportunity to interact with other statisticians and clinicians that work on similar problems specific to a given clinical area. These interactions create opportunities for learning and collaboration. Professional meetings provide an excellent opportunity to meet other researchers (statisticians and others), build relationships with them, creating an important circle of professional friends. Furthermore, professional meetings help statisticians remain current in their knowledge of statistical methods and medical advancements.

References

Chuang-Stein C. 1996. On-the-job training of pharmaceutical statisticians. *Drug Inf J* 30:351–357.
Chuang-Stein C. 1999. Pharmaceutical statisticians in the U.S.: Our future and our direction. *Biopharm Rep* 7(3):1–5.
Chuang-Stein C. 2005. Cultivating non-technical skills of pharmaceutical statisticians. *Biopharm Rep* 13(2):1–4.
Johnson JR. 2007. Some comments on improving collaboration skills of the statistician on a drug development team. *Drug Inf J* 41(5):629–632.
Phillips A. 1999. Guidelines for assessing the performance of statisticians involved in clinical research in the pharmaceutical industry. *Drug Inf J* 33:427–433.
Zelen M. 2006. Biostatisticians, biostatistical science and the future. *Stat Med* 25:3409–3414.

Section II

Scientific and Practical Issues

5

General Considerations in Clinical Trial Design

Most errors in clinical trials are a result of poor planning. Unfortunately, there are no statistical methods that can rescue flawed designs. Thus, careful attention to clinical trial design is essential for ensuring that a clinical trial can address its objectives.

The best strategy for addressing problems is to avoid them. Using foresight, the project team should anticipate obstacles and address them in trial design with logic and creativity. Potential sources of bias should be identified during the design stage.

In this chapter, we discuss general considerations in trial design. We begin by discussing issues relevant to all trials (Section 5.1), then discuss issues specific to controlled or randomized trials (Section 5.2), and discuss design issues specific to trials evaluating biologics, devices, or interventions for rare diseases.

5.1 General Design Issues in Clinical Trials

5.1.1 What Is the Question?

The design of every clinical trial starts with a primary clinical research question. The first requirement for designing a robust and an efficient clinical trial is to clearly define and understand the research question. Clarity of the research question can require much deliberation often entailing a transition from a vague concept (e.g., "to see if the drug works" or "to look at the biological effects of the drug") to a particular hypothesis that can be tested or a quantity estimated using specific data collection instruments with a particular duration of therapy. Secondary research questions may also be of interest, but the trial design usually is constructed to address the primary research question.

Although clinical trials are conducted prospectively, one can think of them as being designed retrospectively. That is; there is a vision of the scientific claim (i.e., answer to the research question) that a project team would like to make at the end of the trial. In order to make that claim, appropriate analyses must be conducted in order to justify the claim. In order to conduct the appropriate analyses, specific data must be collected in a manner suitable to conduct the analyses, and in order to collect these necessary data, a thorough plan for data collection must be developed. This sequential retrospective strategy continues until a trial design has been constructed to address the research question.

Once the research question is well understood, and associated hypotheses have been formed then the project team must evaluate the characteristics of the disease, the interventions, the target population, and the measurement instruments. Each disease and intervention have their unique challenges. For example, in trials to evaluate treatment for pain, researchers should consider the subjective and transient nature of pain, the heterogeneity of pain expression, the placebo effect often encountered in pain trials, and the likely use of concomitant and rescue medications. A design must be customized to address these challenges. The goal of design is to construct the most efficient and robust trial within practical constraints, that will address the research question while considering these characteristics.

5.1.2 Design Efficiency and Robustness

The primary objective of most trials is to estimate the effect of an intervention often relative to a control. The effect of an intervention is estimated most efficiently when extraneous variation is minimized. Thus, it is important to consider minimizing sources of variation in design. Methods for minimizing variation are listed in Table 5.1.

One important method for reducing variation is to construct consistent and uniform endpoint definitions. Ideally endpoints could be measured objectively (e.g., via a laboratory test), however, many endpoints are based on subjective evaluation. For example, the diagnosis of neuropathy or dementia may be an endpoint. These diagnoses are partly subjective, but the variation in these diagnoses can be minimized using clear definitions and consistent evaluations.

A common design feature is the use of central laboratories for quantifying lab parameters to eliminate between-lab variation. For example, the AIDS Clinical Trials Group (ACTG) uses a central laboratory to quantify HIV-1 RNA (ribonucleic acid) viral load in all its studies while the Neurologic AIDS Research Consortium (NARC) uses a central imaging laboratory to quantify

TABLE 5.1

Methods to Reduce Extraneous Variation

- Use of uniform and objective endpoint definitions
- Use of central laboratories
- Use of central evaluation/adjudication committees
- Standardized participant evaluation (potentially enhanced with comprehensive site training)

all images. Similarly, central (common) evaluators can be used to eliminate between-evaluator variation.

Variation can also be reduced with standardization of the manner in which study participants are treated and evaluated. For example, in studies that involve imaging, it is important to have an imaging protocol that standardizes the manner in which images are collected to minimize variation due to inconsistent subject placement. Training modules can be developed to instruct site personnel on the appropriate administration of evaluations. Extensive training on the administration of neuropsychological exams was conducted in the International Neurological HIV Study (ACTG A5199) while the Oral HIV AIDS Research Alliance (OHARA) developed a training module for conducting oral examinations in studies conducted by OHARA, and a training module was developed to instruct sites for the proper administration of the NeuroScreen that is employed in the adult longitudinally linked randomized treatment (ALLRT) trials (ACTG A5001).

It is also desirable for designs to be robust (i.e., valid under minimal assumptions). Every trial is associated with design assumptions. However, many of these assumptions cannot be verified before the completion of the trial. If the assumptions do not hold, the trial results may not be valid or may be uninterpretable. Thus, it is important to understand the assumptions associated with the trial design and ensure that the assumptions are realistic. Trials designed with fewer, or less stringent assumptions are more robust.

5.1.3 Selection of a Population and Entry Criteria

Many factors affect the selection of the entry criteria in a clinical trial. The selection of the entry criteria is based upon consideration of the trial objectives, the characteristics of the interventions and the disease, the intended target population, the availability of potential participants, comorbid conditions, the potential for variable intervention effects (Table 5.2).

The selected entry criteria can depend on the expected toxicity of the intervention. When constructing entry criteria, the safety of the study participant is paramount. Researcher should consider the appropriateness of recruiting *participants* with various conditions into the trial. For example, in

TABLE 5.2

Factors Affecting the Selection of Entry Criteria

- The objectives of the trial
- Potential toxicity of the interventions; participant safety
- The disease and associated severity
- The intended target population to use the intervention
- Availability of participants (e.g., prevalence of disease) and ability to enroll
- Comorbid conditions
- Expected heterogeneity of intervention effects

phase I trials with an objective to evaluate tolerability or measure PK or PD parameters (but not evaluate efficacy), where the interventions are not highly toxic, healthy volunteers (i.e., people without the targeted disease) can be recruited. The use of healthy volunteers, in this case, can still address the objectives of the trial and is considered ethical if the interventions are not too invasive or toxic. However, when evaluating highly toxic interventions, participants with the disease should be enrolled. For example, phase I cancer trials recruit cancer patients since the interventions are often highly toxic.

The selection of entry criteria depends on the objectives of the trial. In some trials, it is desirable to enroll a more homogeneous group of participants as this effectively reduces response variation, enabling easier identification of intervention effects. For example, an objective of some phase II trials is to see if there is a potential signal that should be evaluated in larger phase III trials. Thus, some phase II trials target participants who are likely to benefit from the intervention and use relatively strict entry criteria (e.g., disease symptoms are neither too severe nor too mild, no comorbid or confounding conditions, not taking concomitant medications). Conversely, in some trials it is desirable to enroll a more heterogeneous group of participants to gain a broader perspective on intervention effects. Early-phase trials in oncology often utilize patients with different types of cancers.

Later phase trials frequently target more heterogeneous populations since it is desirable to have the results of such trials to be generalizable to the population in which the intervention will be utilized in practice. It is often desirable for this targeted patient population to be as large as possible to maximize the impact of the intervention. Thus, phase III trials tend to have more relaxed entry criteria that are representative (both in demographics and underlying disease status) to the patient population for which the intervention is targeted to treat.

The ability to accrue study participants is a practical issue that can also affect the selection of entry criteria. Although strict entry criteria may be scientifically desirable in some cases, studies with strict entry criteria may be difficult to accrue particularly when the disease is rare, or alternative interventions or trials are available. Entry criteria may need to be relaxed so that enrollment can be completed within a reasonable time frame.

Researchers may consider restricting entry criteria to reduce variation that might be introduced by comorbid conditions. For example, in a trial evaluating interventions for HIV-associated painful neuropathy, conditions that may confound an evaluation of neuropathy such as diabetes, or a B12 deficiency may be considered exclusionary as these conditions can also affect outcomes.

If the intervention effects are expected to be heterogeneous across participant subgroups, then different intervention effects will be estimated for different subgroups. Project teams must then consider if they wish to exclude particular subgroups of participants so that single common intervention effect can be estimated, or whether to enroll sufficient participants into specific subgroups so that subgroup-specific effects can be estimated with acceptable precision.

5.1.4 Selection of Endpoints

The selection of efficacy endpoints is extremely important and requires a marriage of clinical relevance with statistical reasoning.

5.1.4.1 Desirable Characteristics of Endpoints

The motivation for every clinical trial begins with a scientific research question. The primary objective of the trial is to address the scientific question by collecting appropriate data. The selection of the primary endpoint is made to address the primary objective of the trial. The primary endpoint should be clinically relevant, interpretable, sensitive to the effects of the intervention, practical and affordable to measure, and ideally can be measured in an unbiased manner (Table 5.3).

Ideally, the endpoint is a measure of how a patient feels, functions, or survives. Be wary of endpoints such as "duration of hospital or ICU stay" or treatment duration. These endpoints may be a function of clinician decision rather than patient outcome. This also adds an additional source of variation (due to clinician). Instead, try to capture the patient characteristics that led to the decisions regarding hospitalization or treatment.

Other considerations for selecting a primary endpoint may include the scale of the endpoint (e.g., continuous, binary, nominal, ordinal, and event time), its potential impact on the analyses, and its impact on, power, and cost.

Endpoints can vary greatly in their characteristics. For example, trials of breast cancer require long-term follow-up to capture outcomes 5–15 years into the future, whereas trials treating acute myocardial infarction measure short-term outcomes within hours, days, or weeks.

TABLE 5.3

Considerations for Endpoint Selection

- Clinical relevance to address the objective
- Is it a measure of how a patient feels, functions, or survives?
- Interpretability
- Sensitivity to intervention effects (i.e., assay sensitivity)
- Practicality to obtain (e.g., invasiveness)
- Cost
- Accuracy of measurement (susceptibility to bias)
- Reproducibility of measurement
- Timeliness of result availability
- Susceptibility to manipulation
- Scale of measurement (e.g., continuous, binary, nominal, and event time)
- Impact on required sample size
- Susceptibility to missing data

5.1.4.2 Scales of Measurement

Endpoints can generally be categorized by their scale of measurement. The three most common types of endpoints in clinical trials are continuous endpoints, categorical (including binary) endpoints, and time-to-event endpoints (Table 5.4).

Continuous endpoints are very common. For example, low-density lipoprotein (LDL) cholesterol levels are used in trials for interventions to reduce cholesterol, and weight may be used in obesity trials. Note that, although some measures that are not truly continuous (e.g., LDL), since they can only take on integer values, are generally treated as continuous given the nature of their measurement.

Event-time endpoints are also very common in clinical trials. For example, the time from randomization to death or disease progression may be utilized as an endpoint in an oncology trial. HIV trials often use the time to virologic failure as an endpoint. Event-time endpoints have an advantage when interim analyses are planned since study participants do not have to reach a final visit in order to contribute information to the analyses. If the intervention has a potential to delay the undesirable event, then event time endpoints can be useful for demonstrating and measuring this.

Examples of binary endpoints include death vs. survival, or response vs. no response in an oncology trial. Binary endpoints can also be created from continuous endpoints. For example, in the LDL example described above, LDL response could be dichotomized into two categories (undesirable vs. desirable). It is generally believed that continuous endpoints result in smaller sample sizes than binary endpoints, however, this really depends on the construction of the hypotheses. Dichotomization of a continuous endpoint is often conducted for simplicity but at the cost of precision. Binary endpoints can be created from event-time endpoints as well. For example, a trial measuring overall survival time, "12-month survival" (surviving 12 months vs. not surviving 12 months), can be utilized as a landmark endpoint. The advantage of binary endpoints is their simplicity, but they can result in large sample sizes.

Categorical endpoints are not limited to binary endpoints but can have more than two levels. These levels can be ordinal or nominal, although nominal outcomes are less common. Examples of ordinal endpoints include a physician assessment of disease activity (e.g., "no symptom," "mild,"

TABLE 5.4

Common Scales of Measurement for Endpoints

- Continuous
- Categorical
  - Ordinal
  - Nominal including binary
- Event time

"moderate," "severe," or "very severe") and the end of therapy assessment (e.g., "improved," "no change," or "deteriorated").

5.1.4.3 Objective vs. Subjective Endpoints

Endpoints can also be classified as being objective or subjective. Objective endpoints are those that can be measured without prejudice or favor. Subjective endpoints are more susceptible to individual interpretation. Objective endpoints are generally preferred to subjective endpoints since they are generally less subject to manipulation, bias, or extraneous variation.

Examples of objective endpoints include HIV-1 RNA and CD4 counts in HIV trials. LDL cholesterol was an objective endpoint in the Lipitor trials. Death is an objective endpoint in many oncology trials. Measurements collected from blood chemistry, or mechanical/electrical equipment are objective endpoints.

An example of subjective endpoint is pain measured in arthritis trials. Many Patient-Reported Outcomes (PROs) are subjective in nature such a depression, anxiety, or fatigue. For many of these subjective outcomes, instruments have been developed (e.g., Hamilton Depression Scale; Gracely Pain Scale) to attempt to make quantitative measurement of these outcomes, more objective by developing well-defined scales.

5.1.4.4 Composite Endpoints

An intervention can have effects on several important endpoints. Composite endpoints combine a number of endpoints into a single measure.

Combining component endpoints that are measured on different scales can be challenging. For this reason, composite endpoints are often used in the context of event times where the composite event could be any one of several component events. Examples include composite endpoints in cardio-vascular trials (e.g., time to death, myocardial infarction, or a revascularization procedure) and HIV trials (e.g., time to death, virologic failure, or an AIDS-defining event). ACTG A5087 utilized a binary composite endpoint of "clinical goal" consisting of attaining LDL, HCL, and triglyceride levels within acceptable ranges.

Composite endpoints are also common with PROs. For example, the Western Ontario and McMaster Universities (WOMAC) score (Bellamy et al., 1997) was developed for measuring the severity of osteoarthritis and has been used to create composite endpoints in osteoarthritis clinical trials. The WOMAC consists of 24 questions; five are considered the "pain" domain, two are considered the "joint stiffness" domain, and 17 are considered the "physical function" domain. Composite summary scores are created for each domain and used in data analysis.

There are several pros and cons to the use of composite endpoints (Table 5.5) (Ferreira-Gonzalez et al., 2007; Freemantle et al., 2003; Lubsen and Kirwan, 2002; Montori, et al., 2005). The advantages of composite endpoints

TABLE 5.5

Pros, Cons, and Considerations of Composite Endpoints

- Pros
  - Potentially more complete characterization of intervention effects
  - Multiple effects are measured within-patient
  - More events and thus higher power for event-driven trials
  - May reduce bias from informative censoring in event-time trials
  - Avoids multiplicity concerns with analyses of several component endpoints
- Cons
  - Significance on the composite does not imply significance on the components
  - Significance on the components does not imply significance on the composite
  - Intervention effects on the components can go in difference directions complicating interpretation
  - Interpretation can be challenging when relative importance of the components differ
  - For nonevent-driven composites, power can be reduced if components for which intervention effects are limited are included (intervention effect is diluted by the inclusion of such components)
- Considerations
  - For event-driven trials, consider including events that are more severe (e.g., death) than the events of interest as part of the composite to avoid informative censoring
  - Collect data on all components of the composite to allow for component-specific analyses (i.e., continue to follow participants for all events and not merely a first event)
  - Whether the components have similar importance
  - Whether the components occur with similar frequency
  - Whether the intervention effects is similar across components

are that they may result in a more completed characterization of intervention effects as there may be interest in a variety of outcomes. Composite endpoints may also result in higher power and resulting smaller sample sizes in event-driven trials since more events will be observed (assuming that the effect size is unchanged).

Composite endpoints can also reduce the bias due to competing risks and informative censoring. This is because one event can censor other events, and if data were only analyzed on a single component then informative censoring can occur. For example, death censors observation of the time until cardiovascular disease, but this censoring may be informative. Thus, cardiovascular studies often use the time until the first of stroke, myocardial infarction, or death. Oncology studies often use the time until the first of disease progression or death.

One question that frequently arises is whether all-cause death (or other event) vs. cause-specific death should be used. Since death due to other causes can informatively censor other events of interest, it is usually advisable to use all-cause death (although arguments against this might be made in noninferiority trials). Furthermore, in randomized studies, causality is determined by a contrast of randomized strategies, not by causality adjudication. If death from other causes is truly random, then it has an equal chance of occurring in each arm.

However, composite endpoints have several limitations (Table 5.5). Firstly, significance on the composite does not necessarily imply significance on the components nor does significance on the components necessarily imply significance on the composite. For example, one intervention could have a positive effect on one component but a negative effect on another component, resulting in a nonstatistically significant composite. Another concern with composite endpoints is that the interpretation can be challenging when the relative importance of the components differs. For example, how do we interpret a study in which the overall event rate in arm A is lower than arm B, but the types of events occurring in that arm A are more serious than arm B? For the nonevent-time composites, power can be reduced if there is little effect on some of the components (i.e., the intervention effect is diluted with the inclusion of these components).

When designing trials with composite endpoints, it is advisable to consider including events that are more severe (e.g., death) than the events of interest as part of the definition of the composite to avoid the bias induced by informative censoring. It is also advisable to collect data and evaluate each of the components as a part of secondary analyses. This means that study participants should continue to be followed for other components after experiencing a component event. Other considerations when using composite endpoints are summarized in Table 5.5 (Neaton et al., 2005).

5.1.4.5 Multiple Endpoints

Traditionally, in clinical trials, a single endpoint is selected as primary and is used as the basis for the design of the trial. However, the effects of most interventions are multidimensional, characterized by a set of possibly correlated outcomes. Use of more than one primary endpoint offers an attractive design feature in clinical trials as they capture a more complete characterization of the effect of an intervention. Examples include clinical cure and microbiological cure in infectious disease trials, overall survival and progression-free survival in oncology trials, trials of comorbidities in which there is an endpoint for each comorbidity or use of a primary efficacy endpoint with a primary safety endpoint. For example, a major HIV treatment trial within the ACTG, a phase III comparative study of three nonnucleoside reverse transcriptase inhibitor (NNRTI) sparing antiretroviral regimens for treatment-naive HIV-1-infected volunteers (The ARDENT Study: Atazanavir, Raltegravir, or Darunavir with Emtricitabine/Tenofovir for Naive Treatment) is designed with two coprimary endpoints: time to virologic failure (efficacy endpoint) and time to discontinuation of randomized treatment due to toxicity (safety endpoint).

Regulators have also provided guidelines recommending coprimary endpoints in some disease areas. The Committee for Medicinal Products for Human Use (CHMP) guideline (2008) recommends multiple endpoints to demonstrate symptomatic improvement of dementia associated with in

Alzheimer's disease, indicating that primary endpoints should be stipulated reflecting the cognitive and functional disease aspects. The FDA (2012) recommends the use of two endpoints for assessing irritable bowel syndrome (IBS) signs and symptoms: (1) pain intensity and stool frequency of IBS with constipation (IBS-C) and (2) pain intensity and stool consistency of IBS with diarrhea (IBS-D) in the design of clinical trials evaluating treatments in patients affected by IBS.

Utilizing multiple endpoints may not only provide the opportunity for characterizing the intervention's multidimensional effects but also create challenges for handling multiplicity. Specifically, the control of type I and II error rates when the multiple primary endpoints are potentially correlated is nontrivial. When more than one endpoint is viewed as important in a clinical trial, then a decision must be made as to whether it is desirable to evaluate the simultaneous effects on *all* endpoints (termed *multiple co-primary endpoints*), or *at least one* of the endpoints (termed *multiple primary endpoints*, or *alternative primary endpoints*). This decision defines the alternative hypothesis to be tested and provides a framework for approaching trial design. When designing the trial to evaluate the simultaneous effects for all the endpoints, then no adjustment is needed to control the type I error rate. The hypothesis associated with each endpoint can be evaluated at the same significance level that is desired for demonstrating effects on all the endpoints (ICH E-9 Guideline, 1998). However, the type II error rate increases as the number of endpoints to be evaluated increases. In contrast, when designing the trial to evaluate an effect for at least one of the endpoints, then an adjustment is needed to control the type I error rate. The design of such trials has been discussed (Asakura et al., 2014; Hamasaki et al., 2013; Sozu et al., 2015; Sugimoto et al., 2013).

5.1.4.6 Surrogate Endpoints

In some trials, it may take a very long time to observe the clinical endpoint of interest. In such circumstances, it may be helpful to utilize a surrogate endpoint (or "surrogate"). A surrogate endpoint is a measure that is predictive of the clinical event of interest but takes a shorter time to observe. The purpose of using a surrogate is to draw conclusions regarding the intervention effect on the clinical endpoint without observing the clinical endpoint (i.e., the surrogate is intended to substitute for the clinical endpoint). The clinical endpoint often measures clinical benefit whereas the surrogate tracks the progress or extent of disease. Surrogates can also be used when the clinical endpoint is too expensive or difficult to measure, or not ethical to measure.

An example of a surrogate can be found in the prevention of osteoporosis, a disease that may cause bone loss over time. An important clinical endpoint for osteoporosis is bone fracture. A clinical trial designed to investigate a treatment for osteoporosis would have to follow many participants for many years and examine whether participants experience bone fracture. Such a trial would be very expensive and would not produce results for many years. For this reason, Bone Mineral Density (BMD), a summary measure of bone density obtained from an x-ray image of bones from the lumbar spine, is commonly used as a surrogate. By tracking the BMD changes over shorter periods of time, investigators can use this to predict the likelihood of bone fracture in the future.

Another example can be found in the treatment of HIV. In the 1980s when HIV was first identified as the cause of AIDS, patient survival time was short and thus feasible to measure, and thus the primary endpoint for AIDS trials was the time-to-death. However, medical advances have significantly prolonged the life expectancy of people living with HIV and reduced the incidence of opportunistic infections. As a result, trials with a survival time endpoint would take many years to complete. HIV-1 RNA viral load or CD4 count is now used as a surrogate endpoint in HIV trials.

Other examples of surrogates include immune responses for clinical vaccine efficacy, blood pressure as a surrogate for hemorrhagic stroke, cholesterol level for myocardial infarction, and PSA level for prostate cancer progression.

However, before a surrogate can be confidently used, it must be validated. Unfortunately, there is no agreed-upon definition of a validated surrogate. Many use the criteria proposed by Prentice (1989) to ensure that demonstration of effects on the surrogate implies demonstration of the effect on the clinical endpoint (Table 5.6). A simplified version of the criteria is (1) the surrogate is predictive of the clinical endpoint and (2) the intervention effect on the clinical endpoint manifests itself entirely through its effect on the surrogate. The proportion of the treatment effect explained by the surrogate (PTE) has been proposed as a measure of surrogacy.

It is important to note that the correlation between a potential surrogate with a clinical endpoint does not validate the surrogate. The effect of the intervention on the surrogate must predict the effect on the clinical endpoint (a much stronger criterion than correlation). Ideally the evaluation of the surrogate would result in the same conclusions if the clinical endpoint had been used.

Unfortunately, many researchers have mistaken this correlation as validation, and only a small number of surrogates have been validated. For many years LDL and HDL cholesterol levels have been used as surrogate

TABLE 5.6

Prentice Criteria for Validation of a Surrogate

- Intervention affects the surrogate
- Intervention affects the clinical endpoint
- The distribution of the clinical endpoint conditional upon the surrogate is the same among intervention arms of a randomized trial (i.e., the association between the surrogate and the clinical endpoint is independent of intervention)
- The null hypothesis for the clinical endpoints implies the null hypothesis for the surrogate

endpoints for cardiovascular disease, however a recent study of tredaptive®
showed that the drug increased HDL and lowered LDL but did not improve
clinical outcomes. In HIV, CD4 was commonly believed to be a surrogate for
clinical outcomes but data has contradicted this belief. An example of a surrogate that has been accepted by clinicians and regulators is blood pressure
as a surrogate for cardiovascular outcomes.

Surrogate endpoints are not appropriate primary trial endpoints when
clinical outcomes are present. For example, in many bacterial infections,
microbiological response (e.g., clearance of the pathogen from a body site)
of the bacterial pathogen is often used as a surrogate. However, the clinical
responses of the patient are often observable within weeks. If clinical outcomes are available, then there is no need to use a surrogate such as microbiological response as a primary outcome. Measurement of microbiological
response may still have value in understanding biology, mechanisms of
action, and causal pathways.

5.1.5 Controlled vs. Uncontrolled Single-Arm Trials

The simplest trial design is a *single-arm trial*. In this design, a sample of
individuals with the targeted medical condition is given the experimental
intervention and then followed over time to observe their response. This
design is employed when the objective of the trial is to obtain preliminary
evidence of the efficacy and to collect safety data but is not generally used
as confirmation of efficacy unless the natural history of the disease is very
well understood.

This trial design has several limitations, and despite the design simplicity, the interpretation of the trial results can be complicated. First, there is
an inability to distinguish between the effect of the intervention, a placebo
effect, and the effect of natural history. Responses could theoretically be
due to intervention efficacy, a placebo effect of an inefficacious intervention,
or to a spontaneous or natural history improvement. For a participant that
has responded, it could be argued that he/she would have responded even
without treatment or that the participant responded because he/she thought
that he/she was receiving an efficacious intervention. Furthermore, it is also
difficult to interpret the response without a frame of reference for comparison. For example, if a trial is conducted and no change in participant status
is observed, then does this imply that the intervention is not helpful? It may
be the case that if the participants were left untreated then their condition
would have worsened. In this case, the intervention has a positive effect, but
this effect is not observable in a single-arm design.

Due to these limitations, single-arm trials are best utilized when the natural history of the disease is well understood, when *placebo effects* (discussed
later) are minimal, or when the available patient pool is limited (e.g., rare disease). For example, such designs are common in oncology where spontaneous improvement in participants is not expected, and placebo effects are not

huge. On the other hand, such designs would not be good choices for trials investigating treatments for chronic pain because of the significant placebo effect in these trials.

Ideally trials include a control group to put observations into perspective. Comparisons to a control group help put results into perspective and help control for various biases. The selection of a control group is discussed later in this chapter.

5.1.6 Sample Size

Perhaps the most frequent question asked of a statistician is, "What sample size do we need?" The determination of a sample size and the evaluation of power are fundamental and critical elements in the design of a clinical trial. If a sample size is too small then important effects may not be detected, while a sample size that is too large is wasteful of resources and unethically puts more participants at risk than necessary.

In order to calculate the sample size, the objective of the trial must be clearly defined. The analyses to address this objective must then be envisioned usually via a specific hypothesis to be tested or a particular quantity to be estimated. The planned analyses are then used as the basis for the sample size calculations.

5.1.6.1 Hypothesis Testing versus Precision

Most trials are sized using a hypothesis testing approach. Typically, null and alternative hypotheses are formulated. These must be associated with specific trial endpoints. A statistical test (e.g., t-test) is then envisioned as the analyses. The test-statistic can then be inverted to obtain the required sample size using standard formulas. Standard software can be used for many hypothesis tests.

An alternative method for calculating the sample size is to identify a primary quantity to be estimated, and then estimate it with acceptable precision. Precision is usually defined by the width of a confidence interval. A sample size is derived to ensure that the width of the confidence interval is desirably small. For example, the quantity to be estimated may be the proportion of people that respond to an intervention. A sample size may be calculated to ensure that the width of the 95% confidence interval for the proportion is less than 10%.

5.1.6.2 Choosing an Acceptable Type I Rate

There are generally four important quantities that are used to calculate the sample size: type I error, type II error, estimates of variation, and an effect size to detect (i.e., the minimum clinically meaningful difference in superiority trials) that is used to define an alternative hypothesis. The derivation and selection of these quantities require diligent discussion as the selection of these quantities reflects the assumptions, limitations, and compromises of the trial design. We discuss the selection of each of these below.

Type I error (often denoted α) is the probability of incorrectly rejecting the null hypothesis when the null hypothesis is true. In clinical trials, a type I error often implies an incorrect conclusion that an intervention is effective (since the alternative hypothesis is often that the intervention is effective). In regulatory settings for phase III trials, the type I error is set at 0.025 with one-sided testing. As a result of this regulation, a type I error of 0.025 has become the "standard."

However, in government sponsored trials and some early phase trials in product development, the selection of the type I error rate is at the discretion of the research team, and the "cost" of a type I error can be evaluated given other design constraints. For example, when evaluating a new intervention, project teams may consider lowering the type I error (e.g., to 0.01) when a safe and effective intervention already exists or when the new intervention appears to have significant safety concerns. Alternatively, project teams may consider increasing the type I error (e.g., to 0.10) when a safe and effective intervention does not exist and when the new intervention appears to have low risk.

5.1.6.3 Choosing an Acceptable Type II Error Rate

Type II error (often denoted β) is the probability of incorrectly failing to reject the null hypothesis when the null hypothesis should be rejected. The implication of a type II error in many clinical trials is that an effective intervention is not identified as effective. The complement of type II error is "power", the probability of rejecting the null hypothesis when it is, in fact, false and should be rejected. Type II error and power are not generally regulated. Thus, project teams can evaluate the type II error that is acceptable. For example, when evaluating a new intervention for a serious disease that has no effective treatment, the project team may opt for a lower type II error (e.g., 0.10) and thus higher power, but may allow type II error to be higher (e.g., 0.20) when effective alternative interventions are available. Typically, type II error is set at 0.05–0.20.

5.1.6.4 Choosing the Minimum Clinically Important Difference

When designing a superiority trial, a minimum clinically important difference (MCID) must be specified in order to define the alternative hypothesis needed to calculate sample size for the trial. Conceptually, one could think about this quantity as the minimum effect size that would be considered to be relevant. This quantity is sometimes mandated by the regulatory agency but may also be proposed by project teams based on clinical effects and effects observed in other trials. The selection reflects the project team's thinking and understanding of the disease, the regulatory environment, and the potential characteristics of this intervention.

When selecting the MCID, it is important to distinguish the MCID (based on clinical relevance) from *expected* effects (perhaps estimated from prior studies). Expected effects may not be relevant and designing a trial to detect effects that are less than important would be imprudent. Furthermore, effects smaller than those expected, may still be relevant. In this case, a trial designed to detect the expected effects would be underpowered to detect smaller relevant effects. Thus, it is important to design the trial to detect these smaller relevant effects in case the results are less than expected. Citing expected effects may be important to funders and investors, to help justify that the trial should be conducted.

5.1.6.5 Estimating Variability

When sizing trials, estimates of variability may also be required. For trials with a binary endpoint, an estimate of variation is the function of the assumed proportion. However, when designing trials with continuous endpoints or time-to-event endpoints, then additional information or estimates are needed. Getting estimates of variability can require some homework. When designing late-stage trials, estimates might be obtained from earlier phase trials (although variability in phase III trials can be larger than in phase II since the study populations may be more heterogeneous and evaluations may be taken after a longer treatment duration). Estimates might also be obtained from publications in the medical literature that report results from trials using similar interventions.

If these estimates are incorrect, then the designed trial could be over- or underpowered. Thus, project teams should evaluate the sensitivity of the required sample size and resulting power to reasonable violations of the assumption during trial design. Project teams may also consider interim analyses that evaluate the accuracy of these estimates and potentially make sample size adjustments should the assumptions not hold.

5.1.6.6 Group Sequential and Adaptive Designs

Interim analysis should be considered a design feature since it can affect the required sample size. Thus, when project teams are considering group, sequential or adaptive designs, appropriate planning and adjustments to the sample size is prudent.

If hypothesis testing is to be conducted as part of the interim analyses, then project teams need to be aware of the multiplicity issue (i.e., potential inflation of error rates due to tests of the hypothesis at an interim time point and at final analysis). Appropriate adjustments must be made to ensure error rates are controlled. Often group sequential methods (discussed in Chapter 7) are utilized to control error rates.

Interim analyses may also be used to assess assumptions (e.g., variability) made during the design of the trial. If the assumed variability that was used to design the trial was underestimated, then the trial could be underpowered. Sample size adjustments may need to be considered. However, in order to maintain trial integrity, such adjustments should be planned as a part of the design to ensure control of error rates and to reduce the temptation and perception of manipulation.

Details regarding group sequential and adaptive designs can be found in Chapter 7.

5.1.6.7 Other Issues to Be Considered during Sample Size Calculation

Most standard sample size formulas are "textbook" formulas that apply in a perfectly conducted trial. However, there are operational aspects of the trial that should be considered when calculating sample sizes. For example, poor adherence and participant dropout can reduce the power of the trial. Simulation or adjustments to the standard formulas could be utilized to account for such factors. Lachin (1981) and Donner (1984) provide excellent summaries for such adjustments. We highlight a few of these here.

5.1.6.7.1 Powering for a Per Protocol Analyses

If the "textbook" sample size N for an ITT analysis has been calculated and it is desirable to also have appropriate power for a per-protocol analyses then inflating the sample size to $N/(1-p)$ where p is expected proportion of randomized participants that will not be in the per protocol population. Of course, this does not imply that the analyses of the per protocol population is free from bias as the exclusion from the per protocol population may be nonrandom. When p is 0.05, 0.10, and 0.20, then the inflation factors are 1.05, 1.11, and 1.25, respectively.

5.1.6.7.1.1 Adjustment for Missing Data, the Dilution Effect of Dropouts, and Poor Adherence
An important and often neglected sample size issue is how to account for the loss of power from missing data. A common approach merely inflates the required sample size in the absence of missing data to achieve the same sample size under the anticipated dropout rate, estimated from similar trials. This approach accounts for a reduction in precision of the study from missing data but does not account for bias that results when the missing data differ in substantive ways from the observed data. Consider an extreme case in which the amount of bias from missing data is similar to or greater than the size of the treatment effect. Then detection of the true treatment effect is unlikely regardless of the sample size. When performing power calculations, one should consider sample-size computations for an ITT analysis that uses a hypothesized effect that is attenuated because of the inability of some trial participants to adhere to the treatment. Alternatively, one could develop power analyses for statistical procedures that explicitly account for missing data and its associated uncertainty.

The ITT principle states that all randomized participants should be analyzed even if they drop out of the trial. The implication of an ITT analysis is that the treatment effect is often diluted. Poor adherence can also dilute the treatment effect. A simple adjustment for this dilution effect is to inflate the "textbook" N to $N/[(1-d)^2]$ where d is the expected proportion of participants that will drop out. When d is 0.05, 0.10, and 0.20, then the inflation factors are 1.11, 1.23, and 1.56, respectively.

5.1.6.7.1.2 Adjustment for Unequal Allocation between Groups When comparing groups, power is maximized when the group sizes are equal. There may be occasions when unequal group sizes are desired. For example, it may be desirable to have more participants in the investigational intervention relative to the control, to obtain additional safety data on the investigational intervention. Let N be the required sample size assuming equal allocation between groups and let p and q be the sample fractions such that $p+q=1$. Then the sample size required for unequal allocation is $N\{[(1/p)+(1/q)]/4\}$.

5.1.6.7.1.3 Adjustment for Nonparametric Testing Most sample size formulas are based on distributional assumptions. However, parametric assumptions do not always hold. In such cases, data are transformed, or nonparametric methods must be utilized. However, a trial could be underpowered if sample size calculations were based on parametric methods but nonparametric methods are needed during analyses. An inflation of the sample size is necessary to ensure sufficient power for conducting nonparametric tests. One approach is to calculate the sample size necessary under parametric assumptions. We next inflate these calculations using Pitman's asymptotic relative efficiency that provides a comparison of the efficiency of a parametric test with a nonparametric analog. Using this approach, if N is the required sample size for a parametric test, then a sample size of $N/(0.864)$ is sufficient for a nonparametric analyses.

5.1.6.7.1.4 Multiple Endpoints A trail that utilizes multiple primary endpoints can present challenges to appropriately sizing the trial and maintaining control of error rates given potential multiplicity concerns. Strategies for addressing these issues can be utilized depending on whether the goal is to demonstrate an effect on at least one endpoint and the endpoints can be prioritized by clinical importance, or whether they are considered jointly with the goal of obtaining significance on all endpoints.

If the multiple endpoints can be prioritized according to their clinical importance, then a stepwise test procedure can be applied by testing the most important endpoint first. If this null hypothesis is not rejected, then no further tests are conducted. Otherwise, a test for the next most important endpoint is conducted. The process continues until completion. This gate-keeping strategy preserves type I error and is discussed further in Chapter 8.

Another approach to evaluate the multiple endpoints is to view them as a joint hypothesis where the aim is to demonstrate the significance on all endpoints. In this case, consideration of correlation between the endpoints is needed to size the trial appropriately. Sozu et al. (2015) discuss sizing of trials with multiple endpoints.

5.1.6.8 Simulations

For many years statisticians calculated sample sizes manually or relied on limited available software. Many standard formulas exist when simple analyses are planned (e.g., t-tests) and many statistical software packages utilize these formulas. Such formulas are frequently used as an approximation (perhaps with adjustments) for more complex scenarios (e.g., when distributional assumptions do not hold).

With the development of powerful and fast programmable software (e.g., R), there has been an increase in the use of simulations to calculate sample sizes. Such simulation allows for flexibility, particularly when sample size formulas are not available in closed form due to design complexity or distributional assumptions.

5.1.6.9 Sensitivity Analyses

The sample size evaluation for a trial is a compromise between available resources and trial design assumptions. The output of a thorough sample size analysis should not be a single N but a series of graphs or tables that show the trade-offs as design assumptions vary. Such summaries can help the project team make decisions regarding the design assumptions.

For example, the project team must decide on the acceptable type I and type II error rates. A statistician can provide the required sample sizes for a range of type II errors by plotting the required sample sizes as a function of power. This could help the project team decide between a power of 80% vs. 90% by providing the difference in the required sample sizes for the two designs, allowing evaluation of the differences in the practicality and costs of participant enrollment.

Also, it is important to evaluate the sensitivity of the trial power when the design assumptions are incorrect, for example, a plot of the power as a function of the assumed variation.

5.1.7 Data Management Considerations

Data quality is imperative in clinical trials but often receives inadequate attention from statisticians and clinicians. Even the greatest of statisticians cannot squeeze-out quality results from poor quality data. Trial results are only as reliable as the data that is collected during the trial. Project teams should view the prevention of data errors as a design feature. A useful reference regarding data management in clinical trials is McFadden (2007).

Statisticians need to develop a vision of what data elements will be used in analyses and how the data will be analyzed. Thus, statisticians should consider data collection and management issues during trial design to ensure that the necessary data are collected for analyses and that these data are collected in an appropriate manner to support the analyses. Additional goals are minimizing missing data, minimizing variability, and ensuring the accuracy of the data that are collected.

The development of a statisticians–data manager (DM) relationship helps to ensure a high-quality database. A DM is the person responsible for data management activities (i.e., collecting and organizing data in an ordered fashion so as to facilitate data entry, retrieval, and analysis). Although data management activities can refer to management of paper files, most modern activity revolves around organization and storage of data in electronic form.

Data in clinical trials are collected from many sources including interviews, questionnaires, and clinical examinations. Data may also be derived from tissue samples drawn from the study participants. The samples may come from many sources such as blood, urine, saliva, cerebral spinal fluid, or a biopsy specimen from a bodily tissue. Laboratories then evaluate the samples to quantify important measures (e.g., chemistries, hematology's, CD4 counts, or viral loads) or record qualities of the sample (e.g., whether the sample contains malignant cells). Imaging data such as functional magnetic resonance imaging (fMRI) or computed tomography (CT scans) have also become common.

Data errors inevitably occur for several reasons. Errors may arise from incorrect transcription from source documents or from laboratories that have issues with duplicate specimens. These data need to be cleaned to ensure reliable trial results.

5.1.7.1 Case Report Form Development

Much of the clinical trial data is often collected on case report forms (CRFs), a series of documents often organized by data type. CRFs are distinct from source documents that are usually located at a study site. Historically, CRFs have been in paper forms but electronic CRFs, where data are entered directly into a computing system, have become the standard. Data are often entered onto CRFs at a clinical site by a site nurse or a study coordinator.

When designing CRFs statisticians, clinicians, and data managers should work together to derive the necessary quality control (QC) procedures including missing data checks, range checks, and logical checks to ensure high-quality data. Examples of range checks are checking whether the ages of study participants are consistent with the required entry criteria or whether diastolic blood pressure is ridiculously high (e.g., 200 mm Hg) indicative of a data entry error. Examples of logical checks are ensuring that a male participant does not report a pregnancy and that the data of treatment initiation is not prior to the date of randomization. Electronic CRFs often have QC procedures "built-in" so that incorrect data (predefined by study personnel) cannot be entered into the system. Data problems that are identified via QC methods can then be queried (i.e., sites are contacted and asked to make necessary corrections to the data). It is important that the data is corrected at the site so that there is consistency between source documents and the trial database and so that an audit trail is available. Statisticians and data managers do not correct data themselves as they were not present for the participant evaluations.

Although data managers often lead the development of CRFs, a team effort including input from statisticians and clinicians is imperative for good CRF development. Guidelines for CRF development are provided in Table 5.7.

5.1.8 The Prevention of Missing Data

Missing data is often considered a data analysis problem that concerns statisticians. However, missing data is equally as much a design and conduct issue and should concern all clinical trialists. Excellent discussions regarding the prevention of missing data can be found in The National Research Council's (NRC) "The Prevention and Treatment of Missing Data in Clinical Trials" (2010) and in Little et al. (2012).

Missing data is one of the biggest threats to the integrity of a clinical trial. It undermines randomization, the foundation for statistical inference, and thus can create biased estimates of treatment effects. Although many methodologies have been developed for the analysis of missing data, there is no way to adequately test the robustness of the assumptions required by these methodologies. The belief that complex and fancy data analyses can compensate for missing data is unjustified. The reliance on untestable assumptions to obtain valid results reinforces the importance of *preventing* missing data. Aspects of trial design that limit the likelihood of missing data is an important objective.

Nearly all trials have missing data but ideally the amount of missing data is limited, so as, not to jeopardize trial integrity. For large trials with many trial participants or for trials with a long duration of follow-up, missing data can be particularly problematic. When a trial has substantial missing data, the interpretation of trial results is very challenging and must be viewed with a level of uncertainty.

Prevention of missing data should be high-level priority in all trials during protocol development. The key is to design and carry out the trial in a way that limits the problem of missing data and maximizes the number of participants who are maintained on the trial until the outcome data are collected. Specific ideas for limiting missing data in clinical trials are presented in Table 5.8.

TABLE 5.7

Guidelines for CRF Development

- Ensure that the important data that will be needed for analyses and interim monitoring (i.e., endpoints and covariates) are collected. Construct a table of important data elements that establishes where (e.g., the specific CRF and question), when, and how the data are collected, to help ensure that these data are collected.
- Collect only data that are necessary. Data collection is resource intensive, costly, and can create complexity and data errors. If statisticians are unsure how the data will be utilized in analyses then they should ask the project team for the rationale for collecting the data in question and how the data will be utilized. Furthermore, avoid redundancy in data collection. For example, if the date of birth and the date of the visit are collected from the participant, then it is not necessary to collect age.
- Ensure that missing data can be distinguished from a negative response. Instructions such as "check here if the participant has characteristic X" should be avoided since a negative response cannot be distinguished from a missing response. A better approach is to ask if the participant has the characteristic and get a yes or no confirmation.
- Avoid text fields whenever possible. Text responses are difficult to summarize and use as part of a formal statistical analysis. For example, if a CRF asks for a description of the exercise for which the study participant engages, responses may include, "running," "jogging," or "cross-country running." These are essentially the same response, but data summaries will display these as distinct responses unless they are recoded. Abbreviations (e.g., "run") or misspelled responses (e.g., "running") will display as yet other distinct responses. Text fields cannot be avoided completely but can generally be reserved for specifying unusual responses and explanations. Coded responses can be created for common responses.
- Use consistent coding from trial to trial. This can be helpful in future meta-analyses and reduce future programming.
- Be wary of categorizing responses of characteristics that are measured on a continuous metric. For example, when collecting alcohol consumption data, it is better to collect the number of drinks per week or month rather than simply recording a classification of the participant as a nondrinker or mild, moderate, or heavy drinker. These categories can be derived from the number of drinks during analysis. However, if only the categories were recorded, and there was interest in redefining the definition of a mild, moderate, or heavy drinker, then the data would be insufficient for such recategorization.
- Do not record results of calculations on a CRF. Instead, record the data elements required to make the calculation. Calculations are more reliable when done centrally and electronically.
- Strive for clarity and simplicity. A clear and logical flow to the CRF reduces errors and missing data. Design CRFs with simple questions and clear instructions regarding how the form should be completed. Use skip logic (i.e., allowing the person completing the form to skip unnecessary questions) to reduce time and effort. Strive for CRFs to be simple for site personnel to follow.
- Use consistency of coding across CRFs to make it easier for the CRFs to be completed and to avoid errors. For example, use consistent date formats and consistent coding for yes/no questions.

A common source of missing data is participants who discontinue the assigned intervention because of adverse events, lack of tolerability, lack of efficacy, or due to inconvenience. However, discontinuation of the intervention should not equate to discontinuation from the trial, for example, intervention discontinuation does not imply that outcomes should not be

TABLE 5.8

Ideas for Limiting Missing Data in Clinical Trials

- Trial design
  - Making a "Strategy to Prevent Missing Data" a subsection of the protocol document and engaging the protocol team in considering methods to implement during trial design and conduct
  - Considering run-in periods prior to randomization to identify who can tolerate the intervention and remain in study and randomize only such participants
  - Considering randomized withdrawal designs for long-term trials, in which only participants that have received an intervention without dropping out undergo randomization to continue the intervention vs. switching the intervention (e.g., to placebo)
  - Considering add-on designs in which an intervention is added to an existing effective treatment typically with a different mechanism of action
  - Considering flexible treatment regimens, for example, dose titration
  - Minimizing the length of follow-up (particularly for the primary outcome) and the number of evaluations
  - Allowing rescue medications that are designated as part of the treatment strategy in the protocol
  - Developing simple protocols that are easy to comply with such as having easy instructions (e.g., user-friendly CRFs), having wide data collection windows, and having patient visits that are not too burdensome
  - Defining outcomes that can be easily obtained
- Trial conduct
  - Provide a positive and friendly visit experience when evaluating participants
  - Limit the burden and inconvenience of data collection on participants
  - Provide continued access to effective interventions after the trial
  - Keep contact information for participants up-to-date
  - Select investigators carefully using only those with a strong track-record for retaining trial participants
  - Evaluate potential study participants carefully with respect to information that may indicate a high likelihood of dropping out of the study
  - Monitor missing data and have a query system in place
  - Educating investigators and participants regarding the importance of limiting missing data

measured. Trial participants commit to the study, not merely the study intervention. When a study intervention is discontinued, efforts should be made to obtain the participants' data on outcomes (consistent with the intent-to-treat principle discussed later) as this allows for appropriate evaluation of the intervention *strategy*. Gathering these data after intervention discontinuation preserves the ability to analyze endpoints for all participants who underwent randomization and thus allow intention-to-treat inferences, the desired inference in most clinical trials. It also allows evaluation of whether the assigned intervention affected the use and efficacy of subsequent interventions and provides the ability to monitor side effects that might occur or persist after the discontinuation of initial intervention. In short, all trial participants should be followed after randomization regardless of intervention status or adherence. The scientific benefits of collecting outcomes after participants have discontinued interventions generally outweigh the costs.

5.2 Design Issues in Controlled Clinical Trials

5.2.1 Randomization

Randomization is a powerful tool that helps control intervention assignment bias in clinical trials. Although the randomization cannot ensure between-treatment balance with respect to all participant characteristics, it does ensure the expectation of balance. Importantly, randomization ensures this expectation of balance for all factors even if the factors are unknown or unmeasured. This expectation of balance that randomization provides combined with the ITT principle provides the foundation for statistical inference.

5.2.1.1 Stratification

Trials commonly employ stratified randomization to ensure that treatment groups are balanced with respect to confounding variables. In stratified randomization, separate randomization schedules are prepared for each stratum. For example, baseline immunologic function, as measured by CD4, is a potential confounder for estimating the effects of antiretrovirals to treat HIV (e.g., a between-group imbalance with respect to CD4 could distort the estimate of the treatment effect). Thus, many trials investigating the effects of antiretrovirals employ stratified randomization based on CD4. For example, two randomization schedules may be utilized, one for participants with a baseline CD4 of <200 μL and another for participants with a baseline CD4 of ≥200 μL. Stratified randomization ensures that the number of participants with CD4 <200 μL in each treatment group is similar and the number of participants with CD4 with ≥200 μL in each treatment group is similar.

Stratification has a few limitations. Firstly, stratification can only be utilized for known and measurable confounders. Secondly, although one can stratify on multiple variables, one has to be wary of overstratification (i.e., too many strata for the given sample size). The sample size must be large enough to enroll several participants for each treatment from each stratum.

5.2.1.2 Block Randomization

Randomization can be conducted in several ways. The simplest and the most commonly utilized method is the block randomization. A block randomization is performed by randomly allocating treatment codes equally within a block. Consider a trial with two treatments, A and B. It is common to select a block size of twice the number of treatments when treatments are to be equally allocated. Thus, a block size of four could be utilized, and the first three blocks may look like

A A B B | A B B A | B B A A

In this case, treatment allocation is balanced after every four study participants. Larger block sizes (multiples of 2) could be considered if treatment groups are desired to be of equal size. In a trial with three treatment groups, block sizes of multiples of three can be considered. Block randomization can be applied to most study designs regardless of the number of treatments and the balance between them. For example, a trial may be designed with three treatments (two dosing arms and a placebo) and it may be desirable to allocate twice as many participants to the active doses than the placebo group (i.e., 2:2:1 randomization). In this case, a block size of five or 10 might be selected.

There are advantages and disadvantages of employing small block sizes, as well as large block sizes. Many of the phase II/III clinical trials are multicenter trials where many centers (or investigators) are involved in one study. These centers recruit study participants at varying rates and the enrollment from each center will vary. Randomization in multicenter studies is often stratified by center. It is possible that a center may enroll fewer participants than the block size. For example, suppose a block size of four was used in a two-treatment study but a center enrolled only two participants. It is possible that both participants were assigned to the same treatment group. In this case, a treatment by center interaction becomes nonestimable during data analyses. From this perspective, smaller block sizes are more desirable.

However, there are disadvantages of small block sizes as well. During study conduct, participant safety is monitored. Occasionally, a serious safety event (e.g., toxicity) occurs that requires unmasking of the treatment assignment to treat the participant appropriately. If the study was designed with a block size of two, then once the treatment assignment is unmasked for this study participant, then the treatment assignment of the other participant in the block can also be deduced, jeopardizing the blind. From this perspective, larger block sizes are desirable.

An alternative is to conduct a random block size randomization. In such a design, two levels of randomization are performed. The first level is to randomly select a block size (e.g., four or six for a two-arm trial) while the second level is to populate the blocks with the randomized treatment codes. Although a random block randomization offers a solution to avoid some of the problems mentioned above, it is not frequently used. One reason for this is the added complexity to intervention packaging and shipment.

For this reason, block randomization is the most commonly employed randomization method due to its flexibility, practicality, and its acceptance by regulatory agencies.

5.2.1.3 Adaptive Randomization

An alternative to block randomization is adaptive randomization. In adaptive randomization, the treatment allocation of each new study participant may depend on his or her baseline characteristics or the characteristics of the previously randomized participants. There are two primary types of adaptive randomization: minimization and response adaptive randomization.

Block randomization can be difficult to implement if there are many stratification variables or many levels within the stratification variables. For example, a study designed to stratify by gender (two groups—male and female), disease stage (three groups—mild, moderate, and severe), and use of background medication (two groups—yes and no), creates 12 strata and concern for over-stratification. "Minimization" has been proposed as an alternative to help create a baseline balance between treatment groups on important variables.

In the minimization, the first study participant is allocated to a treatment group with a prespecified probability (e.g., 0.50). Then, for each new participant to be randomized, an algorithm evaluates the distribution of the important prognostic factors of the previously randomized participants for each treatment group, and assigns probabilities for randomization to the treatment groups based on the new participants' baseline characteristics. The probabilities of assignment are weighted in order to "minimize" the imbalances between the treatment groups. Minimization is adaptive but depends only on baseline characteristics of enrolled participants.

Response adaptive randomization allocates participants to treatment groups based on responses observed from participants who are already enrolled. Such trials are only feasible when the primary outcome is known in a relatively short period. For example, in many clinical trials investigating treatments for neurological emergencies (e.g., stroke, traumatic brain injury, and epilepsy) the outcome (e.g., primary endpoint) is known within a relatively short period relative to study entry. Such results may be utilized to adapt the remaining randomization allocation of the trial. For example, in the "play-the-winner" design (Wei and Durham, 1978; Wei, 1978; Yao and Wei, 1996), proportionally, more patients are randomized to the more effective arm. This adaptation is a particularly attractive option in trials studying serious conditions (e.g., possibly fatal). This design feature is attractive to researchers since more patients are assigned to the more effective treatment and may be a more attractive design to participants as there is a higher probability of being assigned to the better treatment. Critics of response adaptive designs state that if there was a clear evidence that one treatment was better than another then a violation of equipoise has occurred, and the trial may need to be terminated.

Adaptive dose–response designs are another example of response adaptive designs. In this case, a study may include several doses. Randomization of new study participants would be based on weights derived from results of study participants that have already been observed for the outcome. More details regarding adaptive randomizations can be found in Rosenberger and Lachin (2002).

Implementing adaptive randomization can be challenging. Randomization schedules cannot be constructed prior to trial initiation as with block randomization. In order to randomize a participant, prompt access to the distribution of baseline factors (in the case of minimization) or responses (in the case of response-adaptive randomization) is needed. Furthermore, researchers need to be careful when there are temporal effects associated with the trial (e.g., surgical trials in which surgeon skill improves with time) as temporal effects with adaptive randomization could confound estimates of intervention effects.

5.2.1.4 Interactive Voice Recognition System

The Interactive Voice Recognition System (IVRS) is a popular and efficient method for conducting randomization and tracking drug supply. Consider a multiple dose clinical trial designed to evaluate a treatment of a chronic disease in which there are participant visits on a regular basis. A typical design would require the participant to go to the site every week for a health assessment and to take a bottle of blinded study medication as the weekly supply. For a 4-week study, each participant receives a total of four bottles of study medication. Based on a block randomization schedule, for every shipment, the drug supply coordinator delivers a block of randomized study medication to the site with four bottles for each participant. Suppose that the block size is four, and then 16 bottles are delivered to the site each time. During the clinical trial, there may only be two or three participants recruited to a particular site with a given study drug shipment, and these participants may not complete the entire 4-week treatment period. When this is the case, many of the drugs shipped to the site would be wasted since each bottle is clearly labeled for participant number X and visit week Y. Such a bottle of medication cannot be used by another participant or by the same participant at a different visit. In order to minimize this type of waste, the IVRS was developed to help with randomization and drug shipment.

The IVRS allows each site to call a central phone number whenever a participant needs study medication. This phone is connected to a computer that asks a series of questions, and depending on the response from the site; the system tells the site which bottle should be dispensed. The basic idea of the IVRS is to provide two levels of randomization—the patient level and the drug bottle level—based on the drug bottles available at each site. Before the trial starts, a set of bottles is delivered to the site. Each bottle has a unique bottle number, and these numbers are stored in the IVRS. When IVRS receives a phone call from the site, it first identifies the trial and the site, and then the system checks if the participant has already been randomized. If this is a new participant, then the IVRS performs the participant level randomization and allocates this participant to a certain treatment group according to the participant characteristics and the stratification requirements. Then, based on treatment allocation and the bottles available at the site, the IVRS performs the bottle level randomization and identifies the bottle number that contains the randomized treatment. The IVRS then directs the site to dispense that given bottle to this new participant. On the other hand, if the IVRS finds the participant has already been randomized, then this system only performs a bottle level randomization and informs the site to dispense the specific bottle to the participant.

In order to implement IVRS, a prespecified number of bottles with study medication must be shipped to the site. The request for drug supply also needs to prespecify the resupply number and the minimum number of bottles to be available at each site. Thus, when the amount of available bottles at the site is running low, a resupply shipment is delivered to the site. The IVRS process reduces the amount of waste from using the traditional drug supply method. When using the IVRS, either the adaptive randomization or the block randomization can be applied to the participant level randomization. IVRS systems are now being replaced by Internet interactive randomization systems that accomplish the same tasks.

5.2.1.5 *Cluster Randomization*

Cluster randomization occurs when clusters of participants (e.g., sites, households) are randomized rather than individuals. The intervention assigned to the cluster is then applied to all individuals in the cluster. Such trials have two units of measurement and levels of inference, for instance, the cluster and the participant.

The rationale for cluster randomization is often one of convenience and feasibility but can also be used to avoid "contamination" effects arising from participants within the same cluster infecting or influencing each other. For example, participants within the same household could infect each other with infectious diseases such as influenza or tuberculosis. Cluster randomized trials are also common with educational, behavioral, or dietary intervention. Tripathy et al. (2010) evaluated the effect of a participatory intervention with women's groups on birth outcomes and maternal depression.

Cluster randomized trials generally have both cluster- and participant-level eligibility criteria and consent. If the trial is unblinded then, cluster randomized trials are subject to the bias of selective entry (i.e., participants within a cluster may know the assigned intervention prior to enrollment). For this reason, although randomized interventions have the expectation of balance with respect to cluster-level characteristics, they do not have the expectation of balance with respect to participant level characteristics.

Cluster randomized trials are statistically less efficient than trials that randomize individual participants. The major issue with cluster randomized trials is that observations (participant responses) within clusters are potentially correlated. A common flaw in the design and analyses of cluster randomized trials is a failure to account for this correlation resulting in an underestimate of the standard errors for intervention estimates and thus spuriously narrow confidence intervals. A key quantity in cluster randomized trials is the *intracluster correlation coefficient* (ρ), defined as the ratio of the between-cluster variation to the total variation. When sizing a cluster randomized trial, one can then calculate the design effect (DE) defined as

$$DE = 1 + (m - 1)\rho$$

where m is the average cluster size. The sample size for a cluster randomized trial can then be obtained by multiplying the sample size required assuming no correlation by DE.

5.2.2 Blinding/Masking

Blinding is a fundamental tool in clinical trial design and a powerful method for preventing and reducing bias. Blinding refers to keeping study participants, investigators, or assessors unaware of the assigned intervention so that this knowledge will not affect their behavior, noting that a change in behavior can be subtle, unnoticeable, and unintentional. Trials utilizing an "open-hidden design" where all participants receive treatment but only some participants are told, have resulted in participants who were told having better outcomes than patients who were not told, illustrating that knowledge of treatment affects outcome.

Some researchers have suggested that objective endpoints can protect against manipulation and thus eliminate the need for blinding. Although objective endpoints help, they do not completely protect against bias induction. For example, participants may drop out of a trial if they are not assigned to their desired intervention or they may not adhere to the intervention. When study participants are blinded, they may be less likely to have biased psychological or physical responses to intervention, use adjunct intervention, drop out of the study, and more likely to adhere to the intervention. Blinding of study participants is particularly important for patient reported outcomes (e.g., pain) since knowledge of treatment assignment could affect their response. When trial investigators are blinded, they may be less likely to transfer inclinations to study participants, less likely to differentially apply adjunctive therapy, adjust a dose, withdraw study participants, or encourage participants to continue participation. When assessors are blinded, they may be less likely to have biases affect their outcome assessments. In a placebo-controlled trial for an intervention for multiple sclerosis, an evaluation was performed by both blinded and unblinded neurologists (Noseworthy et al., 1994). A benefit of the intervention was suggested when using the assessments from neurologists that were not blinded, but not when using the assessments from the blinded neurologists.

Clinical trialists often use the terms *single-blind* to indicate blinding of only study participants, *double-blind* to indicate blinding of both study participants and investigators, and *triple-blind* to indicate blinding of participants, investigators, and the sponsor and assessors. There is not a clear distinction between the double-blind and the triple-blind terminology. Because almost all the double-blind studies are also triple-blind, the term *triple-blind* is not frequently used. Trials without blinding are often referred to as *open-label*.

Successful blinding intervention assignment is a nontrivial task. In a placebo-controlled trial, a placebo must be created to look, smell, and taste just like the intervention. For example, in a dose-response study (Lachenbruch, 1996) that includes four treatment groups: 20 and 5 mg of an investigational drug and a placebo, where there are only 5 and 10 mg tablets, and the two tablets do not look alike, then each participant at each visit would require three bottles of study medication: bottle 1 contains either 5 mg tablets or matching placebo, bottle 2 contains 10 mg tablets or matching placebo, and bottle 3 is the same as bottle 2. If a participant is randomized to placebo, then he or she receives matching placebo tablets from bottle 1, bottle 2, and bottle 3. If a participant is randomized to the 5 mg group, then he/she takes bottle 1 filled with 5 mg tablets, and placebo for bottles 2 and 3. If a participant is randomized to the 10 mg treatment group, then he/she takes placebo from bottle 1 and 3 and the 10 mg tablets from bottle 2. Finally, if a participant was randomized to the 20 mg treatment group, then he/she takes placebo from bottle 1 and 10 mg tablets from both bottles 2 and 3.

Blinding can be challenging or impractical in many trials. For example, surgical trials often cannot be double-blind for ethical reasons. The effects of the intervention may also be a threat to the blind. For example, an injection site reaction of swelling or itching may indicate an active intervention rather than a sham injection in vaccine trials. Researchers could then consider using a sham injection that induces a similar reaction.

In the late phase clinical trials, it is common to compare two active interventions. These interventions may have different treatment schedules (e.g., dosing frequencies), may be administered via different routes (e.g., oral vs. intravenously), or may look, taste, or smell differently. A typical way to blind such a study is the "double-dummy" approach that utilizes two placebos, one for each intervention. This is often easier than trying to make the two interventions look like each other. Participants are then randomized to receive one active treatment and one placebo (but are blinded). The downside of this approach is that the treatment schedules become more complicated (i.e., each participant must adhere to two regimens). Consider a randomized three-arm trial comparing a QD dosing (once-a-day treatment) vs. a BID dosing (twice-a-day treatment) vs. placebo. Participants would receive two sets of bottles for each supply or resupply—the morning bottle and the evening bottle. Each set of these bottles would contain the matching placebo tablets. The participant who is randomized to BID will receive the intervention in both bottles, and the participant randomized to QD will receive one bottle of the intervention and another bottle with placebo. If randomized to placebo, the participant will receive placebos in both bottles.

Having double-blind in the title of a trial does not imply that blinding was successful. Reviews of blinded trials suggest that many trials experience issues that jeopardize the blind. For example, in a study assessing zinc for the treatment of the common cold (Desbiens, *Annals of Internal Medicine*, 2000), the blinding failed because the taste and aftertaste of zinc were distinctive. Creative designs can be utilized to help maintain the blind. For example, OHARA and the ACTG are developing a study to evaluate the use of gentian violet (GV) for the treatment of oral candidiasis. GV has staining potential that could jeopardize the blind when the assessors conduct oral examinations after treatment. A staining cough drop could be given to study participants prior to evaluation to help maintain the blind.

When blinding is implemented in a clinical trial, a plan for assessing the effectiveness of the blinding may be arranged. This usually requires a blinding questionnaire for the participant or the investigator that evaluates the participant for the outcome. Unplanned unblinding should only be undertaken to protect participant safety (i.e., if the treatment assignment is critical for making immediate therapeutic decisions).

Blinding has been poorly reported in the literature. Researchers should explicitly state whether a study was blinded, who was blinded, how blinding was achieved, the reasons for any unplanned unblinding, and the results of an evaluation of the success of the blinding.

5.2.2.1 Selection of a Control Group

The selection of a control group is a critical decision in clinical trial design. The control group provides data about what would have happened to participants if they were not treated or had received a different intervention. Without a control group, researchers would be unable to discriminate the effects caused by the investigational intervention from effects due to the natural history of the disease, patient or clinician expectations, or the effects of other interventions. Researchers should consult ICH E-10 when selecting a control group (Ellenberg and Temple, 2000; Temple and Ellenberg, 2000; ICH E10, 2001).

There are three primary types of control groups (Table 5.9): (1) historical controls, (2) placebo/sham controls, and (3) active controls. The selection of a control group depends on the research question of interest, ethical constraints, and feasibility constraints.

5.2.2.2 Placebos/Shams

In the early stage of clinical drug development (e.g., proof-of-concept [PoC] or a dose–response study), it is a common practice (if deemed ethical) to

TABLE 5.9

Types of Control Groups Used in Clinical Trials

- Placebo/sham control
  - Advantages
    - Can randomize intervention assignment and thus have the integrity of an RCT
    - Control for natural history and placebo effects
    - Allows for potential blinding
    - Entirely prospective
  - Disadvantages
    - Costs (requires recruitment of participants into multiple arms when compared to historical controls)
    - May not be ethical to apply in settings where effective interventions exist
    - Is not a comparison to the best available therapy
- Active control
  - Advantages
    - Can randomize intervention assignment and thus have the integrity of an RCT
    - Allows for potential comparison to best available therapy
    - Potential for use when placebos/shams are unethical
    - Allows for potential blinding
    - Entirely prospective
  - Disadvantages
    - Costs (requires recruitment of participants into multiple arms when compared to historical controls)
    - Does not allow for direct comparison to placebo to demonstrate "any effect" (often the requirement for approval)
- Historical Control
  - Advantages
    - Efficiency (fewer participants to recruit; lower cost)
    - Attractive for rare diseases
    - No ethical constraints
  - Disadvantages
    - Nonrandomized design (subject to biases from observational studies)
    - Nonblinded
    - Partly retrospective
    - Concern for data quality from historical studies
    - Bias from changes in medical practice over time

compare the drug with a placebo control group. Detecting effectiveness beyond that achieved with placebo is an important milestone for continuing development of the drug. In these cases, the placebo group essentially represents a "zero dose." Participant response at a dose of zero provides a foundation for comparison with active doses.

A placebo can be defined as an inert pill, injection, or other sham intervention that masks as an active intervention in an effort to maintain blinding of treatment assignment. Often termed a "sugar pill," it does not contain an active ingredient for treating the underlying disease but is used in clinical trials as a control to account for the natural history of disease and for psychological effects. One disadvantage to the use of placebos is that sometimes they can be costly to obtain.

Although the placebo (from Latin, meaning "I shall please") has no activity for the disease being treated, it can provide impressive treatment effects (negative side effects from placebo treatment are sometimes referred to as "nocebo effects"). This is not to say that you can necessarily "think yourself better." Placebos/shams do not shrink tumors, eliminate viruses, or heal broken bones. However, research indicates that placebos/shams can stimulate real physiological responses (e.g., changes in heart rate or blood pressure; chemical activity in the brain). Cues and associations can also affect how well the placebo works. Thus, when the endpoint is subjective (e.g., pain, depression, anxiety, fatigue, or other patient reported outcomes), significant effects can occur. Figure 5.1 below is extracted from the Viagra label (http://media. pfizer.com/files/products/uspi_viagra.pdf.). The vertical axis represents the percentage of patients reporting improvement. Note that of the 463 patients receiving placebo treatment, 24% report that they experienced improvement (hence, there are 24% placebo responders).

Other studies have also demonstrated placebo effects. Evans et al. (*PLoS Clinical Trials*, 2007) reported a significant improvement in pain in the placebo arm of a trial investigating an intervention for the treatment of painful HIV-associated peripheral neuropathy. Trials evaluating blinded titration/dilution of active treatments have found continued effects (Ader et al., *Psychosomatic Medicine*, 2010). Some studies have shown that proper adherence to placebo results in better outcomes (e.g., Simpson et al., *BMJ*, 2006). This could be a "healthy adherer effect" or a simple case that adherers are different from nonadherers.

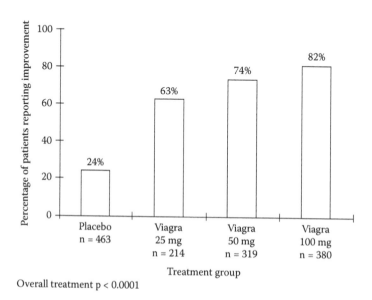

FIGURE 5.1
Percentage of patients reporting improvement by dose.

Several studies have compared different methods of placebos and induced different levels of effect indicating that the method of administration can have effects (e.g., placebo pill vs. sham acupuncture for pain relief; Finniss et al., *Lancet*, 2010). Other studies have compared placebos with no treatment with mixed results (Hrobjartsson and Gotzsche, *NEJM*, 2001). Kaptchuk et al. (2008) reported that a friendly and supportive patient–clinician relationship augmented a placebo effect in a randomized trial for IBS. Kaptchuk et al. (2010) reported the results from a randomized trial comparing responses in IBS sufferers between a "no treatment" group and a group that was given a placebo and was told that it was a placebo, but that placebos sometimes had positive effects. The placebo group reported twice as much symptom relief comparable to improvements found in trials evaluating the best-approved treatments.

A trial that compared albuterol vs. placebo vs. sham acupuncture vs. no intervention for the treatment of asthma (Wechsler, 2011) resulted in albuterol showing superiority to the other three groups with respect to an objective endpoint, the change in forced expiratory volume (FEV). However, three of the interventions (albuterol, placebo, and sham acupuncture) performed similarly with respect to subjective outcomes while the no intervention was inferior.

In certain studies, the use of placebo is not feasible, and a sham control is employed. For example, Macugen®, an injection treatment for age-related macular degeneration, is applied as an injection to the eye in every 6 weeks. The risk from the injection itself (with or without the active treatment) is very high, and thus it is not ethical to inject a placebo treatment to the patient's eye. Hence, in trials evaluating Macugen, a sham control was introduced by having the treating physician use a needle to simply touch the eye, rather than using actual injection.

Many logistic and ethical concerns may prohibit the use of placebos or shams in some clinical trials. The inability to use a sham is common in the development of devices. For example, surgical trials rarely have a sham surgery for ethical reasons.

5.2.2.3 Active Controls

In certain therapeutic areas, there are already very good treatments available and it may not be ethical to run placebo-controlled trials. In this case, an active control could be selected and used as a control. An active control is an intervention that has often shown effectiveness in prior studies to treat the disease.

In some disease areas, there may be many available treatments and thus several options for the active control. Often the active control is selected because it is the standard of care (SOC) treatment or the best available treatment. As the trial is comparing two active interventions, the ability to blind the study via double-dummy strategies, and so on, can play a role in the active control selection.

Active and placebo controls could be simultaneous. For example, it may be of interest to see if a new intervention is superior to placebo but also how the new intervention compares to an effective SOC. If the new intervention was unable to show superiority to placebo, but an active control group was able to demonstrate the superiority to placebo, then this may be evidence that the new intervention is not effective. However, if the active control did not demonstrate superiority to placebo, then it is possible that the trial was flawed, may have been underpowered because of the variability being unexpectedly high, or may be indicative that the effect of the active control has changed.

5.2.2.4 Historical Controls

Historical controls are obtained from studies that have already been conducted and are often published in the medical literature. The data for such controls is external to the trial being designed and is compared with the data collected from the trial being designed.

The advantage of using historical controls is that the current trial will require fewer participants and thus use of historical controls provides an attractive option from a cost and time efficiency perspective. For this reason, historical controls can be considered in trials for evaluating rare diseases (i.e., advanced cancers). Historical controls can also be considered when placebo controls are not ethical (e.g., device trials) particularly when recent historical data is available. It may also be easier to recruit potential study participants into a trial with a single intervention.

The drawback of historical controls is that they are nonrandomized studies and thus subject to considerable bias. Bias can occur if the controls systematically differ from the treatment group in a manner that affects the outcome. Differences can occur due to participant selection (e.g., patients with more favorable prognosis are selected for the new treatment, but poor risk patients are not enthusiastically recruited; differing demographics or disease severity), evaluation, supporting care, concomitant therapies, and follow-up. Some important factors may not have been known or measured in the historical control group. Bias due to factors that could have changed over time (e.g., patient standard of care, diagnostic criteria, or referral patterns) cannot be completely eliminated when using historical controls. Furthermore, historical-controlled trials are not blinded and thus are subject to patient and evaluator bias when assessing eligibility or outcomes. Historical controls are rarely used in clinical trials for late-stage drug development due to the concerns for bias due to lack of the foundation for statistical inference (i.e., randomization). Some research suggests that historical controls tend to produce results that favor the new intervention. Sacks et al. (1982) reviewed 50 randomized clinical trials and 56 historical clinical trials evaluating six therapies and found 76% of the historical trials and 20% of the randomized trials demonstrated superiority to the control with the differences largely due to the performance of the control group in the trials (i.e., the control group in the historical-controlled trials performing worse than in randomized trials).

The validity of the use of historical controls depends on the assumption that study participants who received the old treatment in the historical trial have the same distribution of covariates as study participants in the new trial. When historical data are very recent and reliable, and other diseases and treatment conditions have not changed since the historical trial was conducted, then use of historical controls can be considered. Historical controls have become common in device trials when placebo-controls are not a viable option.

When designing historical-controlled trials, it is important to remember that estimates from prior studies are measured with error (i.e., they are estimates). Uncertainty in these estimates should be considered in trial design. Design and analyses often include stratification and modeling adjustments for potentially confounding variables. Analyses should include a between-group comparison of potentially confounding variables. Propensity scores are often used to adjust for population differences (i.e., study participants are stratified by their propensity score, comparisons within a propensity score are conducted, and results are pooled across propensity categories).

Gehan (1984) outlined desirable characteristics for a historical-controlled trial: (1) the control group has received a well-defined treatment in a recent study, (2) the criteria for eligibility, observation, and evaluation are the same for both groups, (3) no unexplained indications lead to the expectation of different results, and (4) if between-group differences in prognostic variables exist, then they are not of sufficient magnitude to explain observed differences in outcome. Although these characteristics are desirable, they are often unrealistic or challenging to demonstrate. For example, the eligibility criteria, observation, and evaluation is often not exactly the same for both groups. There may be prognostic variables that are unknown or unmeasured. Thus, these characteristics may be viewed as a target but with the realization that all of them may not hold.

5.2.3 Parallel Group vs. Crossover Designs

The design of a clinical trial that compares two or more interventions (or intervention strategies) generally falls into one of the two major categories: parallel group designs or crossover designs. It is important to understand their distinction.

5.2.3.1 *Parallel Group Designs*

Most randomized trials utilize parallel-group designs. Typically a group of participants with the target disease is identified and randomized to one of two or more interventions (e.g., new intervention vs. placebo). A randomized participant only receives one intervention (or intervention strategy) during the duration of the trial. Participants are then followed over time, and the responses are compared between randomized groups. For example, Evans et al. (2007) describes a randomized, double-blind, placebo-controlled, multicenter, dose-ranging study of prosaptide (PRO) for the treatment of HIV-associated neuropathic pain. Participants were randomized to 2, 4, 8, 16 mg/day PRO or placebo administered via subcutaneous injection. The objective was to compare each PRO dose group with placebo with respect to pain reduction. The primary endpoint was the 6-week change from the baseline in the weekly average of random daily Gracely Pain Scale prompts using an electronic diary. The study was designed to enroll 390 subjects equally allocated between groups.

Parallel designs are attractive because when they are utilized with randomization and the ITT principle, they provide valid intervention comparisons. They are also very powerful in that they can be used to evaluate many different research questions and settings. For example, when no standard intervention exists, they can be used to compare an experimental intervention to a placebo/sham. When a standard intervention exists, they can be used to compare an experimental intervention to a standard intervention. They can also be utilized to evaluate the optimal timing of intervention application, whether a patient should switch from an existing intervention to a new intervention, or to compare single intervention to a combination of interventions (Figure 5.2). A disadvantage of parallel designs is that they can require large sample sizes due to the existence of both within- and between-subject variation.

A useful variant of a parallel group design is the randomized withdrawal design that assigns all participants (step 1) to an intervention and follows them for a response (i.e., efficacy, safety, or adherence). Only participants that have a desired response are then randomized (step 2) to continue the intervention or withdraw from the intervention. Outcomes between the continuation arm and the withdrawal arm can then be made to evaluate the effectiveness of the intervention (Figure 5.2). This design has been used in schizophrenia trials, for example, where loss-to-follow-up is a major concern. By only randomizing participants that adhere to study protocol in step 1, the trial "weeds out" participants who are likely to drop out of step 2, thus enhancing the integrity of the randomized comparison (albeit with a less generalizable outcome).

5.2.3.2 Crossover Designs

In a crossover design, each participant is randomized to a *sequence* of interventions that will be consecutively administered during intervention *periods* although the trial objective remains to compare interventions. For instance, in a two-period, two-sequence (i.e., 2×2) crossover trial designed to compare two treatments A and B, a participant is randomized into one of the two sequences: (1) A then B, or (2) B then A. The randomization of the treatment sequence helps to account for temporal trends (such as disease progression or seasonal variation).

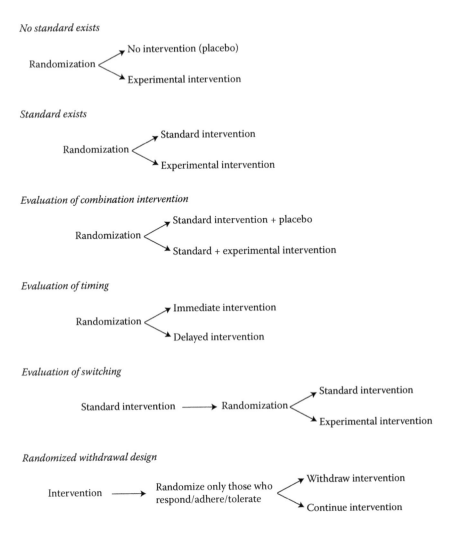

FIGURE 5.2
Examples of parallel group designs.

Crossover trials have several advantages (Table 5.10). Firstly, they generally require fewer participants than parallel designs because each participant serves as his or her own control, thus eliminating interparticipant variation. Thus, a crossover study may reduce the sample size of a parallel group study by 60%–70% in some cases. Also, since each participant is evaluated for each treatment, potentially confounding variables are balanced between intervention groups by design, hence making intervention comparisons "fair." Secondly, researchers can study individual participant responses and examine participant-by-intervention interactions. Lastly, trial recruitment may be

TABLE 5.10

Advantages and Disadvantages of Crossover Trials

- Advantages
  - Smaller sample sizes (i.e., eliminates interpatient variation as each patient serves as his or her control)
  - Balance of confounding variables
  - Can examine participant-by-intervention interaction
  - Potentially enhances recruitment as all participants receive active intervention
- Disadvantages
  - Potential carry-over effect (possibly remedied by wash-out periods)
  - Longer participant follow-up (multiple periods of intervention and washout periods for each participant)
  - More loss-to-follow-up with longer trials and exposure to multiple interventions
  - Usually only feasible with short-term endpoints
- Consider crossover trials when
  - Evaluating interventions for chronic, stable diseases for which no permanent cure exists and for which the risk of death and subject dropout is low
  - Evaluating interventions with a quickly reversible effect with discontinuation
  - Evaluating drugs with a short half-life
  - Endpoints have large inherent intrasubject variation
  - Only short intervention periods are necessary (i.e., effects can be observed quickly)

enhanced as potential participants are aware that they will receive active intervention at some point during the study.

However, crossover trials should be used selectively. The primary concern with crossover trials is the potential *carry-over effect*. If the residual effect of the intervention provided in the first period continues into the second period when assessments of the second intervention are made (despite the discontinuation of the intervention at the end of the first period), then intervention comparisons could be biased since one cannot distinguish between the intervention effect and the carry-over effect. The carry-over effect can make the attribution of observed safety events challenging. For this reason, a *washout period* is often built into the trial design to separate two intervention periods and eliminate carry-over effects. In drug trials, a frequent recommendation is for the washout period to be at least five times the half-life of the drug that has the maximum half-life. Endpoint evaluations can also be made at the end of a period to allow more time for the effects of prior interventions to dissipate.

A second concern with crossover trials is the increased rate of participant dropout. The dropout rate may be high in a crossover trial since they require a longer duration of follow-up for each participant to accommodate for multiple intervention and washout periods. Participants are also exposed to multiple interventions, and thus more potentially harmful interventions than in parallel group designs, and hence may be more likely to drop out due to toxicity. The consequence of dropouts in a crossover trial is a threat to the generalizability of the trial results as analyses are often conducted on only the subset of participants that completed at least two periods. Thus, when conducting crossover trials, it is important to take measures to minimize dropout (e.g., diligent follow-up of participants). A strategy to replace participants who drop out is frequently considered in order to maintain a balance in intervention comparisons.

Period effects can also be a concern in crossover trials. Finally, the evaluation of long-term safety effects is generally not possible. For these reasons, crossover trials are rarely used in confirmatory phase III trials.

Crossover trials may, therefore, be an option when investigating interventions: (1) for chronic, stable diseases for which no permanent cure exists and for which the risk of death and subject dropout is low, (2) with a quickly reversible effect with discontinuation, (3) with a short half-life, (4) with endpoints that have large inherent intrasubject variation, and (5) with short intervention periods (i.e., effects can be observed quickly). Phase I PK trials frequently utilize a (2 × 2) crossover design. Crossover designs are generally appropriate in PK studies because (1) intersubject variability in PK variables is often very large and crossover studies eliminate this source of variation by design and (2) the duration of PK studies is usually short. Crossover trials have been proposed for use in trials for the treatment of drug abuse due to the large intersubject variation in the self-evaluated responses that are used as endpoints. Many treatments are often investigated, but there is difficulty in recruiting participants since drug abusers often do not meet entry criteria due to health complications.

More complicated crossover studies can be designed to include more than two treatments and respective periods. Williams (1949) proposed an algorithm to design balanced crossover studies with multiple formulations. An example of a three period balanced design with three formulations (two test formulations and one reference) looks like

Sequence	Period		
	1	2	3
1	A	B	C
2	B	C	A
3	C	A	B
4	A	C	B
5	B	A	C
6	C	B	A

In this design, each formulation is preceded by each other formulation twice. Studies of 3, 4, or 5 periods with many more formulations can also be designed using Williams algorithm.

Gilron et al. (2005) used a crossover design to evaluate morphine and gabapentin for the treatment of neuropathic pain. The ACTG and NARC utilized a four-period crossover design in a phase II randomized, double-blind, placebo-controlled study (ACTG A5252) of combination analgesic therapy in HIV-associated painful peripheral neuropathy (Figure 5.3). The trial investigated the use of methadone, duloxetine, and their combination (vs. placebo)

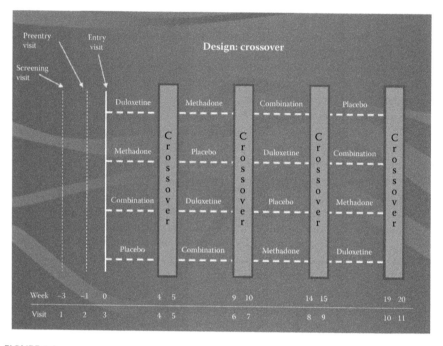

FIGURE 5.3
ACTG A5252 crossover design schema.

for the treatment of neuropathic pain associated with HIV. The design was appropriate since (1) neuropathic pain is chronic, nonlife-threatening, noncurable, and relatively stable over time, (2) pain measurements are often subject to high intrasubject variation, (3) there is considerable concern for a placebo effect in studies of pain, and (4) pain generally returns to baseline levels with discontinuation of the treatments. To address the concern for potential carry over, a washout between each treatment period was implemented, and pain was measured at the end of the treatment period to allow more time for residual effects to dissipate. Measure to minimize dropout included use of rescue medication, follow-up calls to participants, a flexible titration schedule for study medications, and recommendations for the management of treatment-emergent adverse events.

5.3 Special Issues

5.3.1 Design Issues in Biologics

Clinical trials evaluating biologics can pose unique challenges. For example, biologics often address rare conditions (e.g., hemophilia or envenomation from snakebites) and thus the target population can be small and clinical trials can be difficult to accrue. Since the pool of potential study participants is small, uncontrolled or historical-controlled trials may be considered. Biologics also often address needs for which there are no currently existing products and thus there may be no precedent for selecting an appropriate endpoint.

The most widely used biologics are vaccines. Vaccines can be classified as "preventive" or "therapeutic." Preventive vaccines are more common and designed to elicit antibodies in healthy individuals. Examples include childhood vaccines for polio, measles–mumps–rubella (MMR), diphtheria–tetanus–pertussis (DPT), and influenza. Therapeutic vaccines are intended to treat an existing disease such as cancer, or HIV and trials are designed and analyzed similar to other therapeutic interventions.

Trials for preventative vaccines are generally designed to estimate the direct effect of the vaccine on disease incidence. A standardized definition and diagnosis of "disease" should be clearly defined during study planning. "Vaccine efficacy" is usually defined as

$$VE = 1 - IR_V / IR_C$$

where IR_V is the incidence rate in the vaccine group and IR_C is the incidence rate in the control group. Indirect effects of the vaccine may also occur but are more difficult to quantify. For example, the vaccine may induce protection of the population through herd immunity. Since preventive vaccines are given to healthy individuals, the evaluation of safety is critical. There must be assurance that the benefits of potential prevention outweigh the risks.

A major challenge in evaluation preventative vaccines is competing risks. For example, many preventative vaccine trials have long-term follow-up (e.g., follow participants for several years). Death may preclude the occurrence of disease incidence and thus is a competing risk. Treating a death as independent censoring of the primary endpoint of disease incidence may be inappropriate (i.e., when death is not independent of event time). Furthermore, the censoring rate caused by loss-to-follow-up or death may be larger than the incidence rate of the event of interest (e.g., disease for a preventative vaccine) thus dominating the analyses. Designing a study to minimize censoring and development of a plan to handle censoring is crucial.

Behavioral changes can also impact the efficacy assessment of preventative vaccines. For example, individuals who are given preventative vaccines for sexually transmitted diseases such as HIV or HPV, may increase risky sexual behavior due to the belief that they are protected from disease. Thus, counseling is often considered as a tandem intervention in such studies. However, the counseling may help to reduce the incidence of disease in these trials to very low levels, thus resulting in larger sample sizes.

If an existing preventative vaccine has demonstrated effectiveness in reducing incidence to very low levels, then it can be challenging to conduct trials to evaluate new vaccines. Bridging studies may be used to demonstrate the equivalence of vaccine with respect to immune responses or safety measures or the equivalence of vaccines under different conditions of use.

5.3.1.1 Immunogenicity Studies, Field Studies, Lot Consistency Studies

Three types of studies are often considered in the development of biologics in some disease areas: immunogenicity studies, field studies, and lot consistency studies. Consider influenza as an example. Immunogenicity studies are exploratory, often use immune response as the primary endpoint and are conducted prior to field studies. In many cases, an immunogenicity study serves as a PoC study for vaccine development. FDA Guidance (2007, p. 5) indicates, "Immunogenicity evaluations in a substantial number of study participants are important elements of the study design. Characterization of the immune response elicited postvaccination in the clinical endpoint efficacy study may allow for extrapolating the effectiveness to other populations if they have an immune response to vaccination comparable to that observed in the clinical endpoint efficacy study. Furthermore, immune response data collected in the course of a prospectively designed clinical endpoint efficacy study may lead to the establishment of an immune correlate of protection. Such a correlate could greatly facilitate future influenza vaccine development."

Consider the design of an immunogenicity study as part of the development of a trivalent influenza DNA vaccine. Influenza virus is a RNA virus of the orthomyxoviridae family and exists as three distinct virus types: A, B, and C. Influenza viruses A and B are associated with significant morbidity and mortality in humans. Enveloped influenza A and B virions contain two major surface glycoproteins termed hemagglutinin (HA or H) and neuraminidase (NA or N), and these are the predominant antigens of these viruses. Influenza type A virus exists as multiple sub-types with different combinations of these antigens. Only three influenzas A subtypes are currently in general circulation in humans (H1N1, H1N2, and H3N2). Influenza vaccines are usually comprised of antigens from three circulating virus strains. For example, in 2008, the strains were influenza A H1N1 and H3N2, and influenza B. The three strains are selected yearly based on assessment of the influenza variants most likely to represent the circulating strains in that influenza season.

The immunogenicity of the HA of the three viral strains are determined by the titer and frequency of the anti-HA antibody responses against each of the viral strains by hemagglutination inhibition (HAI). Responses are often defined using the geometric mean titer (GMT) or by defining seroconversion using GMT changes with vaccination.

Field trials serve a confirmatory purpose and are generally required by regulatory agencies. These trials use the clinical endpoints and thus generally have large sample sizes and a long duration of follow-up. Each participant receives a single randomized dose of the vaccine and is followed for the clinical endpoint. For example, an influenza vaccine trial may recruit many participants shortly before the influenza season (e.g., September/October in the northern hemisphere, or March/April in the southern hemisphere) and follow participants for influenza symptoms.

Lot consistency studies are designed to demonstrate the consistency of vaccine manufacturing by demonstrating that consecutively manufactured lots of the vaccine, elicit equivalent immune responses (FDA, 2007). Typically, vaccine is manufactured in consecutive lots, and consistency is evaluated with pairwise comparisons of the immunogenicity responses using bio-equivalence methods.

5.3.2 Design Issues in Devices

There are many design challenges in device trials. Many of the statistical issues in drug and biologic trials also apply to devices. Often blinding cannot be implemented and thus resulting biases are a concern. In some cases, a sham can be utilized but this is not always ethical (e.g., in the case of invasive surgeries). Use of historical controls is more prevalent in device trials and such studies thus carry all the concerns associated with nonrandomization trials. As a result, causal inference methods such as propensity scores are often utilized.

Devices are *invented* but drugs are *discovered*. A challenge to the development and evaluation of devices is that the device may change during development. Newer models are developed that improve on older models. Bayesian methods that borrow strength from prior studies have become more common in device trials (Campbell, 2006).

Another challenge to design of trials to evaluate devices is that the success of the device can depend on the skill or experience of the surgeon to implant the device or, in the case of imaging technologies, the skill of the radiologist that reads and interprets the images. Thus, there can be considerable site variation or site-by-treatment interaction. Trials may need to be powered to detect these interactions. The abilities of the surgeons can also change over time. Surgeons often improve over time with greater experience with the device, due to learning effects. Thus "burn-in" periods are frequently implemented as part of the design to allow learning effects to level-off. Training for consistency and uniformity of implementation can also be a critical as part of the design of the trial to reduce variation.

Devices are generally "local" therapies with toxicities confined to the localized area. This can be a significant safety advantage over systemic therapies although implementation of the device is often associated with safety concerns. Some implant devices cannot easily be removed and thus the risk of removal must be weighed against the risk of remaining in the body. Device trials are less likely to have PK/PD studies conducted given the localized therapy.

5.3.3 Multicenter Trials

Many late-phase trials require a large sample size (e.g., several hundred of several thousand participants). In order to enroll such a trial, multiple investigators and associated sites (or "centers") are needed to participate in the trial. These trials are termed "multicenter trials." Investigators and associated centers are often selected by matching the needs from the trial with the expertise of the investigator and the center. For example, investigators for phase I trials often have expertise in PK whereas investigators participating in a phase III trial for rheumatoid arthritis may be rheumatologists.

Another reason for conducting multicenter trials is to ensure generalizability of the trial results. A multicenter trial ensures that the intervention is evaluated in a variety of different geographical areas (noting that medical practices may vary by location) and on participants with varying demographics. Participants recruited from multiple centers tend to be more representative of the broader population. This is particularly important in phase III pivotal trials.

Some centers may recruit participants quickly while others will only be able to recruit a few participants. If some centers do not enroll many participants, then centers may have to be combined during analyses to estimate "center effects" or treatment-by-center interactions when it is desirable to estimate these effects. If a few centers enroll most of the study participants, then generalizability may be limited. Researchers can consider upper limits to center enrollment to help avoid this issues.

5.3.3.1 Multiregional Trials

Multiregional trials may be utilized (1) when it is difficult to enroll a sufficient number of study participants on a single continent within a reasonable time frame, (2) to establish the efficacy and safety of an intervention in different populations, (3) to investigate the effect of ethnic differences in intervention response, and (4) to register the intervention with multiple regulatory agencies. For example, many trials investigating therapies for HIV are conducted in resource-limited countries since the prevalence of HIV is highest in these countries. Many pharmaceutical sponsors have an interest in simultaneous development of an intervention in different regions. A common multiregional study is a bridging study that attempts to investigate the effect of ethnic factors on the response to an intervention.

In multiregional trials, there can be considerable center variation. This variation could occur for many reasons. For example, considerable site variation in neuropsychological responses were observed in ACTG A5199, an international neurological study of HIV-infected participants from Malawi, Zimbabwe, South Africa, Thailand, Brazil, Peru, and India. Differences in responses could be due to differing populations, cultures, socioeconomic factors, HIV resistance patterns, viral subtypes, or variation in test administration. Training is critical in order to standardize test administration and reduce variation when possible. Researchers should consider stratified randomization when conducting multiregional trials to ensure that intervention is balanced within region.

Multiregional trials pose many practical challenges including language barriers, variation in ethics, variation in drug distribution laws, establishing an appropriate infrastructure in each country, and arranging meetings and conference calls with people from varying time zones. For example, the translation of CRFs may be required, and then they will need to be back translated and validated, before they can be used to collect clinical data.

5.3.4 Design Issues in Rare Diseases

Trials involving rare diseases are challenging to conduct since an available pool of study participants are not readily available (Haffner, 1998). Thus, trials investigating rare diseases often have protracted enrollment periods. A trial of itraconazole for the prevention of severe fungal infection in children and adults with chronic granulomatous disease required 10 years to enroll 39 participants (Gallin et al., 2003).

When designing trials for interventions to treat rare diseases, it is advisable to search for designs (e.g., crossovers, matching, or historical controls) that can reduce variation, and thus, effectively the required sample size. Gallin et al. (2003) used a novel design in which participants were randomized to itraconazole or placebo for 1 year and then to alternate between itraconazole and placebo on an annual basis. The number of participants that developed a severe fungal infection while on itraconazole was compared to the number of participants that developed a severe fungal infection while not on itraconazole. The validity of such a design relies on the assumptions associated with crossover trials (Lagakos, 2003).

Since trials of interventions to treat rare diseases often have protracted enrollment periods, the outcome of participants enrolled early into the trial is often known prior to other participants being randomized. This allows for the opportunity to conduct interim analyses and make mid-trial adjustments, if necessary. Exact methods may be needed during the analyses of trials for rare diseases due to the small sample sizes.

A common question with trials of rare diseases is that whether the usual standards of certainty should be used to inform clinical practice. Some argue that when traditionally powered trials cannot be completed for a long period, it may be reasonable to adopt a therapy with less than definitive evidence (particularly when alternative options are limited or when risk is low), by relaxing the usually selected type I and II error rates, and then continually reevaluate the evidence as it accumulates.

5.3.5 Bayesian Designs

p-values and confidence intervals are a product of a traditional "frequentist" approach to statistics. A p-value is the probability of observing data as or more extreme than that observed if the null hypothesis is true (i.e., the probability of the data given a hypothesis being true). Researchers are often more interested in the question, "what is the probability that a hypothesis is true given the data?" Traditional frequentist statisticians view this as asking about the probability of a fact (i.e., either the hypothesis is true or not, and thus the probability is either 0 or 1).

However, an alternative statistical approach, Bayesian statistics, allows for the calculation of the probability of a hypothesis being true (e.g., an intervention is effective) given the data. This concept can be appealing to many researchers. Bayesian approaches are based on the idea that unknown quantities (e.g., a treatment difference) have probability distributions. One potential disadvantage of this approach is that it requires additional assumptions and researchers generally try to move towards having fewer assumptions so that results are robust. However, these assumptions can also be viewed as a mechanism for incorporating prior information about an intervention. The assumptions or beliefs (called *prior distributions* or *priors* in Bayesian terms) incorporate prior beliefs about the hypothesis. Historical data (e.g., all currently available information) or expert opinion can be used to help construct the prior distribution although specification is always partly subjective. This might be an attractive approach when sound prior knowledge based on reliable data is available. For example, use of Bayesian statistics has become more common in the design of clinical trials evaluating devices since there is often a great deal of prior information (e.g., the mechanism is physical and local) in contrast to drugs. After a trial is completed, the prior distribution is then updated to a *posterior distribution* or *posterior* using data collected in the trial. For this reason, a Bayesian approach is often viewed as a formal iterative mechanism for learning (i.e., update prior information based on new information). Researchers can then interpret this posterior as a "degree of belief" that a hypothesis is true (e.g., a positive intervention effect) conditional upon the data. A goal of the trial may then be to show that the posterior probability (i.e., degree of belief) that the intervention is effective exceeds a predetermined value.

A simple example that illustrates the differences between frequentist and Bayesian approaches is in the evaluation of diagnostic tests. Sensitivity (the probability of a diagnostic test being positive when a person is truly diseased) and specificity (the probability of a diagnostic test being negative when a person is truly nondiseased) are examples of frequentist probabilities. However, one may wish to know the positive predictive value (the probability that a person is truly diseased given a positive diagnostic test). This probability can be calculated using Bayes theorem but it requires an assumption about the prevalence of the disease in the population to which the patient belongs. The positive predictive value of a diagnostic test will vary depending on the prevalence of the disease in the population in which the diagnostic is being utilized. The probability that a particular person is diseased might first be estimated using the prevalence of the disease in the general population (the prior). The probability that the person is diseased can be updated when the result of the diagnostic test is known (posterior).

One of the primary challenges to a Bayesian approach to design is the selection of a prior distribution. Some suggest that the selection should be developed using only objective, quantitative prior information (e.g., data from prior trials). However, the available data may be limited or may be subject to selection bias. Others argue that the flexibility to use subjective judgment is an advantage. The prior distribution can theoretically be of any form. The prior is then combined with the observed data via a likelihood function to produce the posterior. The shape of the posterior depends on the prior and on likelihood function. Bayesian analyses are simplified by an appropriate choice of the prior distribution so that the posterior distribution is a member of the same family of distributions as the prior. Priors that satisfy this condition are called *conjugate priors*. The advantage of conjugate priors is that they result in closed-form expressions for the posterior distribution. For example, if both the prior and the likelihood are normal then the posterior is normal. When making inferences about a proportion, if the prior is Beta(a,b), and then successes (s) and failures (f) are observed, then the posterior is Beta(a + s,b + f). Calculating nonconjugate posteriors is computationally intensive but methods include asymptotic approaches (e.g., normal approximations, Laplace's method), noniterative Monte Carlo methods (e.g., direct sampling, weighted bootstrap), and Markov chain Monte Carlo methods (e.g., Gibbs sampling).

Some suggest outlining a family of priors. However, a proper analysis must assess the sensitivity of the resulting interpretation as priors vary. Several common priors include a *noninformative prior* (a prior in which no treatment effect size is considered more likely than any other), a *reference prior* (where there is minimal prior information), the *skeptical prior* (where large intervention effects are unlikely), the *clinical prior* (formed on the basis of clinician opinion), and an *enthusiastic prior* (where large intervention effects are likely).

One advantage of a Bayesian approach to sample size determination is the ability to account for the uncertainty associated with prior information via specification of the prior distribution in contrast to frequentist methods. There is no consensus approach to sample size determination in Bayesian designs, but approaches can be classified into two primary strategies: (1) techniques that treat sample size determination as a decision problem by employing a loss or utility function (i.e., a penalty for errors in estimation and cost of obtaining sampling information) with the strategy of finding the sample size that minimizes the loss function and (2) methods that are solely concerned with inference about a parameter (e.g., size to obtain a reasonable probability of getting a posterior interval less than a specified width). When monitoring a trial, the reason for stopping a trial early does not affect Bayesian inference and thus there is no penalty for multiple looks at the data. There are no p-values in Bayesian inference and thus no error spending concern.

A common dilemma is whether a frequentist or Bayesian approach to trial design is "best" for a particular trial. This has caused some divide in the statistical community. However, the approaches should not be viewed as competing. Frequentist and Bayesian approaches address different research questions and should be viewed as complementary. Bayesian methods should be seen as another approach or tool to help understand the data. It is important that Bayesian methods be used responsibly (e.g., outlined prospectively), and should not be used to "rescue" failed or negative trials that utilize frequentist methods.

References

Ader R et al. 2010. Conditioned pharmacotherapeutic effects: A preliminary study. *Psychosom Med* 72:192–197.

Asakura K, Hamasaki T, Sugimoto T, Hayashi K, Evans SR, Sozu T. 2014. Sample size determination in group-sequential clinical trials with two co-primary endpoints. *Stat Med* 33:2897–2913.

Bellamy N, Campbell J, Stevens J, Pilch L, Stewart C, Mahmood Z. 1997. Validation study of a computerized version of the Western Ontario and McMaster Universities VA 3.0 Osteoarthritis Index. *J Rheumatol* 24:2413–2415.

Campbell G. 2006. The role of statistics in medical devices—The contrast with pharmaceuticals. *Biopharmaceut Rep* 14:1.

Desbiens NA. 2000. Lessons learned from attempts to establish the blind in placebo-controlled trials of zinc for the common cold. *Ann Intern Med* 133(4):302–303.

Donner A. 1984. Approaches to sample size estimation in the design of clinical trials—A review. *Stat Med* 3:199–214.

Ellenberg SS, Temple R. 2000. Placebo-controlled trials and active-control trials in the evaluation of new treatments (Part 2). *Ann Intern Med* 133(6):464–470.

Evans SR et al. 2007. A randomized trial evaluating Prosaptide™ for HIV-associated sensory neuropathies: Use of an electronic diary to record neuropathic pain. *PLoS ONE*, 2(7): e551. doi:10.1371/journal.pone.0000551.

FDA. 2007. Guidance for Industry: Clinical Data Needed to Support the Licensure of Seasonal Inactivated Influenza Vaccines. U.S. Department of Health and Human Services Food and Drug Administration, Center for Biologics Evaluation and Research, May 2007.

Food and Drug Administration. 2012. Guidance for Industry. Irritable Bowel Syndrome: Clinical Evaluation of Products for Treatment, Center for Drug Evaluation and Research, Food and Drug Administration, Rockville, MD, 2012.

Ferreira-Gonzalez I et al. 2007. Problems with use of composite endpoints in cardiovascular trials: Aystematic review of randomized trials. *BMJ* 334:786.

Finniss DG, Kaptchuk TJ, Miller F, Benedetti F. 2010. Biological, clinical, and ethical advances of placebo effects. *Lancet* 375:686–695.

Freemantle N et al. 2003. Composite outcomes in randomized trials. *JAMA* 289(19):2554–2559.

Gallin JI et al. 2003. Itraconazole to prevent fungal infections in chronic granulomatous disease. *NEJM* 348:2416–2422.

Gehan EA. 1984. The evaluation of therapies: Historical control studies. *Stat Med* 3:315–324.

Guidance for Industry. Clinical Data Needed to Support the Licensure of Seasonal Inactivated Influenza Vaccines U.S. Department of Health and Human Services Food and Drug Administration Center for Biologics Evaluation and Research. May 2007.

Haffner ME. 1998. Designing clinical trials to study rare disease treatment. *Drug Inf J* 32:957–960.

Hamasaki T, Asakura K, Evans SR, Sugimoto T, Suzo T. 2015. Group-sequential strategies in clinical trials with multiple co-primary outcomes. *Stat Biopharm Res* 7(1): 1–19.

Hamasaki T, Sugimoto T, Evans SR, Sozu T. 2013. Sample size determination for clinical trials with co-primary outcomes: Exponential event-times. *Pharm Stat* 12:28–34.

Hrobjartsson A, Gotzsche PC. 2001. Is the placebo powerless? *NEJM* 344(21):1594–1602.

ICH E10. 2001. Choice of control group and related issues in clinical trials, International Conference on Harmonization. www.ich.org

International Conference on Harmonisation of Technical Requirements for Registration of Pharmaceuticals for Human Use. February 1998. ICH Harmonised Tripartite Guideline. Statistical Principles for Clinical Trials. E9.

Kaptchuk TJ et al. 2008. Components of placebo effect: Randomised controlled trial in patients with irritable bowel syndrome. *BMJ* 336(7651):999–1003.

Kaptchuk TJ et al. 2010. Placebos without deception: A randomized controlled trial in irritable bowel syndrome. *PLoS One* 5(12):e15591.

Lachin JM. 1981. Introduction to sample size determination and power analysis for clinical trials. *Controlled Clin Trials* 2:93–113.

Lagakos SW. 2003. Clinical trials and rare diseases. *NEJM* 348(24):2455–2456.

Little et al. 2012. The prevention and treatment of missing data. *NEJM* 367(14): 1355–1360.

Lubsen J, Kirwan BA. 2002. Combined endpoints: Can we use them? *Stat Med* 21:2959–2970.

McFadden E. 2007. *Management of Data in Clinical Trials*. 2nd edn. Hoboken, NJ: Wiley.

Montori VM et al. 2005. Validity of composite endpoints in clinical trials. *BMJ* 330:594–596.

National Research Council. 2010. *The Prevention and Treatment of Missing Data in Clinical Trials*. Washington, DC: National Academies Press.

Neaton J et al. 2005. Key issues in end point selection for heart failure trials: Composite endpoints. *J Card Fail* 11(8):567–575.

Noseworthy JH, Ebers GC, Vandervoort MK, Farguhar RE, Yetisir E, Roberts R. 1994. The impact of blinding on the results of a randomized placebo-controlled multiple sclerosis clinical trial. *Neurology* 44(1):16–20.

Prentice RL. 1989. Surrogate endpoints in clinical tirals: Definitions and operational criteria. *Statistics in Medicine* 8(4):431–440.

Rosenberger WF, Lachin JM. 2002. *Randomization in Clinical Trials: Theory and Practice.* New York: John Wiley & sons, 0-471-23626-8.

Sacks H, Chalmers TC, Smith H. 1982. Randomized versus historical controls for clinical trials. *Am J Med* 72:233–240.

Simpson SH et al. 2006. A meta-analysis of the association between adherence to drug therapy and mortality. *BMJ* 333(7557):15.

Sozu T, Hamasaki T, Sugimoto T, Evans SR. 2015. *Sample Size Determination in Clinical Trials with Multiple Correlated Endpoints.* Heidelberg: Springer.

Sugimoto T, Sozu T, Hamasaki T, Evans SR. 2013. A logrank test-based method for sizing clinical trials with two co-primary time-to-event endpoints. *Biostatistics* 14(3):409–421. doi:10.1093/biostatistics/kxs057.

Temple R, Ellenberg SS. 2000. Placebo-controlled trials and active-control trials in the evaluation of new treatments (Part 1). *Ann Intern Med* 133(6):455–463.

Tripathy P et al. 2010. Effect of a participatory intervention with women's groups on birth outcomes and maternal depression in Jharkhand and Orissa, India: A cluster-randomized controlled trial. *Lancet* 375:1182–1192.

Wechsler ME et al. 2011. Active Albuterol or placebo, sham acupuncture, or no intervention in asthma. *NEJM* 365(2):119–126.

Wei LJ. 1978. The application of an urn model in controlled clinical trials. *J Am Stat Assoc* 73:559–563.

Wei LJ, Durham S. 1978. The randomized "play-the-winner" rule. *J Am Stat Assoc* 73:840–843.

Williams EJ. 1949. Experimental designs balanced for the estimation of residual effects of treatments. *Australian Journal of Scientific Research* 2(3):149–168.

Wong WK, Lachenbruch PA. 1996. Tutorial in biostatistics: Designing studies for dose response. *Stat Med* 15:343–359.

Yao Q, Wei LJ. 1996. Play the winner for phase II/III clinical trials (Disc: p2455–2458). *Stat Med* 15:2413–2423.

6

Clinical Trial Designs

The selection and development of the trial design depends upon the objectives of the trial and the specific characteristics of the disease, the medical interventions under study, the population of interest, and the instruments that will be used to measure the effect of the intervention. General considerations in designing clinical trials are discussed in Chapter 5. This discusses common clinical trial designs. Most trials utilize designs that are variants or extensions of the designs described here.

6.1 Phase I

For a drug to enter clinical development and be tested in humans, the drug must pass all the nonclinical challenges and satisfy both the IND and the IRB reviews. Once these milestones have been accomplished, then phase I studies can begin.

There are many types of phase I clinical trials including first in human (FIH), single-dose dose escalation, multiple-dose dose escalation, PK/PD, food effect, drug–drug interaction (DDI), bioavailability (BA), bioequivalence (BE), and other types of studies. Many phase I studies utilize crossover designs. For drugs that are developed to treat nonlife-threatening diseases, phase I studies generally enroll healthy normal volunteers into placebo-controlled designs. However, when developing drugs to treat life-threatening diseases, subjects with the disease are typically recruited. For example, most traditional chemotherapies for treating cancer are "cytotoxic agents." These therapies are not only toxic to the tumor cells but also to normal tissues. Thus, it is not ethical to use healthy volunteers when studying these agents. Therefore, cancer patients are recruited into the phase I studies.

Phase I clinical trials are generally designed to study the PK/PD properties and tolerability (i.e., safety) of a drug, with the intent of providing appropriate information for designing phase II/III studies. Specifically, a primary goal of phase I studies is to estimate the maximally tolerated dose (MTD).

6.1.1 PK/PD Designs

PK is the study of how a drug is processed after it is administered into a human body. This process is often divided into several areas including, but not limited to, the extent and rate of absorption, distribution, metabolism and excretion (ADME). Absorption is the process of a substance entering the body, distribution is the dispersion or dissemination of the drug throughout the bodily fluids and tissues, metabolism is the transformation of the drug and its daughter metabolites, and excretion is the elimination of the drug from the body. In some cases, drugs will irreversibly accumulate in bodily tissues.

PD is the study of the biochemical and physiological effects of drugs, the mechanisms of drug action, and the relationship between the drug concentration and effect. For example, a drug may cause a change in heart rate or respiration. It is often said, "PD is the study of what a drug does to the body, whereas PK is the study of what the body does to a drug."

Figure 6.1 presents an example of a drug concentration-time curve, plotting the concentration of the drug within a compartment of the body versus the time since the drug was administered. Data on drug concentration are collected at discrete time points. Typical variables used for analysis of PK activities include area under the curve (AUC), maximum concentration (Cmax), minimum concentration (Cmin), time to maximum concentration (Tmax), and half-life (Table 6.1). To calculate the AUC, discretely observed points over time are connected, and the AUC is estimated using the trapezoidal

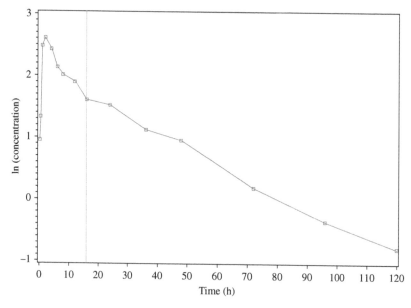

FIGURE 6.1
Example of a time–concentration curve.

TABLE 6.1

Common PK Analysis Variables

- *Area under the curve (AUC)*: Defined as an area under the curve (drug concentration plotted over time after dosing). Often calculated using the trapezoidal rule.
- *Maximum concentration (Cmax)*: The peak concentration of the drug in the body.
- *Minimum concentration (Cmin)*: Trough concentration of a drug after dosing but prior to administration of a subsequent dose.
- *Time to maximum concentration (Tmax)*: The time of dosing to Cmax.
- *Half-life*: Defined as the time required for the drug concentration to decrease by half.

rule. Often the AUC and Cmax are transformed using a natural log for data analyses. The half-life of a drug is the time interval required for the drug concentration to decay to half of its initial value.

In a PK study, samples are generally collected at several prespecified time points from each participant in the study. Observations between participants are independent, but observations within a participant are dependent. Data analysis can be approached from three perspectives: (1) establish a summary measure for each participant and then summarize across participants, (2) construct a summary statistic (e.g., mean) across participants for each time point and then combine these summary statistics over time, or (3) use a mixed effects model incorporating both a (random) participant effect and the time effect (either as a fixed effect or as a random effect). The most common approach is based on the method (1), to analyze summary measures obtained from each participant.

6.1.2 Bioavailability/Bioequivalence

Many phase I trials are designed to study the BA of a drug or the BE among different formulations of the same drug. BA refers to the rate and extent to which the active ingredient is absorbed from a drug and becomes available at the site of action within the body. BE refers to demonstration of the absence of an important difference in the rate and extent to which the active ingredient in pharmaceutical equivalents or pharmaceutical alternatives becomes available at the site of drug action when administered at the same dose under similar conditions in an appropriately designed study (FDA Guidance, 2002). Participants recruited for these studies are generally in good health. BA/BE studies are conducted by measuring drug concentration levels in blood or serum over time from participating subjects.

The objective of a BA study is usually to estimate the amount of drug that is "biologically available" in the body and to study the ADME processes. The objective of a BE study is to demonstrate the "biological equivalence" among two drugs or different formulations of the same drug. BE is important in drug development since various formulations and dose strengths are being developed, tested, and compared during development. BE is also important in the evaluation of the optimal manufacturing process. Often the formulation used in clinical trials is different from the formulation used for marketing. Two 10 mg tablets may not deliver the same amount of active ingredients as one 20 mg tablet and the drug activity of one 10 mg tablet may be different from that of one 10 mg capsule. Thus, BE studies may also be conducted during later stages of new drug development.

A major application of BE studies is the development of generic drugs. After the patent on a brand name drug is expired, generic drugs can be made available to the public at lower prices. In order for a generic drug to be approved for general public use, the generic drug must demonstrate BE to the brand name drug. FDA guidelines to regulate the approval of a generic drug are based on these BE studies. The 2002 FDA guidance (FDA guidance, 2002) is applicable for approving generic drugs, in addition to evaluating different formulations of the same drug.

Conceptually, one needs to show that two formulations are "not different" in order to claim BE. "Not different" does not mean that the two formulations are exactly equivalent, but instead that the magnitude of the differences is not meaningful or relevant. Since one cannot conclude BE with traditional use of hypothesis testing and p-values, BE is generally assessed using a confidence interval (CI), noting whether the bounds of the CI are within the bounds of differences that are considered irrelevant.

The FDA has developed several guidelines to regulate the evaluation of BE. The 2002 guidance discusses average bioequivalence (ABE), where the AUC and Cmax are logs transformed, and then CIs for the ratio of geometric means of the test formulation relative to the reference is calculated. If the entire CI lies in within 80% and 125%, then the drug is considered BE with respect to the parameter in question. In general, two formulations are considered BE if both the AUC and the Cmax satisfy this 80%–125% criteria. In this guidance, the determination of BE is based on the geometric means without consideration of the variability of the PK parameter estimates.

In 2001, the FDA issued a guidance document that discusses approaches to establishing the average population and individual bioequivalences (IBEs). When establishing BE between a test drug and a reference drug (or two formulations of the same drug), the common approach is to compare the two group means of the PK parameters. This is the ABE approach. However, it may also be of interest to establish the similarity of the variances of the PK parameters for the two drugs. Analyses of population bioequivalence (PBE) and IBE include evaluation of both the means and variances of the PK parameter.

One concern with ABE is that it does not address the issue of drug interchangeability. PBE and IBE address drug interchangeability under two different clinical situations. PBE evaluates the interchangeability for a patient that has not previously received either the test or the reference drug, whereas IBE evaluates interchangeability for a patient that switches from a reference drug to a test drug (Anderson and Hauck, 1990).

To guarantee PBE of the test and the reference drugs, the population distribution (mean and variability) of the PK measurements for the test and the reference drugs should be similar when these drugs are administered to a drug-naive patient. To aid the evaluation of "similarity," the FDA 2001 Guidance adopted the concept of the population difference ratio (PDR), which measures the ratio of the expected squared difference between the test and the reference drugs to the expected squared differences between two administrations of the reference drug, when these drugs are given to drug-naive patients. The PDR is defined as

$$PDR = \sqrt{\frac{E(T-R)^2}{E(R-R')^2}}$$

Conceptually, a test drug (T) is PBE to a reference drug (R) if the PDR is within an acceptable limit. The individual difference ratio (IDR) can be defined similarly, and criteria for evaluation of IBE can also be derived.

6.1.3 Estimation of MTD

One of the key objectives of phase I drug trials is to provide a good estimate of the MTD. MTD is considered the maximum dose acceptably tolerated by a particular patient population. Phase II dose-ranging and dose-finding trials are designed using the MTD as an upper limit for doses to be evaluated. If the MTD is over estimated, then phase II studies may use a dose that is too high, and participants may experience intolerable toxicity. However, if the MTD is underestimated, then an efficacious dose may not be identified.

In the development of drugs for treating life-threatening diseases such as cancer, phase I trials are conducted to obtain information on the dose–toxicity relationship (Ivanova, 2006). Toxicity in oncology trials is graded using the National Cancer Institute Common Terminology Criteria for Adverse Events. The dose-limiting toxicity (DLT) is usually defined as treatment related non-hematological toxicity of grade three or higher, or treatment-related hematological toxicity of grade four or higher. The toxicity outcome is typically binary (DLT/no DLT). The underlying assumption is that the probability of toxicity is a nondecreasing function of dose. The MTD is defined as the dose at which the probability of toxicity is equal to the maximally tolerated level, Γ. For example, the MTD can be defined as the dose level just below the lowest dose level where two or more out of six patients had toxicity. In this case, the MTD can be estimated for any monotone dose–toxicity relationship. In phase I oncology studies, Γ ranges from 0.1 to 0.35. An important ethical issue to consider in designing such trials (Ratain et al., 1993) is the need to minimize the number of patients treated with toxic doses. Therefore, patients in oncology dose-finding trials are often assigned sequentially starting with the lowest dose.

For many nonlife-threatening diseases, MTD estimation is not necessarily based on formally designed MTD studies. One main reason for this is that DLT is not a well-defined and widely accepted term in these applications. Here, dose-escalation studies as well as PK/PD studies are used to help estimate MTD without considering DLT.

Whether the estimated MTD is based on formally statistically designed MTD studies or not, there is always a risk of over- or underestimating the MTD, especially for nonlife-threatening diseases. When MTD is estimated in phase I from healthy normal volunteers, this estimate may not necessarily directly translate to MTD in patients with the disease. There are many potential reasons that differences in responses from patients and from healthy volunteers may differ: patients may be older than healthy volunteers, PK/PD profiles could be different, patients with the target disease may also have other diseases, and patients may be taking additional medications. Even when patients were used in phase I with formally designed MTD studies, the MTD is estimated from short-term studies with small sample sizes. Later phase trials are longer, and the true MTD might be different from the MTD estimated from phase I. Drug formulation may differ in early versus later phase studies as well. In order to allow flexibility in phase I dosing, drug formulation in phase I tends to be dry powder formulation, suspension, or other flexible formulations. However, phase II formulation tends to be tablets or capsules to provide convenience for patients. Furthermore, phase I clinical research units usually keep the subjects in the unit overnight (inpatient settings) for data collection, while phase II clinics observe patients only when there is a scheduled clinical visit (outpatient settings). These differences may also cause potentially over- or underestimation of MTD.

6.2 Other Trial Designs Including Phase II and III

6.2.1 Proof of Concept Study

In the development of a new drug for nonlife-threatening diseases, a PoC study is designed in early phase II to evaluate if the drug could show drug efficacy based on the nonclinical and early clinical findings. Typically, phase I trials for these drugs recruit healthy normal volunteers to help establish the PK/PD profile. Since efficacy can only be evaluated in patients with the target disease, there is no available efficacy data at the beginning of phase II. A drug may show efficacy in animal studies and nonclinical experiments, but this does not imply that efficacy will be shown in humans. A PoC study evaluates efficacy to aid decisions regarding future drug development.

A typical PoC study is designed with two treatment groups: a high dose of the drug and a control group. The high dose is usually the MTD or a dose that is a bit lower than the MTD to avoid potential safety concerns. This dose should be high enough so that it will provide the best opportunity for the drug to demonstrate efficacy. The advantage of this design is resource efficiency; a single PoC study can identify a nonefficacious drug and provide sufficient information to stop further development. A disadvantage of a PoC study is that little dose–response information is obtained. Thus, after the concept is proven, the next step is often to conduct a dose-ranging study to evaluate the efficacy and safety at various doses. This dose-ranging study can be viewed as the first dose–response study and it typically includes a high dose, a placebo control, and a few lower doses.

6.2.2 Dose-Finding Study Designs

Dose–response trials are usually carried out in phase II. The MTD is often known from phase I studies, and it is generally assumed that the efficacy is nondecreasing with increasing doses. However, there is often a considerable uncertainty regarding the shape of the dose-response relationship. Under varying assumed dose–response curves, there are differing strategies for allocating doses. For example, consider possible dose–response curves shown in Figure 6.2 to detect the relationship displayed in curve 3, higher doses should be selected. However, to detect the relationship displayed in curve 1, lower doses should be selected. Thus, the dose allocation strategy varies depending on the assumed underlying dose–response curve.

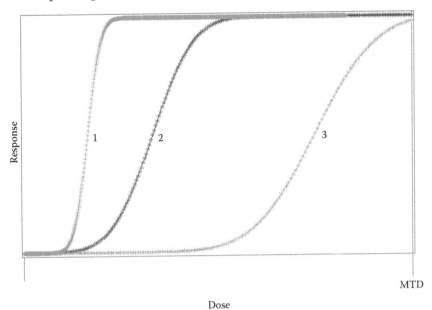

FIGURE 6.2
Several possible dose–response curves.

When designing a phase II dose–response clinical trial, these issues need to be considered: dose frequency, dose range, number of doses, dose spacing, use of a control, sample size for each dose, and fixed-dose versus dose titration.

6.2.2.1 Frequency of Dosing

It is important to determine how often a subject should take the drug (e.g., once a day [QD] vs. twice a day [BID]). Dosing frequency is frequently guided by the results of the phase I PK/PD findings. The PK estimate of the drug half-life is often used to propose a dosing frequency. In certain cases, there may be a need to study more than one dosing frequency in a single study. This may occur, for example, when the PK variability of the drug is high, and it is difficult to identify a single dosing frequency to utilize for further development. In these cases, a factorial design of multiple doses and frequencies could be considered.

It is important to consider simultaneously the PK and PD effects at particular doses. Drug concentrations necessary for PD activity may be different from those required for PK activities. It is possible for PK estimates to indicate insufficient drug in the body, but PD activities can still occur. Conversely, PK parameter estimates may indicate sufficient drug concentrations but may not be enough to elicit PD responses. Thus, the dose frequency derived from PK analyses may overestimate or underestimate the concentration necessary for a PD response.

If it is desirable to have a once daily-dosing treatment, but the drug under development has a twice-a-day PK profile, then a formulation change may be necessary so that the test drug can be used as once-a-day treatment. Figure 6.3 presents the time–concentration curves of a once-a-day versus a twice-a-day dosing. The horizontal line that is above the x-axis represents the efficacy concentration level (often based on PD information). Theoretically speaking, it is desirable to keep the drug concentration above this line at all the times to ensure drug efficacy. Two strategies can be used to achieve this concentration: (1) dose the subject BID with a low dose (the lower curves in Figure 6.3) or (2) dose the subject QD with a high dose (in this case, twice the dosage of the low dose, the higher curve in Figure 6.3). Note that in the first few hours postdosing, the high dose may result in a very high concentration that could potentially cause severe adverse events. When this is the case, reformulation of the drug may be necessary so that when dosed as QD, the Cmax would not be too high, while the efficacy concentration can be maintained throughout the entire 24-h period.

Sometimes drug reformulation is necessary. Possible reasons for this include to help absorption or to change the half-life. Notably, the PK/PD properties of the drug are different from what they were prior to reformulation. Hence, all of the studies to obtain dosing information that were conducted prior to the reformulation need to be repeated.

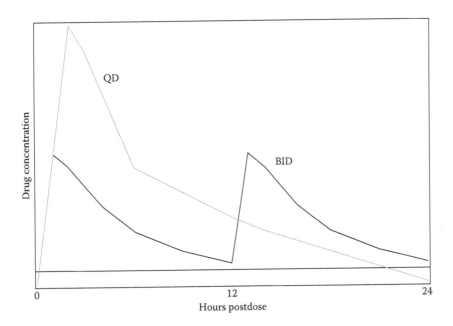

FIGURE 6.3
Once-a-day vs. twice-a-day dosing.

6.2.2.2 Fixed Dose versus Dose Titration Designs

Dose-finding studies can be characterized as a fixed dose design or a dose titration design. For fixed-dose designs, participants are randomized to a dose, and they continue to take the same dose throughout the complete dosing period. Contrastingly, in a dose titration design, a participant is randomized to a dosing strategy such that the dose can change over time. Participants are often randomized to start with a low dose and depending on their response or a predetermined schedule, the dose is gradually increased until a suitable dose level is found. For example, in "titration to response" designs, a participant who responds to a low dose may continue to remain on the low dose, whereas a participant that does not respond to the low dose after a prespecified period will receive the next higher dose of the drug.

Titration designs have some advantages. These studies allow participants to be treated at the optimal patient-specific dose reflecting actual medical practice (personalized medicine). However, titration designs are difficult to analyze and interpret the effects of time and doses are confounded. It is also difficult to distinguish the effects of accumulating exposure to a drug versus the effect of the most recent dose. For example, if a participant responded to the drug after the dose has been escalated, it is unclear whether the higher dose or the accumulation of the lower dose caused the response. Titration designs can have multiple treatment groups with overlapping doses (e.g., one treatment group is 10 mg escalating to 20 mg and another group is 20 mg escalating to 40 mg) which can create challenges in the evaluation of the overlapping dose (i.e., 20 mg). Although drug efficacy may be demonstrated in a titration design, it is very challenging to make valid dosing inferences. Such studies are designed to evaluate the strategy of titration dosing rather than the evaluation of individual doses.

In rare situations, a concentration response study is designed. A concentration response study assesses the efficacy and safety measurements observed from each participant according to the plasma concentration of the drug instead of the dose of the drug. There are many practical challenges in such designs including how to blind the participant and the physician, and when to measure the blood concentration.

Due to the complexities associated with dose titration designs and concentration response designs, the fixed dose designs are more commonly used for assessment of dose–response. Thus, participants are generally randomized to a few fixed dose groups that are compared to one or more control groups.

6.2.2.3 Range of Doses to Be Studied

Drug efficacy can only be studied from participants with the targeted disease. Since phase I trials for nonlife-threatening diseases typically enroll healthy participants, efficacy data is unavailable and thus cannot inform regarding the dose range to be studied in phase II. With the primary goal to describe the efficacy and safety dose–response curves, studies should be designed to estimate the maximal effective dose (MaxED) and the minimal effective dose (MinED). Furthermore, although MTD estimates are available from phase I studies, data from phase II studies is helpful to confirm and adjust the MTD estimates. Strategically the first dose ranging study should cover a wide dose range to identify the "active" doses. A second study can then be designed to capture the dose–response relationship in the sensitive dose range.

Nonclinical data on the drug and possibly clinical and nonclinical data on related drugs may provide an estimate of the minimum drug concentration that is required for efficacy. This information combined with PK data will suggest a target dose range to explore and will further suggest a minimum efficacy dose that may be desirable to include in the trial.

In dose–response designs, the *dose range* is defined as the range between the lowest and the highest dose. The dose range can be expressed as the ratio of the highest dose over the lowest dose. For example, in a design with placebo, 20, 30, and 40 mg, the dose range is 2. As a rule of thumb, in the first dose-ranging study, the dose range should be at least 10.

6.2.2.4 Number of Doses

In order to develop a complete picture of the dose–response relationship, it is desirable to study as many doses as feasible. There are practical constraints in determining the number of dosing groups. In general, increasing the number of doses increases the number of required participants, potentially affecting feasibility and complexity. For these reasons, it may be desirable to set a ceiling on the number of dosing groups.

The selection of the number of doses can be affected by other factors. Often the number of doses that can be used is restricted due to limited dosage form availability in the early stage of clinical development. For example, if the available tablet strengths are 10, 20, and 50 mg, then it would not be possible to study doses of 1 or 25 mg. More dosing groups may also create the need for additional matching placebos in order to maintain the study blind. The complexity of the regimen could also affect whether participants can adhere to the assigned therapy. For these reasons, it is rare to have clinical trial designs with more than 6 or 7 dosing groups, with 3–5 dosing groups being the most common.

With careful and creative planning of capsule sizes (e.g., 0, 5, 25, 100 mg) and the use of multiple capsules per dosing period (e.g., 3), it is possible to construct numerous distinct dosing levels that span a wide range (e.g., $300/5 = 60$-fold). This often requires reasonable knowledge or prediction of the MTD.

6.2.2.5 Dose Allocation and Dose Spacing

Dose allocation depends upon the primary clinical question of interest. This may be detecting an effect, estimating the slope of the dose–response curve near the low doses (or the high doses), or identifying the lowest dose (MinED) with an effect of at least a minimally clinically important difference.

Data from phase I studies are helpful in selecting a high dose (near the MTD) for inclusion. However, limited data is available for guiding the choice of the lower and medium doses due to the uncertainty of the underlying dose–response curve. In these cases, consideration of other information including preclinical data and data from similar compounds and consultation with pharmacokineticists, clinicians, and pharmacologists are important. Wong and Lachenbruch (1996) introduced the strategy of equal dose spacing from low to high. Others have employed log dose spacing (e.g., 1, 3, 10, and 30 mg). At the earlier stages of phase II, it is desirable to test a wide dose range, and thus, the log dose spacing strategy is often preferable. In modern dose ranging trials, equal dose spacing designs are rare.

Hamlett et al. (2002) proposed the use of a binary dose spacing (BDS) design. If the trial includes two dosing groups and a placebo, then the BDS design identifies a midpoint between the placebo and the MTD and allocates one dose above and another below the midpoint. If the trial includes three dosing groups and a placebo, then the BDS design identifies the high dose as described in the two-dose case. A second midpoint is then identified between the placebo and the first midpoint, a low dose is selected below the second midpoint, and the medium dose is selected between the two midpoints. If there are more dosing groups, then the BDS design iteratively follows this strategy. The BDS design employs a wide dose range, helps identify the MinED, employs a log-like dose spacing strategy avoiding allocating doses near the MTD (BDS identifies more doses near the lower end), and is flexible and easy to implement.

One example of the BDS design is described in Ting (2008), where an early phase II dose-ranging clinical trial was designed to evaluate a drug for the treatment of osteoarthritis (OA). Seven treatment groups were included: a placebo control, an active control (AC), and five dose groups. The five doses were expressed as percentages of the MTD (i.e., 2.5%, 5%, 12.5%, 25%, and 75% of MTD). At the design stage of the trial, the available capsule strengths were 2.5% and 12.5% of the MTD. Given the MTD delivered from phase I and the available dose strengths, these five doses were selected based on the BDS algorithm.

6.2.2.6 Adaptive Dose-Finding

Adaptive dose-finding or dose-ranging designs have recently been proposed and have received much attention (e.g., Gaydos et al., 2006). Some authors have recommended use of seamless phase I/II or seamless phase II/III designs (e.g., Schmidli et al., 2007; Inoue et al., 2002). Dose-ranging studies are designed with the objective of identifying a dose or a few doses to be used in future trials and are often conducted as a phase II trial. Adaptive dose-finding designs are an attempt to speed up the development process and to provide an efficient selection of doses.

The fundamental idea is to start a trial with many doses, a placebo control, and possibly an AC. Interim analyses are conducted, and the results are utilized to drop or add doses (e.g., Krams et al., 2003).

There are several cautions to note with adaptive dose-ranging studies. Firstly, they can be challenging to appropriately power. Phase II trials are not generally used for registration purposes. Hence, in most of the cases, type I error is not scrutinized by the regulatory agencies. Thus, some phase II trials are designed as a type I error >0.05 as sample sizes will be smaller. Type II error can be difficult to quantify and often simulations are used to evaluate design characteristics. Careful planning is needed to ensure that error rates are controlled given the interim analyses.

A second concern is that many adaptive dose-ranging trials are designed to achieve several objectives. It is critical to clearly prioritize the objectives (Grieve, 2007). A third concern is that most adaptive dose-ranging designs assume a dose–response model. These assumptions may not hold in practice.

6.2.3 Noninferiority Trials

Noninferiority (NI) trials have become very common in clinical research. One motivation for NI trials is that in order to properly evaluate an experimental intervention (T), a comparison to a control group is necessary to put the results of a T into context. However, for the targeted medical indication, randomization to a placebo (P) is unethical due to the availability of an effective standard of care (SOC) intervention. Thus, a SOC therapy is selected to be the AC group.

However, no longer, is it necessary to show that the T is superior to the control (as generally the case in placebo-controlled trials) but instead it is desirable to show that the T is "at least as good as" or "no worse" than (i.e., NI to) the AC. Hopefully, the T is better than the AC in other ways (e.g., better safety profile, less expensive, more convenient, or less invasive to administer, e.g., fewer pills, shorter treatment duration, different resistance profile, better compliance, better QoL). For example, in the treatment of HIV, researchers seek less complicated or less toxic antiretroviral regimens that can display similar efficacy to existing regimens.

It is important to note that NI cannot be concluded from a nonsignificant test of superiority. The traditional strategy for an NI trial is to decide upon an NI margin (M, to be discussed later), and if treatment differences can be shown to be within the NI margin (i.e., <M), then NI can be claimed. From the hypothesis testing perspective the null and alternative hypotheses are H_0: $\beta_{T,AC} \geq M$ and H_A: $\beta_{T,AC} < M$ where $\beta_{T,AC}$ is the effect of T relative to AC. The standard analysis is to construct a CI for the difference between arms and note if the bounds (entire CI) is within the bounds of the NI margin. For example, if the primary endpoint is binary (e.g., response vs. no response) then a CI for the difference in response rates (T minus AC) can be constructed. If the lower bound of the CI is >–M, then important differences can be ruled out with reasonable confidence and NI can be claimed. Consider Figure 6.4, CIs A–E represents potential trial outcome scenarios. If the trial were designed to evaluate the superiority, then a failure to reject the null hypothesis would result from scenarios A and D (since the CI does not exclude zero). Inferiority would be concluded from scenarios B, C, and E while superiority would be concluded from scenario F. If the trial were designed as an NI trial, then a failure to reject the null hypothesis of inferiority would result from scenarios A, B, and C, but NI could be claimed in scenarios D, E, and F since the lower bound of the interval is >–M. Some confusion often results from scenario E in which inferiority is concluded from a superiority trial, but NI is concluded from an NI trial. This case highlights the distinction between statistical significance (i.e., the CI excludes 0) and clinical relevance (i.e., the differences are less than M) and that inferiority in a superiority trial is not the complement of NI in a NI trial. Scenario A is a case where neither superiority, inferiority nor NI can be claimed because the CI is too wide. This may be due to a small sample size or large variation.

6.2.3.1 Examples

We highlight the use of NI trials with examples from several disease areas in Table 6.2.

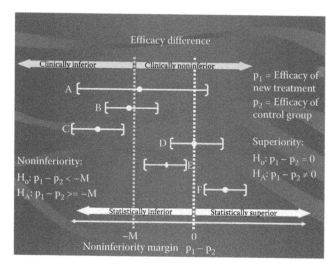

FIGURE 6.4

Interpreting the results of noninferiority trials: Potential trial results.

TABLE 6.2

Examples of NI Trials

- ACTG 116A was designed to compare two antiretroviral (ARV) therapies for the treatment of HIV infection by evaluating if DDI (500 and 750 mg/day) was NI to azidothymidine (AZT). In 1989, AZT was the only approved ARV and had been shown to be better than P in reducing disease progression. More treatments were needed due to resistance development. A trial of DDI versus P would not be ethical due to the known efficacy of AZT and the seriousness of HIV infection. The primary endpoint was the time to an AIDS-defining event or death. It was determined that the NI would be concluded if the upper bound of the 90% CI for the hazard ratio were <1.6. The resulting 90% CIs were (0.79, 1.33) and (0.80, 1.34) for the 500 and 750 mg/day arms, respectively, thus resulting in the NI.
- In 2003, the FDA reviewed an NDA submission based on a randomized, double-blind, multicenter trial evaluating if Piperacillin/Tazobactam (P/T, 4 g/500 mg) was NI to Imipenem/Cilastatin (I/C, 500 mg/500 mg) administered intravenously every 6 h to treat nosocomial pneumonia in hospitalized patients. It was determined that NI would be claimed if the lower bound for the 95% CI for the difference in response rates (P/T – I/C) were > –0.20 (i.e., it could be shown that the response rate for P/T was not more than 20% worse than I/C). The response rates in the I/C and P/T arms were (60/99 = 60.6%) and (67/98 = 67.7%), respectively, resulting in a lower bound for the 95% CI of –0.066 (> –0.20). P/T was judged to be NI to I/C and the FDA approved P/T.
- The PROFESS study (European Stoke Conference 2008) concluded that aspirin + extended-release dipyridamole was not an NI to clopidogrel for stroke prevention. The primary endpoint was recurrent stroke, and an NI margin was set at a 7.5% difference in relative risk. The 95% CI for the hazard ratio was (0.92, 1.11). Since the upper bound of the CI was >1.075, an NI could not be concluded.
- Brodie et al. (Neurology, 2007) was able to demonstrate that Keppra was the NI to Carbatrol for newly diagnosed epilepsy. The primary endpoint was 6-month freedom from seizure and an NI margin was set at a 15% difference. The 95% CI for the risk difference was (–7.8%, 8.2%), and thus NI was concluded.

6.2.3.2 Design Issues

There are several important considerations when designing NI trials. Two important assumptions associated with designing NI trials are constancy and assay sensitivity.

6.2.3.2.1 Constancy

In NI trials, an AC is selected because it has been shown to be efficacious (e.g., superior to P) in a historical trial. The constancy assumption is that the effect of the AC over P in the historical trial would be the same as the effect of P in the current trial if a P group were included. This might not be the case if there were differences in trial conduct (e.g., administration of treatment, endpoints, population) between the historical and current trials. This assumption is not testable in the current trial without a concurrent P group.

The development of resistance is one threat to the constancy assumption. For example, *Staphylococcus aureus* is a bacterial infection that lives on the skin. Historically, "staph" was commonly acquired in hospital settings and was successfully treated with antibiotics such as penicillin, amoxicillin, and methicillin. Recently, staph is more commonly spread outside of the hospital setting (i.e., "community-acquired") and has become resistant to treatment with antibiotics. There is a lack of data to support a claim that these antibiotics are currently efficacious against methicillin-resistant Staphylococcus aureus (MRSA). Thus, use of these antibiotics as an AC in an NI study could result in showing NI to therapies that are no better than P.

FDA guidance released in October 2007 addressed the constancy issue and the use of NI trials to support approval, noting that NI trials are appropriate when there is adequate evidence of a defined effect size for the AC so that an NI margin can be justified. A comprehensive synthesis of the evidence that supports the effect size of the AC and the NI margin should be assembled. They note that for some indications such as acute bacterial sinusitis (ABS), acute bacterial exacerbation of chronic bronchitis (ABECB), and acute bacterial otitis media (ABOM), that data will not support an NI design.

6.2.3.2.2 Assay Sensitivity

The assumption of assay sensitivity is that the trial is designed to be able to detect differences between therapies, if they indeed exist. Unless the instrument that is measuring treatment response is sensitive enough to detect differences, if they exist, then the therapies will display similar responses due to the insensitivity of the instrument, possibly resulting in an erroneously concluding NI. The endpoints that are selected, the way they are measured, and the conduct and integrity of the trial can affect assay sensitivity.

6.2.3.2.3 Selection of the AC

The AC in an NI trial should be selected carefully. Researchers should review ICH E-10 when selecting the AC for an NI trial. Regulatory approval does not necessarily imply that a therapy can be used as an AC. The AC ideally will have clinical efficacy that is (1) of substantial magnitude, (2) estimated with precision in the relevant setting in which the NI trial is being conducted, and (3) preferably quantified in multiple trials. Since the effect size of the AC relative to P is used to guide the selection of the NI margin, superiority to P must be reliably established and measured. Assurance that the AC would be superior to P if a P were employed in the trial is necessary.

Recently, there has been concern over the development of NI studies using ACs that violate the constancy assumption (i.e., AC efficacy has changed over time) or do not have proven efficacy over P. Research teams claim that P controlled trials are not feasible because (1) Ps are unethical because the AC therapy is SOC, (2) patients are unwilling to enroll into placebo-controlled trials, and (3) IRBs question the ethics of the use of Ps in these situations. For example, the Oral HIV AIDS Research Alliance (OHARA) developed a trial for the treatment of HIV-associated oral candidiasis (OC) in resource-limited countries within the ACTG system. Fluconozole is the SOC in the United States but is not readily available in resource-limited settings due to high costs. Nystatin is used as a SOC in many resource-limited settings. Gentian violet (GV), an inexpensive topical agent has shown N excellent *in vitro* activity against OC. A trial evaluating the NI of GV to nystatin was proposed. However, despite the standard use of nystatin, there were no published results from randomized trials that documented the efficacy of nystatin over P. Thus, an NI trial could not be justified and a simple superiority comparison of the two therapies was proposed.

One scientifically attractive alternative design is to have a three-arm trial consisting of a T, an AC, and a P arm. This design is particularly attractive when the efficacy of the AC has changed, is volatile, or is in doubt. This design allows assessment of NI and superiority to P directly, and allows for within-trial validation of the NI margin. Unfortunately, this design is not frequently implemented due to a concern for the unethical nature of the P arm in some settings and the need for higher sample sizes.

6.2.3.2.4 Biocreep

When selecting the AC for an NI trial, one must consider how the efficacy of the AC was established (e.g., by showing NI to another AC vs. showing superiority to P). If the AC was shown to be effective via an NI trial, then one must consider the concern for biocreep. Biocreep is the tendency for slightly inferior interventions (but within the margin of an NI) that were shown to be efficacious via NI trials and to be ACs in the next generation of NI trials (D'Agostino, SIM, 2003). Multiple generations of NI trials using ACs that were themselves shown to be effective via NI trials, could eventually result in the demonstration of the NI of a therapy that is not better than P. NI is not transitive: if A is NI to B, and B is NI to C, then it does not necessarily follow that A is NI to C. For these reasons, when selecting the AC, consider choosing among the best available ACs.

6.2.3.2.5 *Choice of the Endpoints and the Target Population*

To enable an evaluation of the retention of some of the effect of the AC over P, study participants, endpoints, and other important design features should be similar to those used in the trials for demonstrating the effectiveness of the AC over P. One can then indirectly assess the constancy assumption by comparing the effectiveness of the AC in the NI trial and the historical trial.

6.2.3.2.6 *Choice of the NI Margin*

The selection of the NI margin in an NI trial is a complex issue and one that has created much discussion. In general, the selection of the NI margin is done in the design stage of the trial and is utilized to help determine sample size. Although there are no absolute algorithms for determining the NI margin, there are a few guiding principles (EMEA Guideline on the Choice of the NI Margin, FDA Guidance for Industry: Noninferiority Clinical Trials). Defining the NI margin in NI trials is context dependent and highly scrutinized since it plays a direct role in the interpretation of the trial results.

The selection of the NI margin is subjective but structured, requiring a combination of statistical reasoning and clinical judgment. Conceptually, one may view the NI margin as the "maximum treatment difference that is clinically irrelevant" or the "largest efficacy difference that is acceptable to sacrifice in order to gain the advantages of the T." This concept often requires interactions between statisticians and clinicians.

Since one indirect goal of a NI trial is to show that T is superior to P, retention of some of the effect of AC over P needs to be retained (often termed "preserving a fraction of the effect"). Thus, the NI margin should be selected to be smaller than the effect size of the AC over P. Researchers should review the historical data that demonstrated the superiority of the AC to P to aid in defining the NI margin. Researchers must also consider the within and across-trial variability in estimates as well. Ideally the NI margin should be chosen independent of study power, but practical limitations may arise since the selection of NI margin can dramatically affect study power.

One strategy for preserving the estimate of the effect is to set the NI margin to 50% of the estimated AC effect versus P. For example, the STAR trial was designed to evaluate the NI of Raloxifene to Tamoxifen on the primary endpoint of invasive breast cancer. An earlier trial of Tamoxifen versus P was resulted in an estimate of relative risk of 2.12 (95% CI = 1.52, 3.03) favoring Tamoxifen. Thus, one option that was considered for defining the NI margin for the Raloxifene trial was $1 + [(2.12 - 1)/2] = 1.56$. If the upper bound of the 95% CI for the relative risk of Raloxifene versus Tamoxifen were <1.56, then NI would be demonstrated. Note, however, that this method does not consider the fact that the estimate of tamoxifen versus P is subject to uncertainty. To account for the variability of the estimate of Tamoxifen versus P, the "95%–95% CI method" could be used. In this strategy, the NI margin is set to the lower bound of the 95% CI for the effect of P versus Tamoxifen. If the upper bound of the 95% CI for Raloxifene versus Tamoxifen were <1.52, then NI would be demonstrated. This criterion is stringent and depends directly upon the evidence from historical trials.

A poor choice of an NI margin can result in a failed NI trial. The TARGET trial was designed to evaluate the NI of tirofaban to abciximab (two glycoprotein IIa/IIIb inhibitors) for the treatment of coronary syndromes. An NI margin for the hazard ratio (HR) of 1.47 was selected (50% of the effect of abciximab vs. P in the EPISTENT trial). The trial was viewed as poorly designed because an agent with an HR = 1.47 would not have been considered NI to abciximab. In the SPORTIF V trial, ximelegatran was compared to warfarin (AC) for stroke prevention in atrial fibrillation patients. The event rate for warfarin was 1.2%, and the NI margin was set at 2% (absolute difference in event rates) based on historical data. Since the event rate in the warfarin arm was low, the NI could be concluded even if the trial could not rule out a doubling of the event rate. For these reasons, the selection of the NI margin should incorporate statistical considerations as well clinical relevance considerations.

Since the choice of the NI margin is a complex issue, a natural question is whether an NI margin can be changed after trial initiation. In general, there is little concern regarding a decrease in the NI margin. However, increasing the NI margin can be perceived as manipulation unless appropriately justified (i.e., based on external data that is independent of the trial). The EMEA released a Points to Consider document on this issue.

6.2.3.2.7 Sample Size

Sample sizes for NI trials are generally believed to be larger than for superiority trials. However, this really depends upon the selection of the NI margin and other parameters. Required sample sizes increase with decreasing non-inferiority margin (M). Stratification can help since adjusted CIs are generally narrower than unadjusted CIs. Researcher should power NI trials for PP analyses as well as ITT analyses given the importance of both analyses (described later). Researchers need to weigh the costs of type I error (i.e., incorrectly claiming NI) and type II error (i.e., incorrectly failing to claim NI).

The general approach to analyses is to compute a two-sided CI (p-values are not generally appropriate). A common question is whether a one-sided 0.05 CI is acceptable given the one-sided nature of NI. Although researchers can decide upon error rates that they believe are appropriate, ICH guidelines suggest that two-sided CIs are generally appropriate for consistency between significance testing and subsequent estimation. Note that a one-sided 95% CI would lower the level of evidence for drawing conclusions compared to the accepted practice in superiority trials. Most regulatory settings require one-sided 95% CIs for pivotal trials (i.e., one-sided 95% CIs are not generally accepted).

A simple approach to sizing an NI trial is to view the trial from an estimation perspective. The strategy is to estimate the difference between treatments with appropriate precision (as measured by the width of a CI). Then size the study to ensure that the width of the CI for the difference between treatments is acceptable. ACTG 5263 utilized this approach to estimate the difference in event rates (clinical progression of Kaposi sarcoma [KS] or death) between two chemotherapies for the treatment of HIV-associated advanced KS disease. The research team concluded that the widest acceptable width of a CI for the difference in event rates was 20%. Data from the AIDS Malignancy Consortium was used to estimate the event rates at 50% and the team further noted that the width of a CI would be maximized when event rates are 50% in both arms. A sample size was calculated to ensure that the width of a CI for the difference in event rates between two arms would be <20% when event rates were 50% in both arms. If the event rates were different from 50%, then the CI width would be <20%. The advantage of this approach for this trial was that distinct NI margins could be specified for different countries given the multinational nature of the study and the varying perspectives on the NI margin in these countries.

6.2.3.2.8 *Importance of High-Quality Trial Conduct*

Poor trial conduct can damage the validity of an NI trial. Poor adherence, missing data, inadvertent enrollment, misclassification, measurement error, and treatment crossovers can bias a result toward an NI. For illustration, consider an extreme example where two therapies are being compared, and study participants in both arms do not adhere to therapy. Then the treatment arms would appear similar simply because poor adherence in both arms would dilute treatment differences. For these reasons, careful planning of the trial to minimize dropout, maximize adherence, and assuring diligent follow-up of study participants is critical. Trial conduct is highly scrutinized by regulatory agencies for these reasons.

6.2.3.2.9 *ITT versus per Protocol*

In superiority studies, an ITT-based analysis tends to be conservative (i.e., there is a tendency to underestimate true intervention differences). As a result, ITT analyses are generally considered the primary analyses in superiority trials as this helps to protect the type I error rate. Since the goal of NI trials is to show NI or similarity, an underestimate of the true treatment difference can bias toward NI, thus inflating the "false positive" (i.e., incorrectly claiming NI) error rate. Thus, ITT is not necessarily conservative in NI trials. For these reasons, an ITT analyses and a per protocol (PP) analyses are often considered as coprimary analyses in NI trials. It is important to conduct both analyses (and perhaps additional sensitivity analyses) to assess the robustness of the trial result. PP analyses often result in a larger effect size since ITT often dilutes the estimate of the effect, but PP analyses frequently results in wider CIs since it is based on fewer study participants than ITT. It is important that the ITT and PP analyses produce similar qualitative results (CPMP, Points to Consider).

6.2.3.2.10 Interim Analyses

Interim analyses of NI trials can be complicated. It generally takes over-whelming evidence to suggest stopping a trial for NI during interim analy-ses. Also there may not be an ethical imperative to stop a trial that has shown NI (in contrast to superiority studies with which if superiority is demon-strated, then there may be ethical imperatives to stop the study since ran-domization to an inferior arm [or continued follow-up on an inferior arm] may be viewed as unethical). In addition, even if an NI is demonstrated at an interim time point, it may be desirable to continue the study to assess whether the superiority could be shown with trial continuation.

It is not uncommon to stop an NI trial for futility (i.e., unable to show NI). We recommend using the theory of repeated CIs to control error rates with predicted interval plots (described in Chapter 7) to aid DMCs reviewing the interim data.

6.2.3.3 Clarification of the Two Distinct Objectives

Gau and Ware (SIM, 2007) critique traditional approaches to the design and analyses of NI trials by noting that they often fail to distinguish between the two distinct sub-objectives of NI trials: (1) to demonstrate T is NI to AC and (2) T is superior to P taking into account historical evidence. The design of an NI trial can be accomplished by planning to test two separate hypotheses based on these sub-objectives.

To test if T is NI to AC, an NI margin is selected based on clinical consid-erations rather than the effect of AC vs. P, as this will allow an NI claim if supported by the data. Using standard methods for NI trials, one test: H_{10}: $\beta_{T,AC} \geq M$ vs. H_{1A}: $\beta_{T,AC} < M$.

A direct comparison of T to P is not possible. However, if the constancy assumption holds, then the difference between T and P can be estimated as a function of the placebo-controlled trial of AC and the NI trial (i.e., $\beta_{T,P} = \beta_{T,AC} + \beta_{AC,P}$), and test: H_{20}: $\beta_{T,P} = 0$ vs. H_{2A}: $\beta_{T,P} < 0$. Adjustments can be made if the constancy assumption does not hold. For example, suppose that the current effect of the AC vs. P is $(1 - \lambda)$ of the historical effect. Then $\beta_{T,P,\lambda} = \beta_{T,AC} + (1 - \lambda)\beta_{AC,P}$, and we test H_{20}: $\beta_{T,P,\lambda} = 0$.

If the goal of the NI trial is to reject both hypotheses, sample size calcula-tions can be performed for each test, and the larger of the required samples sizes can be selected.

A particular trial may only accomplish one of the two sub-objectives. If T is shown superior to P but fails to demonstrate NI to AC, then use of T may be indicated for patients in whom AC is contraindicated or not available. In contrast, T could be shown to be NI to AC but not superior to P. This may occur when the efficacy of the AC is modest.

Recently, there have been claims that the second of the two sub-objectives outlined by Gau and Ware is the objective of interest in the regulatory set-ting. Industry groups have argued that regulatory approval of new therapies should be based upon evidence of superiority to placebo (demonstration of clinically meaningful benefit) and not necessarily NI to AC. Proponents of this perspective (often termed the "synthesis method") pose several dilemmas and inconsistencies with traditional approaches to NI trials in support of this position. First, T could look better than the AC but not meet the preservation of effect condition. Second, two trials with different ACs have different standards for success. Third, if T is shown to be superior to AC, then a natural question that arises is should the AC be withdrawn from the market? The argument is that the required degree of efficacy should be independent of the design (superiority vs. NI) and that superiority to P is sufficient. Proponents of the synthesis method thus argue that "NI trial" terminology is inappropriate since the superiority of T to P is the true objective.

6.2.3.4 Analyses

When designing NI trials, it is important that the research team has clarity regarding the claims that it desires to make upon completion of the analyses and a vision of the analyses that will support the claim. The choice of the NI margin plays a direct role in the interpretation of the trial (unlike the minimum clinically relevant difference that is often defined in superiority trials). Thus, justification for the NI margin should be outlined in the protocol.

The analyses of NI trials use information outside of the current trial to infer the effect of the T versus P in the absence of a direct comparison. Thus, it is recommended that a comparison of the response rate, adherence, and others of the AC in the NI trial be compared to historical trials that compared the AC to P and provided evidence of the efficacy of the AC. Using historical data, epidemiological methods (e.g., meta-regression) could be used to estimate the difference between the T and P assuming constancy. If the AC displays different efficacy than in prior trials, then the validity of the predefined NI margin may be suspect, and the interpretation of the results is challenging. Researchers can consider modeling the changes in the effect of the AC versus P, but appropriate data may not be available.

6.2.3.5 Missing Data

As with all studies, the most effective way to handle missing data is a solid prevention strategy. This is particularly important in NI studies since missing data can d bias toward NI resulting in a threat to trial integrity.

Despite prevention efforts, missing data occur. In NI trials, researchers may consider strategic imputation techniques for analyses that protect error rates (e.g., protect against falsely claiming NI). For example, consider an NI trial that utilizes a binary endpoint (success vs. failure). Researchers may impute success for the AC and failure for the T (i.e., a very conservative assumption). If NI is shown in this extreme case, then the NI is not affected by missing data, and a result of NI can be confidently concluded. Conversely, researchers may impute success for the T and failure for the AC. If an NI cannot be shown in this extreme case, then futility can be concluded. However, these are extreme approaches, are biased, and may be unrealistic in many cases. Thus, these approaches should be used and interpreted with caution.

An alternative is to consider strategic imputation techniques that are consistent with the hypotheses of the NI trial. For example, when an NI trial has a continuous endpoint, researchers can impute a reasonably expected value (i) for the AC and i–M for the T. Similarly, for a binary endpoint, an expected proportion (p) can be imputed for the AC and p–M can be imputed for the T, and then analyses of means can be performed.

Decisions regarding imputation techniques are complex in NI trials. For registration trials, it is recommended that the researcher consult with the regulator during trial design regarding appropriate imputation techniques.

6.2.3.6 *Switching between NI and Superiority*

If an NI trial is conducted and the NI of T to an AC is shown, then a natural question is whether a stronger claim of superiority can be made. In other words, what are the ramifications of switching from NI trial to a superiority trial? Conversely, if a superiority trial is conducted and significant between-group differences are not observed, then a natural question is whether a weaker claim of NI can be concluded. Can one switch from a superiority trial to an NI trial? Issues to consider when addressing these questions are summarized in the EMEA document, "Points to Consider on Switching Between Superiority and NI."

In general, it is considered acceptable to conduct an evaluation of superiority after showing an NI. Due to the closed testing principle, no multiplicity adjustment is necessary. Although an ITT and PP analyses are both important for the NI analyses, the analyses of superiority should be conducted using an ITT approach. For clarity and to avoid the perception of manipulation, it is also advisable to describe this plan in trial design.

It is more difficult to justify a claim of NI after failing to demonstrate the superiority. There are several issues to consider. First, whether an NI margin has been prespecified. Defining the NI margin post hoc can be difficult to justify and can be perceived as manipulation. The choice of the NI margin needs to be independent of the trial data (i.e., based on external information) which is difficult to demonstrate after data has been collected and revealed. Second, is the control group an appropriate control group for an NI trial (e.g., it has demonstrated and precisely measured superiority over P)? Third, was the efficacy of the control group similar to that displayed in historical trials vs. P (constancy assumption)? Fourth, ITT and PP analyses become equally important in NI trials. Fifth, trial quality must be high. In summary, it may be possible to switch from superiority to NI if (1) an NI margin was prespecified or can be justified based on external information, (2) ITT and PP analyses result in similar interpretations, (3) the quality of trial conduct was high with few dropouts and good adherence, (4) the control group showed similar efficacy to historical trials comparing the control to P (constancy), and (5) the trial was sensitive enough to detect differences in effects (assay sensitivity). Notably one can always use the CI to "rule-out" effect sizes with reasonable confidence.

6.2.4 Futility Designs

An estimated 8% of investigational therapies that are studied in clinical trials are eventually given FDA approval (FDA Critical Path). Given the high negative-trial rate and considering that phase III clinical trials are very expensive and resource intensive, it is important to identify futile therapies as quickly as possible. Phase II futility trials have been developed as a "screening instrument" to identify futile therapies in phase II and prevent them from being studied in expensive phase III trials, thus providing cost and resource savings. Futility trials thus provide a strategy for evaluating a number of candidate therapies before conducting large phase III trials. Tilley et al. (Neurology, 2006) illustrates how a futility study could have saved resources in the DATATOP trial, an 800-participant phase III study that investigated deprenyl and tocopherol to delay Parkinson's disease progression. A phase II futility study could have resulted in early identification of the futility of tocopherol and hence resource savings. Futility trials have been used in cancer and in neurological diseases such as Parkinson's disease, stroke, multiple sclerosis, amyotrophic lateral sclerosis, Huntington's disease, and Alzheimer's disease to investigate new therapies over a short duration in a small number of participants to examine if the therapies are worthy of phase III trials (Huntington Study Group DOMINO Investigators, 2010; Kaufmann et al., 2009; NINDS NET-PD Investigators, 2007; Zhao et al., 2006).

The null hypothesis of a futility trial is "not futile" while the alternative hypothesis is "futile." Specifically, assume that a futility trial is being designed to evaluate intervention x, and the trial will have a binary primary endpoint (e.g., success). Then the hypotheses are

$$H_0 : p_x \geq \Delta \text{ vs. } H_A : p_x < \Delta$$

where p_x is the proportion of successes with intervention x and Δ is "minimum worthwhile success rate," often constructed by determining a minimally worthwhile improvement over a placebo success rate estimated most commonly from a historical trial but could also be estimated from the futility trial (Palesch et al., 2005). Thus, type I error in a futility trial is incorrectly concluding that an effective therapy is futile (often resulting in a decision to decline further development on an effective therapy) while type II error is incorrectly concluding an ineffective therapy is not futile (often resulting in a decision to conduct phase III trials with an ineffective therapy). Researchers must weigh the costs of these errors when designing a futility study. For example, researchers may be willing to risk more type II error for therapies being developed to treat diseases with no known cure (e.g., progressive multifocal leukoencephalopathy [PML] or to prevent disability from hemorrhagic stroke) but unwilling to risk type II error for therapies that are expensive, invasive, or risky when alternative therapies exist (e.g., mild ischemic stroke). In general, futility trials have high predictive value for therapies found to be futile but the predictive value of nonfutile therapies might be modest. Thus, simply because a futility trial does not conclude futility, may not provide sufficient justification for a phase III trial. One must also be careful when using historical data to define the minimum worthwhile success rate since recent medical advances may have occurred.

A futility trial was employed to investigate whether creatine and minocycline should be investigated as therapies for early Parkinson's disease in phase III trials (NINDS NET-PD Investigators, 2006). Two-hundred participants were randomized equally to creatine, minocycline, or placebo. A futility trial was reasonable due to the existence of an appropriate short-term outcome (the Total Unified Parkinson's Disease Rating Scale [UPDRS]) which has been shown to be sensitive to Parkinson's disease tracking in the short-term (Elm et al., 2005). A change of 10.65 on the UPDRS was observed in the placebo arm of a prior trial and "futility" was defined as showing a <30% improvement relative to the historical placebo. Thus, futility would be concluded if the lower bound of the CI for the within-arm mean change in the UPDRS were >7.46 (70% of 10.65). The observed mean changes for the creatine and minocycline arms were 5.6, 95% CI = (3.48, 7.72) and 7.09, 95% CI = (4.95, 9.23), respectively. Thus, futility could not be concluded, and the therapies should be further considered for phase III trials.

6.2.5 Factorial Designs

Often a research team is interested in studying the effect of two or more interventions applied alone or in combination. In these cases, a factorial design can be considered. Factorial designs are attractive when the interventions are regarded as having independent effects or when effects are thought to be complmentary, and there is interest in assessing their interaction (Couper et al., 2005; McAlister et al., 2003; Montgomery et al., 2003).

6.2.5.1 The 2 × 2 Factorial Design

The simplest factorial design is a 2 × 2 factorial (Figure 6.5) in which two interventions (factors) are being evaluated, each at two levels (e.g., intervention vs. no intervention). Each study participant is assigned to one level of each of the factors. Four intervention groups are defined based on whether they receive interventions A only, B only, both A and B, or neither A nor B. Thus, in order to apply the factorial design: (1) you must be able to apply the interventions simultaneously and (2) it must be ethically acceptable to apply all levels of the interventions (e.g., including placebos if so designed).

		Intervention A	
		No	Yes
Intervention B	No	Group 1	Group 3
	Yes	Group 2	Group 4

FIGURE 6.5
Schema for a 2 × 2 factorial design.

The factorial design can be viewed as an efficient way to conduct two trials in one. The factorial design is contraindicated when primary interest lies in comparing the two interventions to each other.

6.2.5.2 The No Interaction Assumption

If one can assume that there is no interaction between the two interventions, that is, the effect of one intervention does not depend on whether one receives the other intervention, then a factorial design can be more efficient than a parallel group design. Since factorial designs are economical under the no interaction assumption, they are often employed when sample sizes are expected to be large as in prevention trials. One must first define the scale of measurement and distinguish between additive and multiplicative interaction. Although there are four groups, there are only two parameters to estimate (the effect of intervention A and intervention B) and all of the groups can be utilized to estimate those parameters. One can estimate the effect of intervention A by comparing groups one versus four and groups two versus four. Similarly, one can estimate the effect of intervention B by comparing groups one versus two and groups three versus four. The effect of the intervention A + B can simply be estimated as the effect of intervention A plus the effect of intervention B.

A limitation of factorial designs is that the assumption of no interaction is often not valid. The effect of one therapy often depends on whether the other therapy is provided. This limits the use of factorial designs in practice. Instances in which a no interaction assumption may be valid include the case when the two interventions have differing mechanisms of action (e.g., drug therapies combined with adjunctive therapies, complementary therapies, behavioral or exercise therapies, diet supplements, or other alternative medicines). For example, Bosch et al. (BMJ, 2002) conducted a factorial trial of ramipril and vitamin E for stroke prevention, Shlay et al. (JAMA, 1998) utilized a factorial design to study the effects of amitriptyline and acupuncture for the treatment of painful HIV-associated peripheral neuropathy, and the first Physicians Health Study (PHS) investigated the use of aspirin and beta-carotene for the prevention of cardiovascular disease and cancer randomizing 22,071 study participants to four treatment groups in a 2 × 2 factorial design.

Interestingly, factorial designs are the only way to study interactions when they exist. They allow direct assessment of interaction effects since they include groups with all possible combinations of interventions. Combination interventions are frequently of interest in medicine particularly when monotherapies are individually ineffective perhaps due to the use of ceiling doses to limit toxicity, but complementary mechanisms of action suggest potential synergistic effects. Quantitative interaction occurs when the effect of the combination intervention of A and B is greater than the effect of intervention A plus the effect of intervention B. Qualitative interaction occurs when the effect of the combination intervention of A and B is less than the effect of intervention A plus the effect of intervention B.

6.2.5.3 Sample Size

A common strategy to sizing factorial trials is to conduct separate calculations based on target effect sizes for each of the interventions and take the maximum of the trial sample size. This approach assumes no interaction and would result in low power to detect interactions if they exist. If a trial were to detect an interaction, then the sample size would need to be adequately increased. This can be costly. For example, to detect an interaction with the same magnitude of a main effect then the sample size would need to be increased four-fold. If the sample size were not increased, then an interaction effect would have to be twice the magnitude of the main effect to have similar power be detected.

On the other hand, having low power to detect interactions could result in incorrect characterization of intervention effects and suboptimal patient care. Researcher should consider whether interactions are possible and appropriately size studies to detect interactions when their existence is unknown.

6.2.6 Factorial Designs with More than Two Factors and Assessment of More than One Outcome

Factorial designs can be considered for more than two interventions. The Women's Health Initiative (WHI) clinical trial utilized a $2 \times 2 \times 2$ factorial design randomizing study participants to a dietary modification (low-fat eating pattern vs. self-selected diet), hormone replacement therapy (vs. placebo), and calcium plus vitamin D supplement (vs. placebo). However, increasing the number of factors will increase the number of groups and associated complexity of the trial. Toxicity or feasibility constraints may also make it impossible to apply a full factorial design, but incomplete factorial designs can be considered although with increased complexity.

Also in factorial trials, the outcomes being studied may vary across interventions. In the WHI clinical trial, dietary modification was studied for its effect on breast and colorectal cancer, HRT was studied for its effect on cardiovascular disease risk, and calcium and vitamin D supplementation was studied for its effect on the risk for hip fractures. In other words, the interventions being studied may be evaluated for different indications. Thus, factorial trials are often considered in preventions trials as they enroll participants free of disease that can be followed for multiple outcomes.

6.2.6.1 Interim Monitoring

Data monitoring of factorial designs can be complicated. Assigning attribution of the effects during the course of a trial can be difficult. It is not uncommon for one component of the trial to be terminated while other components continue, essentially viewing the factorial design as separate trials for each factor. The aspirin component of the PHS I was terminated early when aspirin was shown to reduce the risk of myocardial infarction (MI). The HRT component of the WHI was terminated due to an increased risk for breast cancer and overall health risks exceeding benefits. However, when considering the termination of one component of the trial, an evaluation of the effect on power is critical. The termination of one component will reduce the power to detect interactions and may complicate analyses and subsequent interpretations of main effects and interactions.

6.2.6.2 Recruitment and Adherence

Participant recruitment is more complex in factorial trials and can decrease accrual rates. Potential trial participants must meet criteria for each intervention with no contraindications to any of the possible intervention combinations, and have a willingness to consent to all the interventions and procedures. Protocol adherence can also be more complicated due to the multiple interventions and greater burden on trial participants. For these reasons, it is important to monitor participant enrollment and adherence.

6.2.6.3 Analyses and Reporting

Researchers should be aware of the multiplicity issue given the assessment of multiple interventions. However, it is often considered desirable to control the error rate for the assessment of each factor separately rather than controlling a trial-wise error rate.

When analyzing and reporting factorial trials, interaction effects should be evaluated even if the trial were designed under the assumption of no interaction. Reporting should include a transparent summary of each treatment cell so that potential interaction can be assessed.

When interactions exist, then it is inappropriate to interpret global intervention effects. Instead, one must estimate intervention effects conditional upon the other interventions using subgroup analyses. For example, in a 2×2 factorial trial with interventions A and B, there would be two effects of intervention A: one for patients that receive intervention B and one for patients that do not receive intervention B.

6.2.7 Biomarker Designs

A biomarker is a characteristic that is objectively measured and evaluated as an indicator of normal biological processes, pathogenic processes, or pharmacological processes to a therapeutic intervention. A biomarker may reflect biological processes closely related to a disease mechanism or processes downstream from the primary disease process (Biomarkers Definition Working Group, 2001).

Many theorized biomarkers have failed to be validated. No new major cancer biomarkers have been approved for clinical use for 25 years (Diamardis, JNCI, 2010). A study by Ioannidis and Panagioutou (JAMA, 2011) indicated that many biomarkers are worthless or have very small or overestimated effects. 29 of 35 highly cited studies reported larger effect sizes than subsequent meta-analyses.

A partial cause of these issues with biomarkers is a lack of biomarker qualification (i.e., how it will be utilized). Biomarkers can be used for screening (e.g., tuberculin skin test), diagnosis (e.g., sputum culture for TB), patient monitoring (e.g., HIV-1 RNA in HIV infection), surrogate endpoints (e.g., antibodies for vaccine efficacy), or for signaling mechanisms of action and causal pathways (i.e., and thus a tool for identifying intervention targets). Trial design is linked to biomarker qualification and biomarker validation is in the evaluation of "fitness for purpose."

An important distinction to make regarding biomarkers is whether they are prognostic or predictive. Prognostic biomarkers affect disease outcome independent of an intervention. For example, age is a prognostic factor for peripheral neuropathy outcomes in HIV (Evans et al., AIDS, 2011). However, a prognostic biomarker may not be useful in treatment selection (i.e., the effect of treatment may not depend on the biomarker).

Predictive biomarkers (also called *companion diagnostics*) predict response to particular interventions (i.e., the intervention response depends on the biomarker). Thus, predictive biomarkers can be useful for guiding treatment selection. Examples of predictive biomarkers are

- HIV
  - CCR5 tropism for Maraviroc in HIV
  - HLA-B*5701 allele for hypersentivity to Abacavir
  - Treatment history (e.g., extent of viral resistance to available drugs) for selecting combination therapies among heavily pre-treated patients
- Cancer
  - K-ras status for Vectibix use in colorectal cancer

- Philadelphia chromosome positivity for Dasatinib use in chronic myeloid leukemia
- Her2/neu expression for Herceptin use in breast cancer
- G6PD deficiency for Rasburicase use in leukemia/lymphoma
- UGT1A1 for Irinotecan use in colorectal cancer
- Cardiovascular disease
  - Familial hypercholesterolemia for Atorvastatin use in reducing cholesterol
  - CYP450 2C9/VKORC1 for Warfarin use in blood clotting
- HCV
  - HCV genotype 1 for boceprevir and telapravir
- MS
  - Anti-JC virus antibodies for PML with natalizumab

Consider the following model describing a patient response:

$$F(outcome) = \beta_0 + \beta_1 B + \beta_2 T + \beta_3 T * B,$$

where
  B is an indicator for whether the biomarker is present (B = 1) or absent (B = 0), and
  T is an indicator of treatment (T = 1 for the experimental treatment and T = 0 for the control).

A biomarker is prognostic if $\beta_1 \neq 0$ (i.e., it is prognostic for outcome independent of treatment). A biomarker is predictive if $\beta_3 \neq 0$ since the treatment effect depends on the biomarker (i.e., if the patient does not have the biomarker then the treatment effect is β_2, but if the patient has the biomarker then the treatment effect is $\beta_2 + \beta_3$).

When designing trials to evaluate biomarkers, it is important to understand whether there is an interest in evaluating prognostics versus predictive biomarkers. Some biomarker designs (Freidlin et al., 2010; Simon, 2012) can evaluate whether biomarkers are prognostic, others whether biomarkers are predictive, and some designs can evaluate both. Consider the biomarker-stratified design (Figure 6.6) in which the participant biomarker status is determined prior to randomization and participants are randomized in a stratified manner depending on the biomarker status. The design is very powerful as as it can be used to estimate the intervention in the biomarker positives, estimate the intervention in the biomarker negatives, evaluate whether the biomarker is prognostic (via comparison of outcomes in biomarker positives vs. negatives controlling for intervention), and evaluate whether the biomarker is predictive (via comparison of the intervention effect or contrast

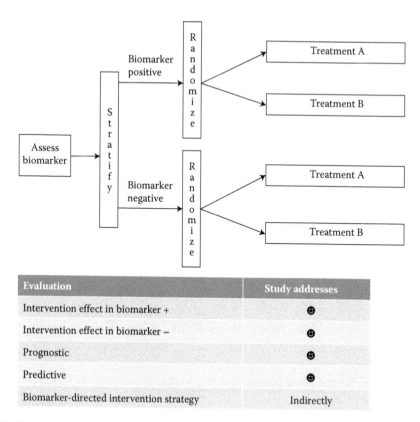

Evaluation	Study addresses
Intervention effect in biomarker +	☺
Intervention effect in biomarker −	☺
Prognostic	☺
Predictive	☺
Biomarker-directed intervention strategy	Indirectly

FIGURE 6.6
Biomarker stratified design.

in biomarker positives vs. negatives). The design may become infeasible if the biomarker has many levels or there several interventions being compared. In order to fully power for a test of the interaction between treatment and the biomarker, large sample sizes may be required.

The enrichment design (Figure 6.7) also evaluates biomarker status prior to randomization but only randomizes participants with the desired biomarker status, thus "enriching" the population with participants who are more likely to have a desired outcome. Participants without the desired biomarker status go from study. The design can be used to estimate the intervention in the biomarker positives but not to estimate the intervention in the biomarker negatives. The design cannot evaluate whether the biomarker is prognostic as there are no outcomes in biomarker negatives to compare to outcomes in biomarker positives. Furthermore, the design cannot be used to evaluate whether the biomarker is predictive as there is no estimate of the intervention effect in the biomarker negatives to compare to the intervention effect in biomarker positives. The efficiency of the enrichment design is greatest (relative to a design enrolling all participants) when there is a low prevalence

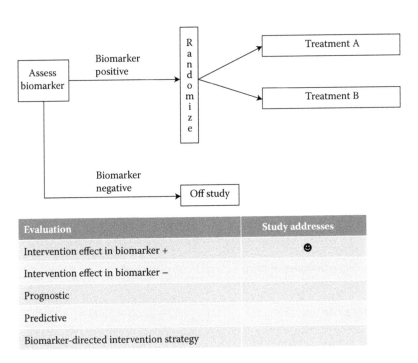

FIGURE 6.7
Enrichment design.

of biomarker positive patients and when the intervention response is vastly better in the biomarker positives than the negatives. The FDA Guidance (2012) has recently released guidance regarding enrichment strategies.

The biomarker strategy design (Figure 6.8) randomizes participants to two treatment strategies: one that is directed by biomarker status while the other is independent of biomarker status. The intent here is to compare selecting treatment based on the biomarker status versus universally treating every-one the same regardless of biomarker status. The design is also able to evalu-ate the effect of the intervention in biomarker positives but not the negatives, and is also able to evaluate whether the biomarker is prognostic but not predictive. Note that the trial does not evaluate the biomarker utility as a positive trial may indicate that intervention A is superior to B rather than the effectiveness of a treatment selection strategy. Mallal et al. (2008) imple-mented a biomarker strategy design to demonstrate that screening for the HLA-B*5701 allele reduces the risk of hypersensitivity to abacavir.

6.2.8 Adaptive Designs

Adaptive designs have become a hot topic in the clinical trial world. Adaptive designs can be defined as a trial design that uses accumulating data from the trial to modify aspects of the trial. Trial adaptation can take several forms.

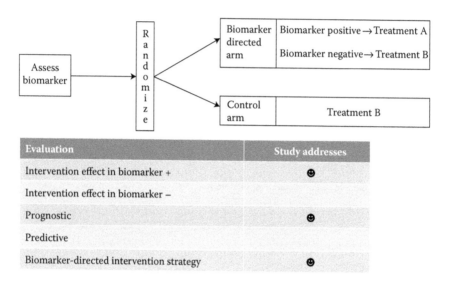

FIGURE 6.8
Biomarker strategy design.

For example, an arm of the trial may be dropped; a trial may be stopped for efficacy, futility, or for safety reasons; sample size may be recalculated; a dose may change; an endpoint may change; or the randomization ratio may change.

Many assumptions are made during the design stage of a clinical trial. Interim analyses can be conducted to evaluate the validity of these assumptions. If the assumptions are invalid, then trial adaptation may be considered. However, trial adaptation must be conducted very carefully to ensure that trial integrity is not compromised.

Implementation of adaptive designs raises unique challenges. Drug supply issues can be more complicated since total sample sizes and duration of follow-up may not be known *a priori*. Thus, adaptive designs may not be ideal for evaluating interventions that have complicated or costly regimens. Two additional concerns are (1) adaptations may convey information about the interim results creating operational bias and (2) some believe that the decision process regarding adaption may require close communications (in a treatment-blinded fashion) between the data monitoring committee and the sponsor given the potential cost implications. To address the first issue, it has been suggested that while the protocol should provide the rationale for the adaptation, the complete details should be withheld from the protocol but fully described in a document with limited availability. This is discussed in further detail in Chapter 7. To address the second issue, it has been suggested that representatives of the trial sponsor be a part of the decision process but (1) these representatives not have other associations with the trial, (2) representation is minimal in terms of the number of representatives and the data made available to the representatives, and (3) protections are in place to ensure knowledge is appropriately limited.

When used appropriately, adaptive designs can be efficient and informative. However, when used inappropriately, adaptive designs can threaten trial integrity. Adaptation is a design feature that requires careful and responsible planning. Adaptive designs are not a rescue medication for poorly designed trials.

Prespecifying adaptive rules can be challenging. Challenges include factoring in safety data, combining interim data with data external to the trial and anticipating all possible contingencies (clinical data, as well as other practical concerns, e.g., funding). Ideally, decisions should be made based on many factors and not simply based on a single efficacy endpoint. However, researchers also need to ensure transparency and prespecify the strategy for the adaptive design. A special protocol assessment (SPA) to review the adaptive design should be considered.

6.2.8.1 Two-Stage Designs

In some instances, the objective of a phase II trial is to investigate whether a new therapy has sufficient promise of efficacy to warrant further investigation with phase III trials. Two-stage designs have been developed to address this objective. The general idea is to enroll a small number of subjects (stage 1) and evaluate them for a response. Accrual may need to be suspended at the end of stage 1 to conduct analyses. If the response is not promising enough for further study, then discontinue the study. Otherwise, continue to enroll subjects (stage 2).

Optimal two-stage designs (Simon, 1989) are based on testing $H_0: p \leq p_0$ vs. $H_A: p \geq p_1$ for a predefined "minimum acceptable response rate" (p_0) and a response rate that is desirable to detect (p_1). Let α and β be the selected desirable type I and II error rates associated with the test of these hypotheses and let n_1 and n_2 be the number of subjects in stage 1 and 2, respectively. The expected sample size is $EN = n_1 + (1 - PET)n_2$ where PET is the probability of early termination. Once α, β, p_0, and p_1 have been selected, the design identifies n_1, n_2, and r_1, which minimizes the expected sample size when the true response rate is p_0. Thus, these designs are optimal in the sense that the expected sample size is minimized with the therapy has little efficacy subject to constraints on the type I and II errors. Early acceptance of efficacy for the therapy is not generally permitted as the ethical imperative is to terminate studies investigating therapies with unacceptable efficacy.

Two-stage designs have been commonly implemented in oncology trials. Oncology trials often employ a dose at or near the MTD (known from phase I trials) to deliver the maximum effect and thus frequently employ single dose trials. Evans et al. (2002) describes a phase II trial evaluating low-dose oral etoposide for the treatment of relapsed or progressed AIDS-related Kaposi's sarcoma (a type of skin cancer) after systemic chemotherapy. The primary endpoint was tumor response defined as at least a 50% decrease in the number or size of existing lesions without the development of new lesions. A two-stage design was employed with the plan for enrolling 41 total participants. However, if there were no objective responses after the first 14 participants have been evaluated, then the trial would be discontinued for futility, noting that if the true response rate was at least 20%, then the probability of observing zero responses in the first 14 subjects is <0.05. Notably, responses were observed in the first 14 participants, the trial continued, and etoposide was shown to be effective (http://www.thebody.com/content/treat/art18814.html).

6.2.8.1.1 Seamless Two-Stage Designs

A seamless design combines objectives traditionally addressed in separate trials in a single trial. This eliminates the time that would have occurred between the two trials and may have additional efficiencies (e.g., reduce total required sample size or total duration of follow-up) thus reducing the development time for new interventions. A typical seamless design has two-stages, one corresponding to each trial that would have been performed sequentially. Decisions regarding trial adaptation are made at the end of the first stage and can be based upon all available information (i.e., not just a single endpoint). Adaptations may include dropping treatment arms (i.e., "drop-the-loser" design particularly useful in Phase II delveopment when there are uncertainties in dose levels) (Posch et al., 2005), recalculating the sample size, or revising the randomization fraction. Final analyses of seamless designs uses data from both stages in a manner that controls the type I error rate and produces parameter estimates with desired precision and coverage probability.

Seamless two-stage designs may be classified by whether the objective and/or endpoints change between stages. One type of seamless design may have the same objectives and the same endpoints analogous to a typical group sequential design. A second type of seamless design may have the same objectives but with different endpoints. For example, a biomarker (e.g., tumor shrinkage) may be used in the first stage, and a clinical endpoint (e.g., survival) may be used in the second stage. A third type of seamless design may have different objectives but similar endpoints between stages. For example, the objective of the first stage may be to select a dose and the second stage would be used to evaluate early efficacy. A fourth type of design may have different trial objectives and different endpoints. For example, the objective of the first stage may be to select a treatment based on safety and the second stage may be to evaluate the efficacy. An important feasibility consideration in designing seamless trials is the amount of time that a trial participant needs to be followed to observe the required endpoint.

6.2.8.1.2 Seamless Phase II/III Designs

Consider a situation in which early phase II data has identified three possible doses. One option for the next step of development is to run a single phase III trial with four arms (three dose arms and a placebo) with appropriate adjustments for multiplicity (e.g., Bonferoni). A second option is to run two separate trials; the first being a phase II trial to select a dose and the second a phase III trial to compare the selected dose versus placebo. The two trials are analyzed separately. A third alternative is to run a seamless Phase II/III trial beginning with four arms (i.e., three doses and a placebo). This alternative provides a seamless transition from *learning* in Phase IIb (learning) to *confirming* in Phase III. An appropriate number of participants for selecting a dose are enrolled into the first stage. Enrollment is then paused until all randomized participants have completed the treatment, and interim analyses are conducted to select the "best" dose. Ineffective or unsafe treatments are eliminated after the first stage. Inclusion of a control group in the first stage since it is used to help pool data from the two stages during final analyses. Enrollment then continues into the second stage (i.e., phase III portion) of the trial, randomizing participants to either the selected dose or placebo. The final analysis then combines data from both stages (Maca et al., 2006).

If analyses of the seamless phase II/III trial was conducted at the nominal one-sided 0.025 rate, then the actual type-I error is >0.025 as dropping doses can inflate the type-I error. The type I error inflation increases with the number of doses that are selected out and the longer it takes to make the selection. Simulation can be utilized to make adjustments for the type I error inflation although simulation does not provide strong type-I error control (i.e., strong control of type I error implies that the probability of selecting a dose that is no better than placebo and obtaining statistical significance in favor of the dose is controlled by any response rate configuration for the other doses). Different adjustments are needed for different choices of sample sizes in each stage and the number of treatments. There is also limited flexibility to make data-dependent changes to the sample size or dose selection algorithm at the interim analyses.

A general two-stage method was described by Posch and Bauer (2005) that provides strong-control of type I error (important for regulatory submissions). The Posch–Bauer strategy is to (1) compute p-values from each stage and then combine p-values and (2) use the closed testing principle to ensure strong control of type-I error. The method permits early stopping, data-dependent sample size adjustment, and dose selection. For example, the total sample size may be modified based on nuisance parameters (e.g., variability of the primary endpoint or control group response rate). Adaptive rules do not have to be prespecified and can be modified at the interim and more than one experimental arm can be carried into stage two since the multiplicity is handled by the closed testing principle. A seamless phase II/III trial utilizing the Posch and Bauer method can result in smaller total sample sizes than the alternatives of a single trial with multiple dose arms and a placebo or two separate trials and is shorter in duration than running two separate trials.

6.2.8.2 Changing Endpoints

A fundamental tenant in the design of randomized trials involves prespecification and clear definition of the endpoints that will be assessed in the trial (ICH E9, 1998), as failure to prespecify endpoints can introduce bias into a trial and creates opportunities for manipulation. However, sometimes new information may come to light that could merit changes to endpoints during the course of a trial. This new information might include, for example, results from other trials or identification of better biomarkers or surrogate outcome measures. Such changes can allow incorporation of up-to-date knowledge into the trial design. However, changes to endpoints can also compromise the scientific integrity of a trial.

Several studies have reported discrepancies between endpoints that were identified in trial protocols compared to publications of the trials in the medical literature. Chan et al. (2004a) compared published articles with protocols for 102 randomized trials approved by the Scientific-Ethical Committees for Copenhagen and Frederiksberg, Denmark in 1994–1995, and reported that 62% of the trials had at least one primary endpoint that had been changed, introduced, or omitted. Chan et al. (2004b) compared published articles with protocols for 48 randomized trials approved for funding by the Canadian Institutes of Health Research in 1990–1998, and reported that primary endpoints differed between protocols and publications in 40% of the trials.

Evans (2007) identified and discussed the principle considerations when evaluating whether to modify an endpoint. A primary consideration is whether the decision is independent of the observed treatment effect obtained from the trial to date. If the decision to revise endpoints is independent of the observed treatment effect from the trial, then such revisions may have merit. In fact, Wittes (2002) encourages consideration of changes in long-term trials, as medical knowledge evolves or when assumptions made in the design of the trial appear questionable. Wittes further argues that researchers, "may consider changes to the primary endpoint when the trial has airtight procedures to guarantee separation of the people involved in making such changes from data that could provide insight into treatment effect" (pp. 2792–2793).

Some trials have successfully changed endpoints after the trials began by maintaining independence between the decision and the trial data. For example, the randomized Post-CABG (Post-coronary artery bypass raft) trial (Campeau et al., 1997) compared two lipid-lowering regimens in patients who had coronary artery bypass surgery. The investigators explicitly did not identify a primary endpoint when they designed the trial. An angiogram to assess lipid deposition in the coronary arteries was conducted at entry and then again 5 years later. The researchers planned to compare changes in lipid deposition over the 5-year interval between the two regimens. Because by design no endpoint would be available for 5 years after randomizing the first participant, the protocol team used this period to define the endpoint and to develop methods for analyses. Although the endpoint was not prespecified in the design phase, a practice that is not generally recommended, trial leadership ensured that the selection of the endpoint was independent of data from the trial.

Another example occurred in ACTG A5265, a randomized trial comparing GV to nystatin (both oral mouthwashes) for the treatment of OC in HIV-infected adults. The primary endpoint in the trial was changed based on a DSMB recommendation. GV that often causes staining of the mouth was diluted as part of the protocol to protect against staining. The original primary endpoint was binary defined as improvement based on clinical evaluation. It was believed that this evaluation would be blinded to the evaluator of the improvement endpoint. However, staining occurred at a significant rate, thus unblinding the evaluator. The DSMB recommended that the primary endpoint be revised to "cure" with the rationale being that an assessment of cure would be more objective and less subject to bias from an unblended evaluator.

If, however, the decision to change the endpoint is not independent of the observed treatment effects, then "cherry-picking" is a serious concern. New endpoints may be selected because they displayed a trend toward significance, while other candidate endpoints may have been examined but not selected or reported because they failed to display a desirable trend; this increases the chance of false positive errors. In the Physicians' Health Study (Young et al., 1988; Cairns et al., 1991), the trial's DSMB recommended termination of the study because interim data seemed unlikely to show any benefit of aspirin with respect to the primary endpoint, total mortality. At the time this decision was made there was evidence of benefit with respect to MI. However, the FDA did not approve an indication for aspirin for the prevention of MI because this was not the prespecified primary endpoint.

Evans (2007) noted that in order to evaluate whether a change in endpoint is independent of data from the trial, investigators and reviewers should ask three important questions (Table 6.3).

Since the decision to revise endpoints should be independent of the trial data, a DMC that has reviewed interim data may not be appropriate for making decisions regarding endpoint revisions. Even DMC review of pooled data can suggest treatment effects (e.g., in a two-group comparison study of response rates, a very high-pooled response implies a relatively high response rate for both groups). In this case, trial leadership may wish to convene an external advisory committee that has not reviewed data from the trial to assess the potential impact on the integrity of the trial and to make recommendations regarding endpoint revision.

It is also important to consider the scientific relevance of the endpoints in question. Does the current state of knowledge make the results of the current trial uninformative or inefficient? Is the trial now scientifically uninteresting or irrelevant? If so, then changing endpoints may be constructive, and perhaps even ethically necessary, to ensure that the study generates a scientific contribution. For example, new scientific questions may arise after recently

TABLE 6.3

Questions to Evaluate When Considering Endpoint Changes

- What is the source of the new information that elicits consideration of the change in endpoints? If the source is external to the trial in question, for example, arising from results from another trial, then the revision of endpoints may be credible.
- Have interim data on the endpoint (or related data) from a trial been reviewed? If trial data have not been reviewed, then the revision of endpoints may again be credible.
- Who is making the decision regarding endpoint revision (e.g., trial sponsors or an independent external advisory committee)? Appropriate decision makers should have no knowledge of the endpoint (or related trial data) results. In particular, if interim analyses have been conducted, the decision makers should not have knowledge of those data. Note, however, that even if no formal interim analysis has been conducted, any impressions that the investigators may have of the trial to date may influence decisions regarding changes in endpoints. For example, investigators may have a "sense" of the endpoint result or a related variable even though formal analysis of the endpoint has not been conducted. An investigator may notice changes in certain patients at his or her site and may attribute these changes to the investigational medication. This can be particularly problematic in unblinded trials. For these reasons, study sponsors, investigators, and DMCs may not be appropriate decision makers for endpoint revisions.

completed trials have already answered the original question of interest. Also, better biomarkers or surrogates may have been identified, or there may have been changes in regulatory oversight.

One should be cautious of potential operational bias induced by the revision of endpoints. Operational bias is created when the conduct of clinical investigators or participants is changed by knowledge (or perceived knowledge) of trial data. Knowledge of revisions to endpoints may influence the actions of clinical investigators or participants as they anticipate the reasons for such revisions. For example, if a decision to change the primary endpoint is made, then participating clinicians and patients may believe that such a change was made due to a lack of efficacy of the intervention. This belief may affect their willingness to participate, affecting accrual and retention.

If the trial leadership decides to modify endpoints, then appropriate documentation is crucial. Changes should be described in amendments to the protocol and the analysis plan. The registry record for the trial should also be updated. Changes in endpoints should also be declared when submitting a manuscript to a journal, so that the results can be properly evaluated. Evans (2007) outlines reporting criteria for a clinical trial with any modified endpoint (Table 6.4) to help ensure clarity and transparency of the analyses, enable the evaluation of the independence of the endpoint revision and trial data as well as the potential for selective reporting, allow assessment of the ramifications of the endpoint revision, and help avoid over interpretation of the data. Researchers may further consider focusing on descriptive analyses using CIs rather than hypothesis testing to avoid overstating the significance of the results.

Hawkey (2001) suggests that journals should require submission of the protocol alongside manuscripts describing clinical trial results, to help ensure

TABLE 6.4

Items to Include When Reporting Results from Trials with Modified Endpoints

- A clear statement describing the fact that information obtained after trial initiation led to the change in the endpoint
- A description of the reasons (e.g., whether the endpoint was suggested by the data) and decision procedure (e.g., who made the decision and whether data were unblinded)
- A discussion of the potential biases induced by the change of the endpoints
- If warranted, (i.e., if the decision to add endpoints was not independent of the data), a disclaimer that the results should be interpreted with caution and should be confirmed in future trials
- A report of the reasons for excluding endpoints from the analyses and whether this was independent of trial data

that the reported endpoints indeed reflect what was defined at the start of the trial. Several journals have adopted this policy, including *The Lancet* and the *British Medical Journal.* Other journals are considering a requirement to submit raw data. For industry-sponsored studies, the *Journal of the American Medical Association* requires that analyses be conducted by an independent statistician at an academic institution, in part to protect against post hoc endpoint revisions.

A particularly common and problematic situation occurs when endpoints are defined vaguely or ambiguously. For example, a protocol designed to study the effects of 24 week of a new investigational drug on immune function might specify "CD4 count" as an endpoint. This endpoint could be interpreted in many different ways, including, for example: (1) CD4 count at week 24; (2) changes from baseline in CD4 count at week 24; (3) the occurrence of a doubling of CD4 count from baseline; or (4) the occurrence of at least a 50-cell increase in CD4 count from baseline. If a precise definition and analysis for each endpoint are not specified in advance, it is possible for many different versions of the endpoint to be examined, followed by selection and reporting of the most desirable result. This form of "cherry-picking" inflates the false positive error rate and leads to an under-reporting of negative evidence. Thus, it is critical to prespecify the precise definition of the primary endpoint together with the method of statistical analysis that will be applied (ICH E9, 1998).

In certain cases, it may be appropriate to change or identify endpoints after initiation of a trial, even when the decision is based on data from the trial. For example, if a trial is very large and of long duration, then investigators may divide the trial into two stages: a hypothesis-generating stage in which endpoints are identified, and a subsequent hypothesis-testing stage. In this case, statistical testing would be based only on data collected after the first stage was complete.

Revisions to endpoints (particularly primary endpoints) should be uncommon. If not appropriately evaluated, such revisions lead to misguided research and suboptimal patient care. If, however, important scientific knowledge has been gained after a trial begins, then this knowledge should be carefully and responsibly evaluated for incorporation into the trial.

6.2.8.3 Sample Size Recalculation

Sample size calculations are conducted during the design of a trial. Such calculations require assumptions regarding parameters such as a control group response rate (for binary endpoints); the variability of the outcome (for continuous endpoints); the accrual interval, the follow-up interval, and the median event time (for time-to-event endpoints); adherence; and the treatment effect to be detected. Assumptions are often made based on historical data. However, these assumptions can be inaccurate due to the novelty of the therapy under evaluation, advancing medical practices or differences between modern versus historical trial design. If the assumptions are inaccurate, then the trial may be under- or overpowered. Given the uncertainly of the validity of the assumptions used in trial design, it may be prudent to evaluate the assumptions used in calculating the sample size using interim data, particularly if power is very sensitive to assumption variation. If the assumptions are erroneous, then in-flight adjustments can be made to ensure that the trial can efficiently address its objective.

Planning for sample size recalculation is important to maintain trial integrity. A careful plan should be outlined in the trial protocol including the timing, methodology, the decision-making process, and implementation plan. Restrictions on trial duration or sample size should be discussed.

Sample size recalculation based on nuisance parameters (e.g., control group response rate or the variance of a continuous outcome) can usually be addressed in a noncontroversial manner and can be considered for many trials although review frequency should be limited to minimize complexity. If resizing based on a variance, for example, researchers can estimate the number of patients needed to estimate the variance with an acceptable level of precision and then plan an interim analyses to adjust the sample size after gathering the appropriate data. The timing of the recalculation will depend on the accrual rate, follow-up time and level of uncertainty in the initial assumptions. The respective analyses should be planned so as not to delay trial progress, however, unless it is so desired.

It is usually preferred that recalculation based on nuisance parameters be conducted in a blinded manner (i.e., without knowledge of intervention assignment) since procedures using blinded data generally have good operating characteristics and the use of blinded data tends to minimize type I error inflation and avoids the potential bias of having someone reviewing unblinded data.

Blinded sample size recalculation is generally accepted by regulators. Blinded data review can often be conducted by trial personnel. However, there are some cases where pooled data can be informative about treatment effects. For example, consider a trial with a binary endpoint of response. If

TABLE 6.5

Challenges with Unblinded Sample Size Recalculation

- Control of statistical error rates
  - Combination tests are often used to control the type I error rate. For example, when comparing two sample means, differences in means from each stage of the trial are combined as independent incremental results using weights proportional to the sample size of each stage. Methods for combining p-values from different stages are often applied in such cases. An alternative to designing trials with combination tests is to prespecify the sample size with superiority/futility rules for the first interim analysis (i.e., assign a weight to the first two stages of the trial), and then iteratively at the time of each interim analysis, define the sample size with superiority/futility rules for the next stage (i.e., divide the weight for the subsequent stage into two parts for the following two stages until a point is reached where all the remaining weight is assigned to a particular stage).
- Control for operational bias
  - People who are aware of the methodology utilized in the sample size calculation may be able to infer treatment effects based on the sample size revisions, jeopardizing trial integrity. Information that is distributed should be limited such that the trial results cannot be inferred, and results must remain confidential. Project teams may opt to withhold the exact recalculation methodology from people involved with the trial and the "public." A confidential document can be created to outline the details of the methodology (discussed further in Chapter 7).
- Evaluation of clinical relevance of observed treatment effects
  - Observed treatment effects may not be clinically relevant, and trials should be powered to detect relevant effects. Researchers should consider the clinical relevance of the observed effects before sample size adjustment is undertaken.
- The integrity of the blind must be meticulously maintained
  - Review of the unblinded data be conducted by individuals not involved with the trial.

---

the pooled data suggest 100% response, then it is clear that all therapies have 100% response.

Sample size recalculation may also be based on assumptions regarding treatment effects. During the conduct of a trial, data from other trials may change the perception of what constitutes a clinically relevant effect or new safety data may change the perception of the benefit:risk profile affecting how the trial should be powered. If the sample size is recalculated based on external data only, then there are no statistical or practical concerns.

However, when utilizing unblinded data and resizing a trial based on observed treatment effects, then extreme care must be taken to minimize the risk to trial integrity. Several challenges and concerns with unblinded sample size recalculation based on observed treatment effects are outlined in Table 6.5. For these reasons, recalculation based on observed treatment effects is generally avoided.

6.2.9 Dynamic Treatment Regimes

The treatment of patients for many complex or chronic diseases is not based on a single decision regarding an intervention but is a sequence of decisions made over time. Clinicians manage a patient's ongoing illness and rou-tinely adjust, change, add, or discontinue treatment based on progress, tox-icity, adherence, and quality of life (QoL). Treatment decisions are tailored to the individual patient based on patient characteristics and prior patient response. A dynamic treatment regime (or adaptive treatment regime) is a set of rules that governs the assignment of time-varying treatment based on observed covariates and intermediate response. Treatment choices are made sequentially as patients make the transition from one health state to another. The goal is to find the best treatment regime that produces the best terminal outcome. The idea of adaptive treatment regimes appeals to clinicians since it is consistent with clinical practice.

Sequentially randomized trials are often used to investigate the effect of adaptive treatment regimes. For example, suppose there are two treatment options A1 and A2 at the first stage and two treatment options for both responders (B1 and B2) and nonresponders (B1' and B2') at the second stage. In sequentially randomized designs, patients are randomized to A1 or A2 fol-lowed by another randomization at the second stage, to B1 or B2 if the patient is a responder or to B1' or B2' if the patient is a nonresponder. With this design, one can investigate a total of eight treatment regimes, namely, $A_jB_kB_l'$, j, k, l = 1, 2 where $A_jB_kB_l'$ stands for "Treat with A_j followed by B_k if respond, by B_l' if otherwise." The number of treatment options in the second stage may vary between responders and nonresponders. For example, a patient that responds in stage 1 may not require a change in therapy at stage 2.

For illustration, consider a trial evaluate treatments for PML and HIV (Figure 6.9). PML is a rare brain disease with a high mortality rate with no proven effective therapy. A patient that is diagnosed with PML and HIV might initiate combination antiretroviral therapy (cART) in hopes of treat-ing the disease. Some data suggests "super-cART" (a boosted cART regi-men) may produce better PML outcomes (e.g., survival). However, cART and super-cART may also induce a life-threatening immune-reconstitution inflammatory syndrome (IRIS). It is also theorized that steroids can help to treat IRIS but can also have toxic effects. A trial randomizing HIV PML patients to cART versus super-cART, followed by randomization to steroids versus steroid-placebo if IRIS occurs, may be designed to compare four treat-ment regimes: (1) cART plus steroids if IRIS occurs, (2) cART without steroids if IRIS occurs, (3) super-cART with steroids if IRIS occurs, and (4) super-cART without steroids if IRIS occurs. A particular challenge to the analyses is that data from an individual patient can contribute to effect estimates from multiple strategies. For example, a patient that is randomized to cART and never has IRIS contributes to regimes 1 and 2.

It is important to realize the distinction between the treatment regime (i.e., the algorithm or strategy that dictates patient treatment over time) versus the possible realized experiences following the regime. Patients on the same treatment regime can have different treatment experiences. For example, two patients may be randomized to a regime "treat with A, then with B if no

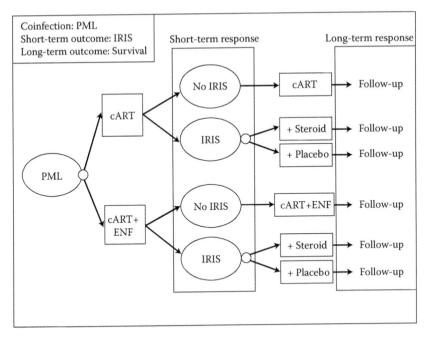

FIGURE 6.9
Schema for example of adaptive treatment regime trial for the treatment of HIV-associated PML.

response, but continue on A otherwise." However, one patient may receive treatment A and respond and thus remains on treatment A. The second patient may receive treatment A, fail to respond, and then receive treatment B. Although the two patients receive difference treatment experiences, they are both part of the same regime. Furthermore, data from individual participants can contribute to multiple treatment strategies.

Several clinical trials have adopted these sequentially randomized designs. An example is the Sequenced Treatment Alternatives to Relieve Depression (STAR*D) study (Rush et al., 2004). STAR*D is a four-level multisite randomized clinical trial that evaluated strategies for "treatment-resistant" depression. At level 1, all participants received Citalopram (CIT), and if they responded to CIT, the treatment with CIT continued. If they failed to respond to CIT, they moved to level 2. At level 2, the participants had a choice to select either to add another treatment to CIT, or to switch to a different treatment. If they selected to add another treatment, they were randomized to one of the three options (bupropion, buspirone, and cognitive therapy); if they selected to switch to a different treatment, they were randomized to one of the four options (sertraline, bupropion, venlafaxine, and cognitive therapy). If they still did not respond by the end of level 2, they moved to level 3. Similar to level 2, participants had a choice to add or switch treatments. If they selected to add a treatment, they were randomized to one of the two options (lithium or thyroid hormone); if they decided to switch to a different treatment, they were randomized to one of the two options (mirtazapine or nortriptyline). Finally, at level 4, nonresponders from level 3 were randomized to one of the two options (tranylcypromine or the combination of mirtazapine and venlafaxine).

Another example was the Cancer and Leukemia Group B (CALGB) 8923 (Stone et al., 1995), a randomized, double-blind, placebo-controlled two-stage trial evaluating therapies for elderly patients with acute myelogenous leukemia (AML). In the first stage, participants were randomized to standard chemotherapy or standard chemotherapy plus granulocyte-macrophage colony-stimulating factor (GM-CSF) for the treatment of elderly patients. In stage 2, participants with complete remission were then randomized to one of the two intensification therapies. Participants that did not display complete remission then further chemotherapy was deferred. The goal of the trial was to compare the four treatment regimes with respect to disease-free survival. Other trials utilizing dynamic treatment regimes include (1) the Veterans Affairs Non-Q-Wave Infarction Study in Hospital (VANQWISH: Stone et al., 1995) that evaluated immediate diagnostic catheterization and angiography followed by revascularization as indicated by evidence of reversible occlusion of the coronary arteries versus noninvasive exercise testing and postponement of catheterization until the patient displayed signs of cardiac ischemia, for the management of mild heart attacks and (2) the Clinical Antipsychotic Trials of Intervention Effectiveness (CATIE) study (Schneider et al., 2001) for Alzheimer's disease.

The design, conduct, and analyses of the trials using adaptive treatment regimes can be challenging. For example, participants that consent to randomization in stage 1 may not consent to randomization in a later stage. Consent to an *adaptive strategy* should be obtained prior to the trial. Furthermore, the ITT principle is more complex in dynamic treatment regimes as it is unclear how to assign treatment in a stage following patient dropout from a previous stage. Estimation of a mean response based on observational longitudinal data is proposed in Murphy et al. (2001) and Murphy (2003). Methods for survival outcomes were considered in Lunceford et al. (2002), Wahed and Tsiatis (2004, 2006), Guo and Tsiatis (2005), Hernan and Robins (2006), and Lokhnygina and Helterbrand (2007). Thall et al. (2007) adapts a Bayesian approach to model the time to failure and compare two-stage adaptive treatment regimes.

6.2.10 Diagnostic Device Trials

Diagnostic tests are an important part of medical decision-making. In practice, many tests are used to screen or diagnose disease or injury. For example, a pap smear is a screening test for cancer of the cervix whereas digital rectal examination (DRE) and prostate-specific antigen (PSA) tests are used for prostate cancer screening.

Developing diagnostics need to be evaluated for accuracy (e.g., how well they identify patients with disease, how well they identify patients without disease, and once a test is administered, what is the likelihood that it is correct). This evaluation requires a comparison to a "gold standard" diagnosis (i.e., a diagnoses that can be regarded as the "truth" or near approximation) which often requires costly, time-consuming, or invasive procedures (e.g., a biopsy). Evaluation consists of examination of sensitivity (the probability of a positive test given a true positive), specificity (the probability of a negative test given a true negative), positive predictive value (the probability of a true positive given a positive test), and negative predictive value (the probability of a true negative given a negative test). The interpretation of these accuracy measures is relative to the disease being studied, implications of therapy upon diagnoses, and alternative diagnostics. For example, if a disease is very serious and requires immediate therapy, then a false negative is a very costly error, and thus high sensitivity is very important. However, if a disease is not life threatening but the therapy is costly and invasive, then a false positive error is very costly (i.e., high specificity is necessary). If a diagnostic device can be shown to have good accuracy relative to the gold standard diagnoses and has other advantages (e.g., reduced costs, faster results, less invasive, practical to administer), then the diagnostic device will be valuable.

Important considerations in design include whether blinding can be implemented and whether test evaluations (e.g., laboratory results on collected specimens) should occur centrally (i.e., a single laboratory for the entire study) or locally (with each site having its own laboratory). If retrospective diagnostic testing is done (e.g., via laboratory evaluation of collected specimens), then operators of the test should be blinded to the reference standard results to protect from bias. Central testing is best at eliminating technical variations and thus is usually preferred. Local laboratories introduce another source of variation but can be cheaper and faster, and may more closely represent results as they occur in practice.

When the outcome of a diagnostic test is positive vs. negative (binary), then the calculation of sensitivity and specificity can be performed directly. Most diagnostic trials are designed to estimate sensitivity and specificity with a prespecified precision where precision is usually measured by the width of a CI. Under assumptions regarding the *observed* sensitivity and specificity and specification for maximum desired CI widths associated with estimates of these quantities, sample sizes can be easily determined (Figure 6.10). Note that you would calculate the number of necessary participants who are disease positive and disease negative separately. If the true disease status of patients is known then disease-status specific numbers can be targeted. If the true disease status is unknown, then an estimate of disease prevalence may be obtained to guide the total number of patients that will be necessary to appropriately size the trial. Disease prevalence can be monitored to make adjustments if necessary.

One can extend these methods to size a trial under the assumption of a *true* sensitivity and specificity and specification for the desired *average* CI width. Such calculations are often performed using simulation. However,

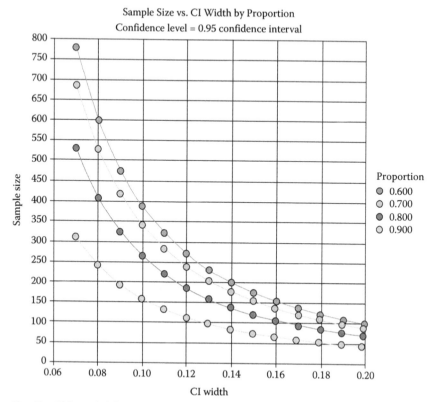

Sample Size vs. CI Width by Proportion
Confidence level = 0.95 confidence interval

Proportion
● 0.600
○ 0.700
● 0.800
○ 0.900

Note: E.g., If the probability of correctly predicting susceptibility is observed to be 0.80, then −100 specimens are required to have a CI width of 0.16.

FIGURE 6.10
Required sample size as a function of desired CI width when the probability of correctly predicting susceptibility (or nonsusceptibility) is 0.6, 0.7, 0.8, or 0.9 (from PASS®).

many diagnostics have an outcome that is measured on a continuum and the identification of a "cut-off" that will discriminate between positive vs. negative diagnoses must be conducted. Evaluation of such diagnostics can be conducted in a trial with two stages. Stage I is used to identify an appropriate cut-off and stage II used to validate the accuracy of the diagnostic using the cut-off identified in stage I. We illustrate this strategy with an example.

6.2.10.1 Example

Stroke is a common cause of death and a major cause of long-term disability. However, stroke is a treatable disease if diagnosed early. Approximately 80% of strokes are ischemic, and 20% are hemorrhagic. The treatment of ischemic stroke is time sensitive and requires intravenous (IV) administration of thrombolytic therapy. However, thrombolytic therapy is contraindicated for hemorrhagic stroke. Thus, it is important to be able to distinguish the two types of stroke as quickly as possible. Current diagnostics include imaging modalities, but imaging is often unavailable in a timely fashion. Additional diagnostics are needed for which timely results can be available.

The NR2 peptide is released into the bloodstream during cerebral ischemia and can be detected and quantified (via a blood sample) quickly after ischemic onset. A clinical trial to evaluate the NR2 peptide as a diagnostic for ischemic events is being planned. It was decided that the minimum acceptable sensitivity and specificity be 80%, and thus a goal is to demonstrate that the sensitivity and specificity of the NR2 peptide are simultaneous >80%. The NR2 peptide level will also depend on the time of the blood sample relative to ischemic onset. Thus, evaluation was conducted in four-time windows (i.e., 0–3, 3–6, 6–12, and 12–24 h after ischemic onset).

The primary objective of the trial was to investigate if the NR2 peptide measurement can be used to discriminate ischemic vs. accurately nonischemic events. The trial is designed with two stages. The intent of stage, I is to estimate optimal cut-off values for each of four-time windows from which the blood sample for the NR2 peptide level quantification is drawn for discriminating ischemic vs. nonischemic events. The intent of stage II is to validate the diagnostic using the cut-off values identified in stage I.

6.2.10.1.1 Stage I

In stage I, an ROC curve (sensitivity vs. 1–specificity) for each time window by varying the cut-off value for distinguishing ischemic events from nonischemic events, will be constructed. The gold standard used for constructing the ROC curve is based on MRI imaging. The partial area under the ROC curve for the region with specificity between 0.75 and 0.95, denoted by $pAUC_{0.75}^{0.95}$ is used as a global measure of the diagnostic effectiveness of NR2 peptide. The partial area rather than the commonly used total AUC is considered since only the region with high specificity is of clinical interest. The standardized partial area $spAUC_{0.75}^{0.95} = \frac{1}{0.20}pAUC_{0.75}^{0.95}$ can be interpreted as the average sensitivity level when the specificity level ranges from [0.75, 0.95]. To test the global null hypothesis that the NR2 peptide does not have adequate accuracy at any of the four periods, evaluation of whether the average $spAUC_{0.75}^{0.95}$ across the four time windows is at least 0.8 is conducted.

The sample size calculation for stage I is based on the following assumptions:

1. The prevalence rate of ischemic events in the enrolled cohort is 80%.

2. A transformation of the NR2 peptide measure is normally distributed to trial participants with and without ischemic events.

Without loss of generality, it is assumed that the standard deviation for the nonischemic participant is one and the counterpart for the ischemic participants is d_0. With n participants per time window in the first stage, Table 6.6

TABLE 6.6

Standard Errors for the $pAUC_{0.75}^{0.95}$ and the Average $pAUC_{0.75}^{0.95}$ (Displayed in the Parenthesis) Estimates When the True $spAUC_{0.75}^{0.95}$ is 0.90

	n = 100 per Time Window	n = 150 per Time Window	n = 200 per Time Window
$d_0 = 1.0$	0.066 (0.033)	0.052 (0.026)	0.047 (0.023)
$d_0 = 1.5$	0.049 (0.024)	0.038 (0.019)	0.033 (0.016)
$d_0 = 2.0$	0.046 (0.022)	0.035 (0.017)	0.031 (0.015)

provides the standard error for the $pAUC_{0.75}^{0.95}$ estimate for each time window and the standard error of the average $pAUC_{0.75}^{0.95}$ (displayed in the parenthesis) when the true $spAUC_{0.75}^{0.95}$ is 0.90.

Results summarized above suggest that a sample size n = 100 per time window is adequate for stage I with a power of no less than 86% for the above settings. With n = 100 per time window, the 95% CI for the average $spAUC_{0.75}^{0.95}$ across the four time windows is expected to be (0.84, 0.96), which is sufficiently narrow to indicate acceptable clinical precision.

At the end of stage I, the empirical estimate for the ROC curve and its standard errors will be estimated via the method by Hsieh and Turnbull (1996). The perturbation technique to construct the confidence band for the ROC curve may also be used (see, e.g., Lin et al., 1993). If the CIs or band for the estimated ROC curves are not narrow enough for n = 100 per time window, then another 50 participants can be enrolled into stage I. Since the adjustment is only made considering the sampling variability and is not for time window comparisons, no statistical adjustment for type I error is necessary even if the sample size is increased.

To evaluate the overall diagnostic accuracy of the NR2 peptide in discriminating ischemic events, the CI estimate for $spAUC_{0.75}^{0.95}$ is constructed. If at the end of stage I, there is no cut-off value for NR2 peptide that provides both specificity and sensitivity of 0.80 or higher in any given time window, then the study may be terminated without proceeding to stage II. The time windows for which no cut-off value for NR2 peptide would yield specificity and sensitivity 0.80 or higher would be excluded from the stage II assessment. For time windows that potentially yield acceptable sensitivity and specificity, cut-off points for the NR2 peptide for discriminating ischemic events will be identified. The cut-offs points do not have to be the same for each time window.

6.2.10.1.2 Stage II

Stage II is designed to validate the cut-off values derived from stage I and establish the diagnostic performance of the NR2 peptide in discriminating ischemic events.

Since this is the validation stage, it is important to control the type I error. Using the data from stage II, we will assess whether the two-sided lower bounds of the 97.5% CIs for sensitivity and specificity are above 0.8,

TABLE 6.7

Expected Lower Bound of 97% Confidence Interval Estimates for Sensitivity and Specificity for Various True Sensitivities, Specificities, and Sample Sizes

			Lower Bound of 97.5% CI	
True Sensitivity	**True Specificity**	**Sample Size**	**Sensitivity**	**Specificity**
Lower Bound Set to Be 0.80				
0.85	0.85	1286	0.83	0.80
	0.90	322	0.80	0.82
	0.95	322	0.80	0.89
0.90	0.85	1286	0.88	0.80
	0.90	231	0.85	0.80
	0.95	58	0.80	0.81
0.95	0.85	1286	0.93	0.80
	0.90	231	0.91	0.80
	0.95	56	0.88	0.80
Lower Bound Set to Be 0.75				
0.85	0.85	326	0.80	0.75
	0.90	106	0.76	0.75
	0.95	82	0.75	0.83
0.90	0.85	326	0.86	0.75
	0.90	106	0.83	0.75
	0.95	31	0.76	0.75
0.95	0.85	326	0.92	0.75
	0.90	106	0.90	0.75
	0.95	31	0.85	0.75

indicating that the NR2 peptide is an effective diagnostic test for identifying participants in ischemic events. Positive and negative predictive values will also be plotted as a function of assumed prevalence.

Assuming that the true sensitivity and specificity are 0.85 and 0.95 for the stage II participants and that the prevalence of ischemic event is 80%, then with n = 322, the expected lower bound of two-sided 97.5% CIs for sensitivity and specificity are 0.80 and 0.82, respectively (i.e., greater or equal to 0.80) indicating that n = 322 for stage II is adequate (Table 6.7).

6.3 Phase IV

Phase IV clinical trials are also known as postmarketing trials. Typically these studies are designed and carried out after the intervention is approved by regulatory agencies. Objectives of phase IV trials greatly vary and thus studies may be initiated by industry, the NIH, hospitals, research/academic institutes, or individual physicians. If a drug has an important public health impact, but questions remain regarding long-term safety and efficacy, then the NIH may support a large phase IV trial to evaluate long-term safety effects.

The objectives of industry-sponsored phase IV trials can be divided into two categories: safety surveillance and label extension. For safety surveillance phase VI trials, some are designed based upon postapproval commitment (PAC). PAC may occur when at the time of initial FDA approval, the FDA has concerns regarding safety signals (e.g., concern for long-term cumulative effects). Some safety signals may take a long time to arise (due to cumulative effects) or may be so rare (e.g., bone fracture, stroke) than many patients need to be followed. Instead of delaying approval while waiting for the long-term study results and delaying public use of a useful intervention, regulators may decide to approve the drug and request or mandate that a large-scale, long-term-follow-up phase IV trial be conducted. A better understanding of drug safety can then be established with long-term follow-up on a large number of patients.

One type of phase IV study often recommended by FDA is a large, simple trial. These trials can be single-arm uncontrolled studies or randomized controlled studies. They are designed for a large number of patients followed for only a few endpoints and thus are useful if there is interest in only a few specific safety events of interest. Procedures at each study visit are greatly simplified encouraging patients and clinicians to participate and remain in study.

Phase IV studies designed for label-extension purposes include at least two categories: (1) label improvement of the approved indication and formulation and (2) new formulation or new use for additional indications. Usually, a drug is initially approved for a single specific indication, formulation, and dose. However, there may be potential to extend the use to include other indications or formulations. In this case, Phase IV trials may be run, and trial results can be included as part of a supplemental NDA (sNDA) for the primary approval. For example, the primary approval may be based on an IV formulation, but a phase IV trial may be run to evaluate an oral formulation.

In some disease areas, a drug may be effective for many indications. For example, in antibiotics, a drug that is effective in treating an upper respiratory tract infection caused by a specific bacteria, may also be effective for the treatment of a urinal tract infection from another type of bacteria. Similarly, an oncology drug that is effective for treating breast cancer may also be effective in treating head and neck cancer. Industry sponsors tend initially to evaluate an intervention for a primary indication and then follow up in phase IV development to pursue additional indications.

Many phase IV label extension studies are designed to compare the intervention versus a standard intervention in patients with the initial indication to see if superiority can be demonstrated (often with respect to secondary endpoints, QoL or health economic (HE) endpoints). Other phase IV trials may evaluate the intervention in combination with a standard "background" intervention, often termed "background studies" or "combination studies." Placebo-controlled phase IV studies may be designed with an objective to demonstrate the superiority in endpoints that were too challenging to evaluate in phase III trials (e.g., long-term endpoints or rare events). Even if a primary objective of a phase IV trial is to evaluate the efficacy, safety is always evaluated.

One special type of phase IV trial evaluates the intervention in a pediatric population. If a drug can potentially be indicated for pediatric use, then a pediatric investigational plan (PIP) may be prepared and submitted to regulatory agencies. If the PIP is accepted, then a phase IV pediatric trial can be conducted to evaluate the safety and efficacy in the pediatric population.

References

Anderson S, Hauck WW. 1990. Consideration of individual bioequivalence. *J Pharmacokinet Biopharm* 18:259–273.

Biomarkers Definition Working Group. 2001. Biomarkers and surrogate endpoints: Preferred definitions and conceptual framework. *Clin Pharmacol Ther* 69(3):89–95.

Bosch J et al. 2002. Use of ramipril in preventing stroke: Double-blind randomized trial. *BMJ* 324:1–5, 699–702.

Brodie MJ, Perucca E, Ryvlin P, Ben-Menachem E, Meencke H.-J, and for the Levetiracetam Monotherapy Study Group. 2007. Comparison of levetiracetam and controlled-release carbamazepine in newly diagnosed epilepsy. *Neurology* 68(6):402–408.

Cairns J et al. 1991. Issues in the early termination of the aspirin component of the physicians' health study. Data monitoring board of the physicians' health study. *Ann Epidemiol* 1:395–405.

Campeau L, Knatterud GL, Domanski M, Hunninghake DV, White CW, Geller NL, Rosenberg Y. 1997. The effect of aggressive lowering of low-density lipoprotein cholesterol levels and low-dose anticoagulation on obstructive changes in saphenous-vein coronary artery bypass grafts. The post artery bypass graft trial investigators. *N Engl J Med* 336:153–162.

Chan AW, Hrobjartsson A, Haahr MT, Gotzsche PC, Altman DG. 2004a. Empirical evidence for selective reporting of outcomes in randomized trials. *JAMA* 291:2457–2465.

Chan AW, Krleza-Jeric K, Schmid I, Altman DG. 2004b. Outcome reporting bias in randomized trials funded by the Canadian Institutes of Health Research. *Can Med Assoc J* 171:735–740.

Couper DJ, Hosking JD, Cisler RA, Gastfreind DR, Kivlahan DR. 2005. Factorial designs in clinical trials: options for combination treatment studies. *J Stud Alcohol Suppl* (15):24–32; discussion 6–7.

D'Agostino RB et al. 2003. Non-inferiority trials: Design concepts and issues—The encounters of academic consultants in statistics. *Stat. Med.* 22(2):169–186.

Diamardis EP. 2010. Cancer biomarkers: Can we turn recent failures into success? *JNCI* 102(19):1462–1467.

Elm JJ et al. 2005. A responsive outcome for Parkinson's disease neuroprotection futility studies. *Ann Neurol* 57(2):197–203.

Evans SR. 2007. When and how can endpoints be changed after initiation of a randomized clinical trial?. *Public Library of Science (PLoS) Clin Trials* 2(4):e18. doi:10.1371/journal.pctr.0020018.

Evans SR, Testa MA, Cooley TP, Krown SE, Paredes J, Von Roenn JH. 2002. A phase II evaluation of low-dose oral etoposide for the treatment of relapsed or progressed AIDS-related Kaposi's Sarcoma: An ACTG clinical study. *J Clin Oncol* 20(15):3236–3241.

Evans SR et al. 2011. Peripheral neuropathy in HIV: Prevalence and risk factors. *AIDS* 25:919–928.

FDA Guidance. 2002. Statistical approaches to establishing bioequivalence. U.S. Department of Health and Human Services. Food and Drug Administration. Center for Drug Evaluation and Research (CDER). January 2001.

FDA Guidance. 2012. Guidance for industry: Enrichment strategies for clinical trials to support approval of human drugs and biological products. U.S. Department of Health and Human Services. Food and Drug Administration. Center for Drug Evaluation and Research (CDER). Center for Biologics Evaluation and Research (CBER). Center for Devices and Radiological Health (CDRH). December 2012.

Freidlin B, McShane LM, Korn EL. 2010. Randomized clinical trials with biomarkers: design issues. *J Natl Cancer Inst* 102:152–160.

Gau P and Ware JH. 2007. Assessing non-inferiority: A combination approach. *Stat. Med.* 27(3):392–406.

Gaydos B, Krams M, Perevozskaya I, Bretz F, Liu Q, Gallo P, Berry D, Chuang-Stein C, Pinheiro J. 2006. Adaptive dose-response studies. *Drug Inf J* 40(4):451.

Gilron I, Bailey JM, Tu D, Holden RR, Weaver DF, Houlden RL. 2005. Morphine, gabapentin, or their combination for neuropathic pain. *N Engl J Med* 352(13):1324–1334.

Grieve A. 2007. Discussion of the "White Paper of the PhRMA Working Group on Adaptive Dose-Ranging Designs". *J Biopharm Stat* 17(6):997–1004.

Guo X, Tsiatis AA. 2005. A Weighted risk set estimator for survival distributions in two-stage randomization designs with censored survival data. *Int J Biostat, Berkeley Electronic Press* 1(1):1–15.

Hamlett A, Ting N, Hanumara C, Finman JS. 2002. Dose spacing in early dose response clinical trial designs. *Drug Inf J* 36(4):855–864.

Hawkey CJ. 2001. Journals should see original protocols for clinical trials. *BMJ* 323:1309.

Hernan MA, Robins JM. 2006. Estimating causal effects from epidemiological data. *J Epidemiol Community Health* 60(7):578–86.

Hsieh F, Turnbull B. 1996. Nonparametric and semiparametric estimation of the receiver operating characteristic curve. *Ann Stat* 24(1):25–40.

ICH E9. 1998. Statistical principles for clinical trials, International Conference on Harmonization.

Inoue LYT, Thall PF, Berry DA. 2002. Seamlessly expanding a randomized phase II trial to phase III. *Biometrics* 58:823–831.

International Conference on Harmonisation of Technical Requirements for Registration of Pharmaceuticals for Human Use (ICH E9). 1998. ICH harmonized tripartite guideline: Statistical principles for clinical trials, E-9. Available: http://www.ich.org/LOB/media/MEDIA485.pdf. Accessed March 6, 2007.

Ioannidis JPA and Panagioutou OA. 2011. Comparison of effect sizes associated with biomarkers reported in highly cited individual articles and in subsequent meta-analyses. *JAMA* 305(21):2200–2210.

Ivanova A. 2006. Dose *Finding in Oncology: Non-parametric Methods Dose Finding in Drug Development*. New York: Springer, pp:49–58.

Kaufmann P et al. 2009. Phase II trial of CoQ10 for ALS finds insufficient evidence to justify phase III. *Ann Neurol* 66:235–244.

Krams M, Lees KR, Hacke W, Grieve AP, Orgogozo JM, Ford GA, and for the ASTIN Study Investigators. 2003. ASTIN: An adaptive dose-response study of UK-279,276 in acute ischemic stroke. *Stroke* 34:2543–254.

Lokhnygina Y, Helterbrand JD. 2007. Cox regression methods for two-stage randomization designs. *Biometrics* 63(2):422–428.

Lunceford JK, Davidian M, Tsiatis AA. 2002. Estimation of survival distributions of treatment policies in two-stage randomization designs in clinical trials. *Biometrics* 58:48–57.

Maca J, Bhattacharya S, Dragalin V, Gallo P, Krams M. 2006. Adaptive seamless phase II/III designs: background, operational aspects, and examples. *Drug Inf J* 40(4):463–473.

Mallal S. et al. 2008. HLA-B*5701 screening for hypersensitivity to abacavir. *NEJM* 358:568–579.

McAlister FA, Straus SE, Sackett DL, Altman DG. 2003. Analysis and reporting of factorial trials: A systematic review. *JAMA* 289:2545–2553.

Montgomery AA, Peters TJ, Little P. 2003. Design, analysis, and presentation of factorial randomized controlled trials. *BMC Med Res Methodol* 3:26.

Murphy SA. 2003. Optimal dynamic treatment regimes (with discussion). *JRSS-B* 65:331–366.

Murphy SA, vander Laan MJ, Robins JM. 2001. Marginal mean models for dynamic regimes. *JASA* 96:1410–1423.

Palesch Y, Tilley BC, Sackett DL, Johnston KC, Woolson R. 2005. Applying a phase II futility study design to therapeutic stroke trials. *Stroke* 36:2410–2414.

Posch M, Bauer P. Adaptive two-satge designs and the conditional error function. *Biometrical J* 1999, 41:689–696.

Posch M, Koenig F, Branson M, Brannath W, Dunger-Baldauf C, Bauer P. 2005. Testing and estimation in flexible group sequential designs with adaptive treatment selection. *Stat Med* 24:3697–3714.

Ratain MJ, Mick R, Schilsky RL, Siegler M. 1993. Statistical and ethical issues in the design and conduct of phase I and II clinical trials of new anticancer agents. *J Natl Cancer Inst* 85:1637–1643.

Rush AJ et al. 2004. Sequenced treatment alternatives to relieve depression (STAR*D): Rationale and design. *Controlled Clin Trials* 25(1):119–142.

Schmidli H, Bretz F, Racine-Poon A. 2007. Bayesian predictive power for interim adaptation in seamless phase II/III trials where the endpoint is survival up to some specified timepoint. *Stat Med* 26:4925–4938.

Schneider LS et al. 2007. National Institute of Mental Health clinical antipsychotic trials of intervention effectiveness (CATIE): Alzheimer disease trial methodology. *Am J Geriatr Psychiatry* 9(4):346–60.

Simon R. 1989. Optimal two stage design of phase II clinical trials. *Control Clin Trials* 10:1–10.

Simon R. 2012. Clinical trials for predictive medicine. *Stat Med* 31(25):3031–3040.

Shlay JC et al. 1998. Acupuncture and amitriptyline for pain due to HIV-related peripheral neuropathy: A randomized controlled trial. *JAMA* 280(18):1590–1595.

Stone RM, Berg DT, George SL, Dodge RK, Paciucci PA, Schulman P, Lee EJ, Moore JO, Powell BL, Schiffer CA. 1995. Granulocyte-macrophage colony-stimulating factor after initial chemotherapy for elderly patients with primary acute myelogenous leukemia. *N Engl J Med* 332(25):1671–1677.

Thall PF, Wooten LH, Logothetis CJ, Millikan R, Tannir NM. 2007. Bayesian and frequentist two-stage treatment strategies based on sequential failure times subject to interval censoring. *Stat Med* 26:4687–4702.

Tilley BC et al. 2006. Optimizing the ongoing search for new treatments for Parkinson disease: Using futility designs. *Neurology* 66:628–633.

Ting, N. 2008. Confirm and explore, a stepwise approach to clinical trial designs. *Drug Information Journal* 42(6):545–554.

The Huntington Study Group DOMINO Investigators. 2010. A Futility Study of Minocycline in Huntington's Disease. *Mov Disord* 25(13):2219–2224.

The NINDS NET-PD Investigators. 2006. A randomized, double-blind, futility clinical trial of creatine and minocycline in early Parkinson disease. *Neurology* 66:664–671.

The NINDS NET-PD Investigators. 2007. A clinical trial of coenzyme Q10 and GPI-1485 in in early Parkinson disease. *Neurology* 68:20–28.

Wahed AS, Tsiatis AA. 2004. Optimal estimator for the survival distribution and related quantities for treatment policies in two-stage randomization designs in clinical trials. *Biometrics* 60:124–133.

Wahed AS, Tsiatis AA. 2006. Semiparametric efficient estimation of survival distribution for treatment policies in two-stage randomization designs in clinical trials with censored data. *Biometrika* 93:163–177.

Wittes J. 2002. On changing a long-term trial midstream. *Stat Med* 27:2789–2795.

Wong WK, Lachenbruch PA. 1996, Tutorial in biostatistics: Designing studies for dose response. *Stat Med* 15:343–359.

Young F, Nightingale S, Temple R. 1988. The preliminary report of the findings of the aspirin component of the ongoing Physicians' Health Study. The FDA perspective on aspirin for the primary prevention of myocardial infarction. *JAMA* 259:3158–3160.

Zhou X et al. 2006. A two-stage design for a phase II clinical trial of coenzyme Q10 in ALS. *Neurology* 66:660–663.

7

Interim Data Monitoring

Interim data monitoring is an important part in the life of a clinical trial. Interim data monitoring has (1) an ethical attractiveness as it can result in fewer patients being exposed to potentially harmful and inefficacious interventions, (2) economical advantages in that research questions may be able to be answered sooner, and (3) a public health impact since trial results may be communicated to the medical community faster. Thus, appropriate thought should be given to how a trial should be monitored.

Monitoring considerations include the potential use of Data Monitoring Committees (DMCs) also called Data Safety Monitoring Boards (DSMBs), and interim data monitoring statistical methodology to ensure control of error rates. Interim analyses and the DMC/DSMB process should be considered a design feature to be planned during protocol development. Ideally procedures and methods are established before trial initiation. Unplanned interim analyses are possible but can create difficulties with error control and threaten trial integrity. We discuss data monitoring committees and interim monitoring methodology. We also briefly mention an evolving risk-based approach to monitoring which will be an important part of monitoring future trials.

7.1 Data Monitoring Committees/Data Safety Monitoring Boards

A DMC/DSMB is an independent group of experts who are appointed to monitor the safety, efficacy, and scientific integrity of a clinical trial. The DMC/DSMB makes recommendations regarding trial design and conduct by reviewing accumulating data from the ongoing clinical trial. Industry has adopted the DMC terminology while National Institutes of Health (NIH) uses DSMB terminology. Useful references regarding issues with DMCs include DeMets et al. (2006), Ellenberg et al. (2003), and Herson (2009).

In industry settings, DMCs can be internal (e.g., consisting of members within the same sponsor) or external (e.g., consisting of independent members from outside of the sponsor). Advantages of internal DMCs include first-hand knowledge of the intervention; a better understanding of regulatory implications, company needs, and dynamics; and easier logistics (e.g., scheduling meetings). Advantages of an external DMC include more scientific objectivity, less biased recommendations, and potential recruitment of expertise that is not available within the sponsor. External DMCs are also consistent with FDA guidance. Thus, we restrict remaining discussion to issues with external DMCs.

7.1.1 Membership

Dr. Jeffrey Drazen, editor of the *New England Journal of Medicine*, said of DMC/DSMB membership: "It's probably the toughest job in clinical medicine. Being on a DSMB requires real cojones." DSMB service can be one of a statistician's greatest contributions to medicine. Thus, DMC/DSMB members should be selected carefully to ensure that the members are knowledgeable and dedicated to the position.

DMC/DSMB membership is multidisciplinary with generally 3–12 members consisting of clinicians that have expertise in the disease and intervention area, statisticians, possibly ethicists, epidemiologists, basic scientists, or pharmacologists. DMCs are generally chaired by a clinician with expertise in the disease area and with clinical trials experience or by statisticians with experience in interim monitoring and clinical trials with complex designs. Members should be free from conflict of interest so that recommendations will be objective. Conflicts can be financial, intellectual, professional, or regulatory-based. Members are often paid, but the payment is not linked to trial outcome. It should not include company stock, should be reasonable, and documented. Contracts with DMC members now often include indemnification clauses for the protection from liability claims.

An ongoing concern with DMCs has been the paucity of experienced DMC members. Although there are many people who are clearly experts in various disease areas, many are not experienced with DMC service and processes, and may not fully appreciate the consequences of DMC actions and recommendations. There is a population of statisticians who are experts in the DMC process because they serve as DMC members in many different disease areas and thus on many DMCs, whereas clinicians tend to focus on a single disease area. Several ideas regarding the education and training of DMC members have been proposed. These include the development of apprentice programs in which a junior person would attend DMC meetings as a nonvoting member but would learn from observation and watching experienced DMC members. Obstacles to this proposal include that confidential and blinded data is being made available to someone who is not a part of the official process, and there can be costs associated with the apprentice attending the meeting. Other proposals have included the development of formal classes, possibly with certification, that could be offered at academic institutions or as short courses at professional meetings. Thus far, however, training has not been adopted on a large scale.

7.1.2 When Are DMCs Needed?

There are several considerations when determining whether a DMC is needed including the seriousness and medical importance of the indication, the potential for harm from the intervention, the endpoints of interest, the populations being studied, the trial duration and size, the trial complexity, and the uncertainty of responses. Since 1979, NIH policy has required DSMB oversight for all phase III multicenter trials. DMCs are more likely to be necessary in trials with

- High-risk interventions, when very little is known about the safety and efficacy profile of the intervention (e.g., novel compounds or devices), or when the intervention is very invasive (e.g., requiring surgery)
- Conditions that are life threatening or have a high mortality or morbidity (e.g., cancers, HIV, counter-terrorism agents)
- Ethical dilemmas
- Larger sample sizes and longer duration as accumulating medical information can influence the utility and ethics of ongoing trials and interim analyses may not be feasible in short trials
- Trials that require a "waiver of informed-consent" (e.g., emergency treatment trials of stroke, status epilepticus, or traumatic brain injury; or psychiatric conditions)
- Trials with vulnerable populations (e.g., pediatric, elderly, pregnant women, mentally ill, and people of poverty)

7.1.3 Roles

The role of the DMC includes stewardship of the trial. DMC members may be viewed as officers of the trial that preserve the scientific integrity of the trial and monitor participant safety. DMC members should aim to be objective and proactive. The DMC mission consists of

- Protecting patients (both enrolled into the trial and outside of the trial) and evaluating equipoise (e.g., Is it still ethical to randomize or follow trial participants on current interventions?)
- Preserving credibility of the trial by preserving the blind, confidentiality, protecting statistical error rates, minimizing operational bias, and maintaining consistency with the protocol
- Ensuring that the research objectives can be appropriately addressed
- Ensuring that reliable results are available to the medical community in a timely fashion (although this role is not universally viewed as an appropriate role particularly in industry-sponsored trials)

It is advisable to have DMCs review the trial protocol prior to trial initiation. DMC members are experts in the disease area and can have important insights into trial design (e.g., appropriate control group, endpoints, or analysis plan). DMC members should ensure that the trial design is appropriate so that the desired monitoring of the trial can be conducted (e.g., that the appropriate data are being collected, and data reviews are scheduled frequently enough to ensure appropriate monitoring). Review of the trial design is common in government-funded trials but has been less common inindustry highlighting the distinction that government-sponsored trials are research-based whereas industry-sponsored trials are development oriented.

DMCs evaluate safety and may recommend stopping a trial (or an arm of a trial) if an intervention is unacceptably toxic or may recommend making dose adjustments. DMCs need to be particularly aware of the multiplicity issue when evaluating safety as between-arm comparisons may be made for many labs (e.g., chemistries and hematologies), diagnoses, vital signs, and signs and symptoms. Statistical significance is neither necessary nor sufficient for the DMC to take action. Clinical relevance and the severity of events need to be carefully considered. Safety is often interpreted within the context of displayed efficacy.

DMCs may be asked to evaluate efficacy and may recommend stopping early for benefit (e.g., accelerate development of treatments when efficacy is greater than expected). DMCs should consider whether such a positive trial should continue to collect important safety data. DMCs may also evaluate futility and recommend stopping a trial with an ineffective intervention. However, DMCs should evaluate whether stopping a study early for futility (i.e., a positive trial is unlikely with trial continuation) should continue to ensure a conclusive result (i.e., that relevant effects can be ruled out with reasonable confidence).

Each time a clinical trial is designed, assumptions (e.g., variability, control group response rate, and event rates) have to be made. However, these assumptions are only educated guesses and often turn out to be incorrect. The DMC may review the design assumptions to evaluate whether the sample size should be recalculated or the duration of follow-up should be modified. An example where the DSMB recommended sample size adjustment occurred in an NIH-funded trial designed to detect a 7.5% difference in response rates (i.e., 97.5% vs. 90% [control arm]). 486 participants were required to have 90% power. However, interim data revealed that these rates were overestimated by ~10%. With 486 participants, there is 56% power to detect the 7.5% absolute difference between 80% and 87.5%. More than 1000 participants are required to have 90% power to detect a difference between 80% and 87.5%.

The DMC may also assess the following:

- Trial progress and feasibility (e.g., enrollment rates and representativeness of enrolled participants)
- Continued relevance of the trial objective

- Data quality (e.g., missing endpoint data and accuracy of important measurements)
- Whether results could change with trial continuation
- Consistency of effect across important subgroups
- Consistency of effect for different endpoints
- External consistency of results with other studies
- Comparability of the intervention groups with respect to baseline factors
- Comparability of the intervention groups with respect to concomitant therapy use, adherence, dropout, protocol violations, length of follow-up, and missing data
- Factors that predict participant response (benefit, harm, or loss-to-follow-up)
- Benefit:risk evaluation
- Impact of external information (e.g., results from other studies) on trial conduct

7.1.4 Organization

A typical organization flow includes the DMC, a data coordinating center with an "independent statistician" (discussed later), the sponsor, and a steering committee associated with the sponsor. A modern industry model is displayed in Figure 7.1. For NIH-funded trials, the steering committee often consists of the NIH program officer, possibly the trial chair, and other consultants. In industry-sponsored trials, the steering committee may consist of members of senior management and possibly academic consultants with expertise in the disease area. The DMC receives the protocol from the sponsor and interim reports from the independent statistician. The DMC does not generally receive electronic data or conduct statistical analyses. After reviewing and discussing the data, the DMC makes a recommendation to the steering committee. DMCs may recommend a protocol modification, early trial termination, or a temporary hold until issues have been resolved. The most common DMC recommendation is to simply continue the trial as planned. However, the role of the DMC is advisory. The Steering Committee can filter the recommendation before providing the trial team with a recommendation. The project team then must decide to accept or reject the recommendation. This can be a difficult decision because the recommendations may conflict with the interests of the sponsor. Recommendations can have significant implications in terms of the future of the trial or intervention development as well as regulatory activities. Most DMC recommendations are carried forward, however.

There has been debate regarding who should prepare DMC reports and present data to the DMC. FDA guidance suggests that this should

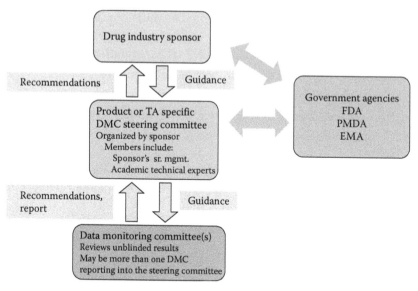

FIGURE 7.1
An industry model of the DMC organizational structure.

be done by an "independent statistician" to protect the blind and avoid potential bias. Thus, the general practice in industry-sponsored trials is that the (unblinded) *independent statistician* prepares the DMC report and presents the data at DMC meetings. The independent statistician is often contracted to contract research organizations (CROs), academics, or a designated unblinded statistician from the sponsor. Although the independent statistician may be paid by the sponsor, it is advantageous to view the independent statistician as working for the DMC. The independent statistician may iteratively interact with the DMC to ensure that they prepare appropriate data summaries for effective DMC evaluation and to assure that the analyses are truly "independent." The protocol statistician remains blinded.

One concern that can arise with the independent statistician model is that an independent statistician may not be knowledgeable enough about the trial to prepare a high-quality report that will be adequate for quality DMC evaluation. The DMC also frequently asks questions about the data and trial conduct. Thus, some clinical trial networks funded by the NIH do not utilize the independent statistician model. For example, in the ACTG, the protocol statisticians create the DMC reports and present interim results in closed session to the DMC. However, some design revisions may require the assistance of an independent statistician in special cases. For example, the recommendation from the DMC for a recent ACTG trial was to change the primary endpoint. Thus, it would follow that the study be resized based on the new endpoint. Since the protocol statisticians had seen the data, an independent statistician was tasked with working with the other members of the protocol team to size the trial, to ensure that independence between the trial data and construction of the alternative hypotheses.

Given the concerns with creating a quality DMC report and the ability to address DMC questions, it is important that the independent statistician be very engaged and knowledgeable about the trial design and conduct, capable of answering questions, and discussing issues. Good independent statisticians understand the roles of the DMC, can anticipate potential questions, can lead the DMC through the unblinded report, and carry out unanticipated analyses that can arise based on unexpected results. Characteristics of an excellent independent statistician include the following:

- Knowledgeable regarding the trial design and conduct
- Knowledgeable regarding the disease and interventions being studied
- Understands DMC needs
- Interacts with DMC Chair and other members to ensure that they received analyses needed to make optimal recommendations
- Predicts and proactively address DMC concerns/questions
- Proactively pursues important issues arising from the data even if not a part of preplanned analyses
- Prepared to walk the DMC through the report (possibly via presentation)

7.1.5 Charter

Whenever a DMC is utilized, a charter should be drafted. A charter is a document that outlines the DMC responsibilities and processes. Typical charters include a description of

- DMC responsibilities
- The anticipated DMC meeting schedule
- DMC membership
- DMC conflicts of interest
- Procedures for replacing DMC members
- Data security procedures
- Procedures for taking meeting minutes and documentation
- The flow of information (i.e., how reports will be distributed to DMC members; collection of DMC reports after the meeting)
- The meeting format and attendees

- The voting process
- Contents of the DMC report (i.e., safety and efficacy data analysis plan)
- Decision rules or guidelines for DMC recommendations
- The recommendation process

The charter is often reviewed and discussed at the first DMC meeting.

7.1.6 Data Monitoring Plan

It is advisable for a project team, often led by the project statistician, to develop a data monitoring plan (DMP). A DMP may be viewed as an analog to an SAP that provides a plan for how the trial will be (data) monitored. The project statistician and the medical monitor should interact with DMC members regarding particular safety concerns and generate a plan regarding when and how often reports need to be constructed and what they will contain. Typical elements of a DMP include the following:

- A description of the reports that will be generated during trial conduct
- What the reports will contain (e.g., the analyses that will be preformed)
- The rationale for the report
- Who prepares the report
- Who receives the report
- When the reports will be generated and their frequency
- Whether the report contains unblinded data

The DMP can be reviewed and discussed at the first DMC meeting.

7.1.7 Meetings

The first meeting of a DMC is generally an organizational meeting to review the charter, the trial design and protocol, the monitoring plan, and proposed data summaries that will be presented as part of the DMC reports. Subsequent DMC meetings typically consist of three sessions. The first session is generally an open session where often a summary of baseline data and trial issues are presented by the study team to the DMC. Unblinded data is not discussed. The second session is a closed session with only the DMC members and the independent statistician. The independent statistician may present highlights from the DMC report, and the DMC members discuss the unblinded or masked results. Questions or recommendations are prepared by the DMC. The third session is generally a session with the steering committee where the DMC provides their recommendations.

DMC's may meet a prespecified number of times per year (e.g., 1–4 times per year) or may meet based upon a trigger (e.g., a trial enrolling a particular number of participants, a particular number of participants having reached a particular time point, or a specified number of events have been observed in an event-driven trial.)

Occasionally, an emergency DMC meeting may be scheduled, for example, if an unexpected death occurs potentially indicating an immediate safety concern or if other data has become available that may influence the ethics or science associated with the trial. An example of a DMC meeting held on an accelerated schedule occurred in the adolescent trials network (ATN), an NIH funded clinical trials group evaluating interventions to treat HIV-infected adolescents. ATN 082 was a randomized controlled-trial evaluating preexposure prophylaxis (PREP) versus placebo to prevent HIV transmission. The first participant was enrolled August of 2008. On November 22, 2010 the results of the iPREX trial (Grant et al., 2010) a similar randomized trial sponsored by the Gates Foundation, were released and indicated that PREP reduced HIV transmission in adults (44% reduction in HIV incidence, 95% CI = (15%, 63%), p = 0.005). A DMC call was scheduled the next day November 23, 2010. The DMC discussed whether equipoise persisted and whether it was ethical to continue randomizing participants to and following participants on the placebo arm. The DMC recommended (1) immediate notification of trial participants and IRBs of the iPREX results, (2) unblinding of the trial participants, (3) discontinuation of randomization into the control arms; enrollment in the PREP arms can continue, and (4) participants on the control arms should be offered the option to rollover onto active PREP.

Two sets of DMC meeting minutes are prepared: one set of minutes for the open session which can be distributed to team members, and another set of minutes for the closed session which is not distributed but is archived in case they are needed as part of a regulatory submission.

7.1.8 Reports

When preparing DMC reports, only the independent statistician should be unblinded. The DMC may need to be unblinded in order to make a thorough evaluation and an informed recommendation. Some independent statisticians mask treatment with arbitrary labels (e.g., "Treatment A") but may provide a sealed envelope to DMC members with the DMC report to be opened at their discretion (i.e., when DMC members feel that they need to know the identity of the treatment groups).

Good planning is an important part of DMC report generation. Planning to have important endpoints adjudicated, batching participant samples for analyses in a timely fashion so that lab data are available, data entry, and cleaning important endpoint data help to ensure that the DMC has the most complete and up-to-date data as possible. Reporting of important data (e.g., deaths, SAEs) can lag resulting in under-reporting. *Sweeping* can be performed whereby each investigator contacts each participant at a particular time point prior to a DMC meeting, essentially fast-tracking important data into the database and ultimately the DMC report. A data freeze is typically conducted as recent as possible (typically 4–6 weeks) prior to the DMC meeting. Reports are typically delivered to DMC members 1–2 weeks prior to the meeting to allow time for a thorough review. DMC reports are generally collected at the end of the meeting and are destroyed to protect trial integrity. Interim data and DMC discussion must remain confidential as leaks can create operational bias by affecting recruitment, adherence, follow-up, or participant/clinician outcome assessment.

To create a DMC report, typically, clean formatted/labeled SAS datasets are sent to the independent statistician along with other important information (e.g., protocol, CRFs). There has been debate as to whether canned programs that produce standard output should also be provided to the independent statistician. The advantage of doing so is that the preplanning provides efficiency and eliminates surprises and analyses that the sponsor views as undesirable. The disadvantage is that the independent statistician and the resulting report that he/she generates is not truly independent of the sponsor and that such preplanned analyses may not address issues raised by review of the data itself (e.g., additional data may be needed to understand an unexpected result or event such as a participant death).

DMC reports should be informative, thoughtful, and efficient. The goal of the report is to inform the DMC members regarding the status and health of the trial participants and summarize the intervention effects. A DMC report should begin with a brief schema that reminds the DMC members of the trial design. A summary of the trial history including minutes from prior DMC meetings is helpful. Historically, there have been issues with the quality of DMC reports. Many are too voluminous (i.e., several hundred or even thousands of pages) and difficult to digest with hundreds of tables and AE listings that span many pages. Examples of poor report preparation are provided in Table 7.1. Tables with zeros in most of the cells, too many decimal places in summaries, or case summaries of all patients are also generally undesirable although patient case summaries may be helpful in very small trials. Creative graphics such as a CONSORT chart, Kaplan–Meier plots, forest plots, and side-by-side boxplots are useful displays that are easy to interpret. Special cases (e.g., deaths, unexpected outcomes) should be investigated and reported in patient narratives. Long listings and other voluminous information can be put into an appendix. DMC reports should include text summaries of important findings (facts and not opinions). Suggestions for DMC report preparation include the following:

- Plan well
  - Clean important data
  - Consider sweeping (e.g., mortality)

TABLE 7.1

Real Examples of Poor DMC/DSMB Report Preparation

- *Example 1:* One day prior to a scheduled DMC meeting for a large pharmaceutical company, DMC members are delivered a disk with the report. The disk contained 49 files. The first file contained 1600 pages of tables.
  - *Lesson:* Reports should be delivered 1–2 weeks prior to a meeting. Reports should not be so voluminous such that it cannot be digested. Text is needed to highlight important results. A table of contents should map the report so that results are easy to find.
- *Example 2:* DMC members receive a 300-page report but only four participants had been randomized.
  - *Lesson:* Automated programming cannot replace critical thinking. A case summary of each participant may be more helpful here.

- Have important endpoints adjudicated
- Schedule to batch participant samples for lab analyses
- Deliver report with sufficient review time (e.g., 1–2 weeks prior to the meeting)
- Protect the blind and trial integrity; minimize access to data to only DMC members
- Use data as recent as possible. Updates to important results can be brought to a face-to-face meeting.
- Use a Table of Contents so that results are easy to find
- Construct a text to highlight important results (facts but not opinions)
- Strive to answer important questions rather than provide extensive listings
- Be aware of the length of the report (strive for quality over quantity)
- Use graphics to display the data (as it is easier to digest)
- Use appendices for long data summaries (e.g., AE listings)
- Predict DMC questions and address them in advance
- Use rehearsal meetings
- Prepare slides to present important results

Preparation for a DMC meeting begins several months prior to the meeting and includes time for data cleaning, endpoint adjudication, a rehearsal meeting, and ample allowance for review time. A timeline that is often used in the ACTG is provided in Table 7.2.

DMC reports may include by-treatment summaries of accrual, important baseline variables and demographics, study status and duration of follow-up, participants off-study with reasons, participants off-treatment with reasons, protocol violations, adherence, concomitant medication use, deaths and AEs, endpoint evaluability and data completeness of important variables, and possibly efficacy endpoint summaries. Whether efficacy endpoint

TABLE 7.2

Typical Timeline for DSMB Review Preparation

Key Task	Relative to DSMB
Memo sent to sites alerting of DSMB review	13 weeks before
First clinical endpoint review[a]	12 weeks before
Last date of clinic visits/specimens included in the review	9 weeks before
Second clinical endpoint review[a]	8 weeks before
All CRFs up through the last date of clinic visits keyed	7 weeks before
Final data delinquency report run	7 weeks before
Final queries/discrepancies to sites	6 weeks before
Deadline for last corrections from sites	5 weeks before
Analysis files frozen	5 weeks before
Draft reports due	3.5 weeks before
Rehearsal meeting	3 weeks before
Final DSMB reports due	2 weeks before
Reports distributed to DSMB members	2 weeks before
DSMB meeting	Day 0

[a] Materials for review are provided 1 week in advance of the scheduled call.

summaries are provided to DMCs when there is no formal efficacy analysis planned has been an issue of debate. Some DMC members feel that even when charged with monitoring safety-only, that efficacy data is important to review because safety data should be interpreted within the context of efficacy.

Open versions of the DMC report are often provided to project team members. Open reports typically contain aggregate summaries (i.e., not by intervention). However, aggregate summaries of efficacy data can be informative regarding treatment effects and should only be reviewed by DMC members. Thus, many independent statisticians limit the open report to only baseline and possibly safety summaries.

7.1.9 Recommendations

The DMC makes recommendations to the Steering Committee. The Steering Committee can then filter the recommendation and decides what to recommend to the project team. Some Steering Committees (frequently with NIH-sponsored trials) prefer not to filter the recommendation and have the DMC make recommendations directly to the project team (e.g., the ATN utilizes this model). DMC recommendations should be brief and must be constructed so that operational bias is not induced, and that blinding is maintained. Operational bias can be induced when participants, clinicians, or others involved in the trial change their actions due to inferences that they have made based on the DMC recommendations. Such actions may be intentional or unintentional. Operational bias may manifest in participants dropping out of the trial or not adhering to the protocol or investigators choosing not to enroll participants or rating patient responses differently. Thus, extreme care must be taken when making recommendations. The DMC recommendation should be limited to a recommendation and is not necessarily a statement of the results particularly if the trial is to continue. Statements such as "the DMC recommends that the trial continue as planned" are generally sufficient.

Summary letters of the DMC recommendations are often distributed to IRBs. Since 1999, the NIH has required that local IRBs be notified of the outcome of all DSMB reviews, even when no major change has been recommended, to document that appropriate data and safety monitoring is occurring as expected.

7.1.10 DMCs of the Future

Concerns have arisen that DSMB decisions are often not truly independent but are driven by marketing considerations rather than the health of trial participants and the public. A recent example where DMC actions have been, a question occurred in a randomized placebo-controlled trial evaluating Zytiga for the treatment of prostate cancer. Interim results of the trial leaned heavily towards a positive trial but were not statistically significant ($p = 0.08$). The DMC recommends participants on placebo be offered Zytiga. Since the initial randomized version of the trial was stopped, there is a perception that the DMC decision was motivated by business rather than scientific and ethical concerns. Another example where DMC operations have been questioned is in a phase II trial evaluating a combination treatment for the treatment of cystic fibrosis. Initial interim analyses showed positive results although the trial continued. Company stock soared upon the news and company executives sold stock for millions of dollars in profit. However, shortly thereafter it was revealed that there was an error made during the initial interim analyses, and the results were not as promising as previously reported. The stock price fell, and questions regarding the DMC practices arose. A security exchange commission investigation ensued.

As a result, proposals for revisions to the DMC process are being considered including making the DMC more independent or separated from the sponsor of the trial and that the DMC reports to a public body rather than a trial sponsor. Another model that was discussed at a workshop held at the Harvard School of Public Health (HSPH) in 2011. The academic "mutual fund" proposal (see the schema in Figure 7.2) consisted of the creation of a not-for-profit organization for DMC coordination. The organization would have an executive committee with academic, industry, and DMC members for each trial and would communicate with industry within a legal boundary. The organization would keep confidential records, have a modern reporting system, develop educational programs for training DMC members, fund research projects and symposia on DMC issues, and handle indemnification.

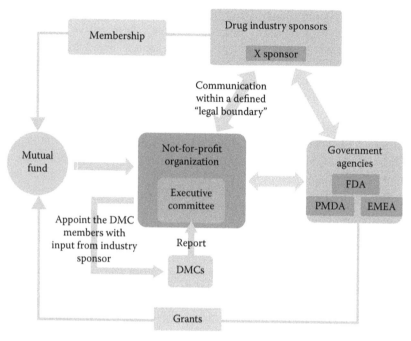

FIGURE 7.2
Proposed nonprofit mutual fund model for DMCs.

Industry, regulators, and others would contribute to a mutual fund to fund the organization.

The increasing use of adaptive designs has also generated discussion regarding how such trials affect the roles and operations of DMCs. Given the statistical complexity of adaptive trials and the experience of some expert statisticians, it is worth considering to have more than one statistician on these DMCs and having a statistician serve as the DMC chair. Care should be taken when drafting a DMC charter in an adaptive trial. It is important to carefully consider the role of the DMC in such trials. Some adaptations should be made independent of the observed treatment effects in the trial. In this case, the DMC is not the appropriate body to make decisions regarding or implementing adaptations as they see these data. Other adaptations can only be made after observing the interim treatment effects. In this case, the DMC may be an appropriate body to review these data and recommend or implement an adaptation. For example, in a recent randomized controlled trial evaluating a treatment of lymphoma, a plan was made to calculate the conditional power (CP) during the interim analyses. A prespecified sample size (actually the required number of events given the event-time primary endpoint) rule was developed (i.e., if very low then stop trial for futility; if very high then continue as scheduled; but if in the middle then adjusts the number of required events based on the observed effect size). Details of the rule were put into a "closed-protocol" and made available only to the DMC but not primary members of the protocol team. To protect trial integrity from operational biased induced back-calculation if a revised number of required events is announced to the project team, the DMC did not reveal the results of the calculation. Instead the DMC was kept apprised of the number of events, and indicated "STOP" when the required number of events was observed.

7.2 Interim Monitoring Methods

7.2.1 Evaluating Efficacy

It is desirable to answer research questions of interest as quickly and efficiently as possible for ethical, resource, and public health reasons. Thus, clinical trial teams frequently employ sequential methods to evaluate early evidence of efficacy. However, statistical tests for sequentially evaluating efficacy can inflate the trial-wise α error. For example, if a trial has one interim analysis and final analyses, then the trial-wise α error is closer to 0.08 while a trial that has four interim analyses and a final, has an α error rate of 0.14. Thus, appropriate adjustments for multiplicity are needed.

Group sequential methods (GSMs) were developed to allow sequential evaluation of efficacy while controlling α error. GSMs identify decision boundaries that preserve the desired α error rate during sequential monitoring. The methods specify how much α will be spent at each interim look and at the final analysis and require that the number of interim looks be prespecified. The decision rules are binding else error rate control is lost. Several GSMs have been developed.

Pocock (1977) and O'Brien and Fleming (1979) proposed the first two formal GSMs to evaluate the mean difference between two intervention groups when the responses are normally distributed with known variance. Both methods were designed to test between the null hypothesis (H_0) that the mean intervention difference is zero and the alternative hypothesis (H_A) that it is nonzero. Using Pocock's test, all interim testing is conducted at a constant and fixed α level (e.g., 0.0294 for one interim and a final; 0.0221 for two interims and a final). This may be attractive when evaluating safety outcomes (e.g., mortality and morbidity) that need to be identified early. O'Brien and Fleming (OF) is more conservative (spending a very small α) at early interim analyses and more aggressive at the final analyses (e.g., for a trial with one interim, the interim test is conducted at $\alpha = 0.015$ and the final analyses are conducted at 0.042). OF is more commonly applied relative to Pocock's approach for efficacy analyses as it saves most of α for the final analyses when significance is most important or desirable. Subsequently, researchers have extended their work to offer stopping boundaries with alternative shapes (Wang and Tsiatis, 1987) to adapt to other response distributions (Harrington et al., 1982; Tsiatis, 1982) through justifying normal approximations to the joint distributions of the test statistics; and to allow unequal and unpredictable increments of group sizes or other generalizations to fit realistic settings (Slud and Wei, 1982; Lan and DeMets, 1983; Kim and DeMets, 1987). Whitehead developed a continuous monitoring procedure permitting unlimited analyses as a clinical trial progresses.

7.2.2 Evaluating Futility

Clinical trial teams also strive to stop trials evaluating ineffective therapies as soon as possible. Inability to predict failures in early clinical trials escalates costs. It has been estimated that a 10% improvement in predicting failures could save $100 million in development costs. Futility analyses (FA) have been developed to assess if a clinical trial is destined to reach inconclusive results or to conclude that an investigational product is not superior to standard therapy (or not noninferior in noninferiority trials) as continuation of such studies is an inefficient use of resources and potentially unethical. Theoretically there is no associated α cost (i.e., the only error that can be made is β error) with FA and strict FA should increase the significance level for the final analysis. However, regulatory authorities may require that some small amount of α be spent since treatment comparisons may trigger study discontinuation for efficacy under extreme circumstances even if discontinuation for efficacy is not planned. FA can decrease power unless appropriate sample size adjustments are made, although a minimal loss of power has been reported in many cases.

There are two primary approaches to FA: informal methods based on CP and formal approaches that use stopping boundaries that are built into the design. Using the CP approach, the probability of statistical significance at the trials end is computed, conditional upon observed data and assumptions regarding future data (e.g., current trends continue, H_0 or H_A is true) usually assuming that the alternative hypothesis is true. Ware et al. (1985) similarly describes the "futility index" (i.e., the complement of CP).

A more formal approach to FA is to construct futility stopping boundaries built into the trial design. The process is similar to GSMs that control α spending with multiple efficacy looks, except sequential methods for futility are applied to β rather than α. Futility boundaries are constructed such that the probability that futility is concluded is β under nonfutility (i.e., H_A is true). Multiple tests have been proposed (Gould and Pecore, 1982; Pampallona and Tsiatis, 1994; Pampallona et al., 2001) to permit early stopping under the null hypothesis by specifying a set of lower stopping boundaries. That is, if the test statistic crosses the lower stopping boundary, the trial will be terminated, and the null hypothesis will be accepted. Lower stopping boundaries are constructed such that the type II error rate of wrongly accepting the null hypothesis is well preserved below a nominal level.

Whitehead and his collaborators also proposed the triangular test and the double triangular ("Christmas Tree") test for one-sided and two-sided group sequential tests respectively (Whitehead and Stratton, 1983; Whitehead, 1997).

7.3 Limitations and Extensions

The advantage of these existing methods for data monitoring is that error rates (α and β) can be preserved using appropriate sample size adjustments during trial design. However, most of the existing methods provide information regarding the statistical significance of effects but do not convey information regarding the clinical relevance, and none use prediction of subject responses to convey information regarding potential effect size estimates and associated precision with trial continuation.

Each of these GSMs is based on hypothesis testing. Therefore, they are subject to limitations including (1) the significance levels (i.e., p-values) that do not convey information regarding precision and effect size and (2) they have inflexible decision rules that do not allow incorporation of other data (e.g., other endpoints, safety data) into the decision process.

Jennison and Turnbull (1984) proposed use of repeated confidence intervals (RCIs) for monitoring clinical trial addresses the first limitation. RCIs are a sequence of confidence intervals (CIs) (one at each interim analysis) with specified simultaneous (joint) coverage probability, constructed such that the confidence of each sequential CI varies using principles of group-sequential α-spending. The attractive features of RCIs are that they provide an estimate of effect, associated precision of the estimate, and they provide flexibility in decision making. The width and the location of the CIs explicitly convey the precision of the estimate as well as the estimated effect size. Simultaneous coverage probability is guaranteed regardless of decisions that are made.

The stochastic curtailment approach addresses the second limitation by defining the criterion for early stopping for futility based on a measure of the probability of rejecting the null hypothesis at the final analysis (Lan et al., 1982; Choi et al., 1985). The CP in the frequentist paradigm and the predictive power in the Bayesian paradigm are two such measures, which can be combined with other data such as the safety data to aid decision-making. Neither approach addresses both limitations. All the aforementioned methods derive inference for the parameter of interest using the observed interim data. None evaluates the potential gain in precision and power with trial continuation to give the investigators a better understanding of the pros and cons associated with the decision to continue or terminate the trial.

7.3.1 Predicted Intervals

Motivated by these limitations, Evans et al. (2007a) proposed a new flexible monitoring tool for clinical trial using predicted intervals (PIs), a prediction of the CI at some future time point t (e.g., the end of study). The concept of PIs is intuitive. One simply predicts the CI at a future time point (e.g., end of the study) conditional upon (1) data observed to date and (2) assumptions regarding data yet to be collected. Reasonable assumptions regarding data yet to be collected include (1) that the observed trend continues, (2) the alternative hypotheses is true (or various alternatives are true), (3) the null hypothesis is true, and (4) best or worst case scenarios are true (often useful for binary data).

7.3.1.1 Binary Endpoints

PIs for binary endpoints can be constructed in a very straight forward manner.

If a trial has a single arm with a binary primary endpoint (e.g., response vs. no response), then one may construct a PI for the response rate for the single arm. If the upper bound of the PI for the response rate is smaller than the lowest "acceptable" response rate, then it may be prudent to discontinue the trial for futility. To construct a PI for the response rate then one only needs the (already known) total targeted sample size (N) and the expected number of total responses. The expected number of responses is

$$e_r = x_1 + (N - n_1)p_f$$

where x_1 is the number of responses observed at the time of the interim analysis, n_1 is the sample size at the time of the interim analysis, and p_f is the expected future response rate (e.g., if the expected future response rate is the same as that observed to date, then $p_f = x_1/n_1$). One then constructs an interval (exact or normal approximation) for the response rate in the usual manner. A normal approximation to the PI is

$$PI = p \pm z_{\forall/2}\text{sqrt}((p(1-p))/N)$$

where $p = e_r/N$.

A PI for the difference between two response rates can be similarly constructed by calculating the expected number of responses for each group and using standard methods to calculate an interval for the difference between two proportions. Assume that N_i and N_c are the total targeted sample sizes and e_i and e_c are the expected number of total responses for the investigational treatment (i) and control (c) arms, respectively. Then the PI for the difference between response rates is

$$(p_i - p_c) \pm z_{\forall/2}\text{sqrt}(((p_i(1-p_i))/N_i) + ((p_c(1-p_c))/N_c))$$

where $p_i = e_i/N_i$ and $p_c = e_c/N_c$.

7.3.1.2 Continuous Endpoints

PIs for continuous endpoints can also be constructed in a very straight forward manner.

If a trial has a single arm with a continuous primary endpoint (e.g., change in a continuous measure, post minus pretreatment), then one may construct a PI for the mean (e.g., mean change) for the single arm. If the upper bound of the PI for the response rate is smaller than the lowest "acceptable" mean change, then it may be prudent to discontinue the trial for futility. To construct a PI for the mean, then one only needs the (known) total targeted sample size (N), the observed mean change ($\bar{x}$), and the predicted standard deviation (s). Various assumptions regarding the predicted standard deviation can be made keeping the objectives of the analysis in mind. For example, the predicted standard deviation may be estimated by the observed standard deviation or using variability estimates from prior studies. The (large-sample) PI for the mean (assuming the current trend continues) is

$$\bar{x} \pm t_{N-1} \frac{s}{\sqrt{N}}$$

A PI for the difference between two means (e.g., for an investigational treatment (i) and a control (c)) can be similarly constructed using the (known) total targeted sample sizes (N_i, N_c), the observed means ($\bar{x}_i, \bar{x}_c$), and the predicted standard deviations (s_i, s_c). The PI for the difference between two means (assuming the current trends continue) is

$$(\bar{x}_i - \bar{x}_c) \pm t_{df} \sqrt{\frac{s_i^2}{N_i} + \frac{s_c^2}{N_c}}$$

where $df = \min(N_i, N_c) - 1$.

Simple adjustments can be made to incorporate other assumptions regarding unobserved data (e.g., under H_A or assuming a different variability).

7.3.1.3 Time-to-Event Endpoints

When constructing PIs for time-to-event endpoints, researchers must consider the distinction between censoring due to loss-to-follow-up versus. administrative censoring (i.e., due to the timing of the interim look). For administratively censored values, simulation of the residual time-to-event after censoring is needed. Once event times have been generated, PIs for a parameter of interest (e.g., hazard ratio or median survival time) can be constructed via standard methods using the generated observations.

Two types of PIs for time-to-event data can be constructed. The first type of PI is based on the study's designed length, that is, censoring the prediction at the time of the planned study completion. This PI is a prediction of what would be seen at the end of the trial if it were continued as currently designed (and thus may be more appropriate for answering the question regarding whether to continue the trial as currently designed). A second type of PI is based on predicting data at a time point other than the planned end of the study. Researchers can evaluate the increase in precision with extension of the duration of follow-up or evaluate the decrease in precision with a shortening of follow-up by comparing a PI of this type to a PI based on the planned study length. This analysis is then useful in helping to determine the ramifications of revising the duration of follow-up.

7.3.1.4 Example: NARC 009

Neurologic AIDS Research Consortium (NARC) 009/Savient c0603/Adult AIDS Clinical Trials Group (AACTG) A5180 (Evans et al., 2007b) was a prospective, randomized, double-blind, placebo-controlled, multicenter, dose-ranging study of Prosaptide™ (PRO) for the treatment of HIV-associated neuropathic pain. Participants were randomized to 2, 4, 8, or 16 mg/day PRO or placebo administered via subcutaneous injection. The primary endpoint was the 6-week change from baseline in the weekly average of random daily Gracely pain scale prompts collected using an electronic diary. The study was originally designed to randomize 390 subjects equally allocated between groups. The study was sized such that the 95% CI for the difference between any dose arm and placebo with respect to changes in the 13-point Gracely pain scale was no wider than 0.24 assuming a standard deviation of Gracely pain scale changes of 0.35.

An interim analysis was conducted after 167 patients completed the 6-week double-blind treatment period. Mean changes in pain and respective 95% CIs for each arm were constructed (Table 7.3) with negative changes indicating a decrease in pain. Pain decreased in all arms including placebo. CIs and PIs were calculated for the between-group difference (active arm minus placebo) in 6-week mean changes in pain scores. Notably, each PI (constructed assuming that current trends continue) straddles zero (except the 8 mg arm in which the PI indicates an advantage for placebo). A comparison of the width of the current CI to the PI for each arm reveals that continued enrollment would not provide a substantial increase in precision (e.g., the width of the CI for the 2 mg arm is 0.29, whereas the PI width assuming that the current trend continues is 0.22). Required differences between means in remaining patients for the CI for the difference in mean changes to exclude zero were also inconsistent with (i.e., much larger than) observed changes (i.e., the required change is not contained in the CI for the mean change). The robustness of the prediction may again be examined with sensitivity analyses using other assumptions regarding future data. The DMC of the NARC recommended termination of the study.

TABLE 7.3

Interim Analysis Results for NARC 009

Treatment	N	95% CI for Mean Change	95% CI for Diff[a]	95% PI for Diff[b]	95% PI for Diff[c]	Required Diff[d]
Placebo	31	(−0.35, −0.11)				
2 mg	34	(−0.21, −0.04)	(−0.04, 0.25)	(−0.01, 0.21)	(−0.16, 0.06)	−0.54
4 mg	34	(−0.38, −0.12)	(−0.19, 0.16)	(−0.14, 0.10)	(−0.23, 0.01)	−0.45
8 mg	32	(−0.18, −0.02)	(−0.01, 0.28)	(0.03, 0.23)	(−0.15, 0.05)	−0.56
16 mg	36	(−0.34, −0.09)	(−0.16, 0.19)	(−0.11, 0.14)	(−0.21, 0.04)	−0.54

[a] 95% CI for the difference in mean changes vs. placebo.
[b] 95% PI for the difference in mean changes vs. placebo assuming full enrollment, assuming the current trend.
[c] 95% PI for the difference in mean changes vs. placebo assuming full enrollment, assuming per protocol, $\mu_{placebo} = -0.17$ and $\mu_{drug} = -0.34$.
[d] Difference in mean changes needed in the remaining patients for the CI for the difference in mean changes to exclude zero (in favor of active treatment) at the end of the trial.

7.3.2 Predicted Interval Plots

An extension of the PI approach is predicted interval plots (PIPs) (Li et al., 2009). Instead of assuming a specific observation for future data, a model generating the data can be assumed. PIPs evaluate the sampling variation associated with the assumed model using simulation, plotting the simulated PIs in a concise, intuitive graphical summary (PIPs) that can be used by DMCs for evaluation.

The construction of the PIP proceeds as follows: (1) a parametric assumption for the unobserved data is imposed (note that this can be estimated using prior data or can be selected, e.g., consistent with a hypothesis), (2) future data is simulated under the assumed model, (3) the observed data is combined with the simulated data to construct a "final dataset," (4) a PI is constructed using standard methods, (5) the first four steps are repeated many times (e.g., 500) to obtain many PIs, and (6) the PIs are aligned and plotted similar to a forest plot. The horizontal axis of the plot shows the effect sizes. The vertical axis shows the distribution of point estimates using a pseudo-box-plot approach.

PIPs have many appealing features. The gain of precision with trial continuation and the associated increase in sample size will be reflected in the length of the PIs. The possibility of yielding a significant test (i.e., CP), if the trial continues can be adequately evaluated under different assumptions. Sensitivity analyses can be conducted by varying the data-generating assumptions. This information can then be used with other data such as safety data to help make decisions.

7.3.2.1 Example

The NARC conducted a randomized, blinded, placebo-controlled single site (Kampala, Uganda) trial evaluating the effect of minocycline on cognitive impairment in HIV-infected people. The primary endpoint was the 24-week change from baseline in the mean of several standardized neuropsychological tests, and the trial was designed to detect an effect size of 0.5. The targeted sample size was 100 participants (50 per group), and an interim analysis was conducted after the results for 41 participants were known. The mean (SD) changes in the minocycline and placebo arms were 0.34 (0.84) and 0.49 (0.78), respectively (p = 0.59, 95% CI for the between-group difference = −0.67, 0.38). Three PIPs (Figure 7.3) were constructed under the assumptions of (1) the observed trend continues, (2) the null hypothesis (i.e., no difference between arms), and (3) the alternative hypothesis (minocycline is superior by 0.5).

The median width of the PIs is 0.63, 0.68, and 0.67 under the current trend, the null hypothesis, and the alternative hypothesis, respectively. The increase in precision can be seen when compared to the width of the current CI (1.05).

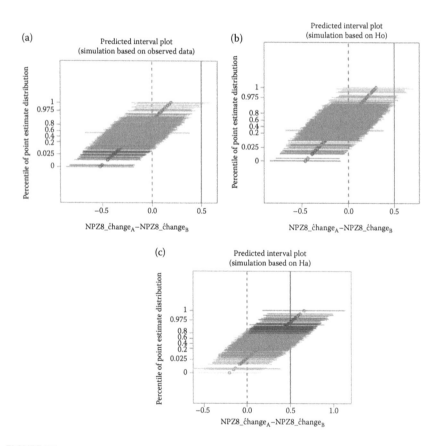

FIGURE 7.3
Predicted interval plot assuming: (a) observed trend continues, (b) the null hypothesis, and (c) the alternative hypothesis.

A summary of the PIP analyses reveals that the probability of rejecting in favor of minocycline with trial continuation is ~0% under the current trend, ~0% under the null hypothesis, and ~19% under the alternative hypothesis. The probability of rejecting in favor of placebo with trial continuation is ~8% under the current trend, ~2% under the null hypothesis, and ~0% under the alternative hypothesis. After evaluating secondary endpoints and noting that most important effects (i.e., >0.5) could already be ruled out, the NARC DMC recommended stopping the trial for futility (i.e., a low likelihood of observing an effect with trial continuation).

7.3.2.2 The Utility of PIs

PIPs convey information regarding the effect size, thus providing a quantitative assessment in addition to qualitative assessment of the effect. Modest p-values (and low estimates of CP) from traditional methods of interim data monitoring, imply either (1) no effect or (2) not enough data. PIs can help distinguish between these two. Low p-values (and high estimates of CP) imply statistical significance but do not convey information regarding clinical relevance of the effect. PIs allow assessment of clinical significance as well as statistical significance. PIs can be used with repeated CIs when monitoring to control error rates.

One may also assess the potential improvement in precision with continued enrollment by comparing the width of the current CI (based on interim data) to the width of the PI. If the width of the PI is not sufficiently narrower than the CI, then the cost of trial continuation may out-weigh the benefit of the minimal improvement in precision. CP can be obtained directly from the plot as the percentage of PIs excluding the null value in favor of the alternative hypothesis. Nonstatistician members of DMCs can easily interpret the PIPs.

PIs are flexible and invariant to trial design (i.e., superiority, noninferiority, or equivalence). PIs can also be useful for adaptive designs (e.g., as a tool for sample size reestimation). For example, if the PI is wider than desired, then an appropriate increase in sample size can be made to improve precision.

It is only reasonable to consider all available information (i.e., data other than the primary endpoint) when monitoring a clinical trial. PIPs are monitoring tools that allow incorporation of such additional data into the data monitoring process. PIs allow assessment of the sensitivity of effect estimates to various assumptions regarding future data, whereas traditional GSM only allow assessment of the sensitivity of the statistical significance of effect estimates.

The concept of PIs can be extended to include Bayesian PIs (i.e., predicting credible intervals). Frequentist and Bayesian analyses can be conducted simultaneously to aid decision making. One may envision an interim analyses consisting of (1) three frequentist PIPs (one for each of assumed trend continues, null hypothesis, and alternative hypothesis), (2) three Bayesian PIPs (one for each of assumed trend continues, null hypothesis, and alternative hypothesis), (3) a probability density at the interim, and (4) three predicted probability densities (one for each of assumed trend continues, null hypothesis, and alternative hypothesis). The concept can also be extended to coprimary endpoints (Evans et al., 2011).

7.4 A Centralized Risk-Based Approach to Monitoring

Traditionally, many clinical trials have implemented extensive on-site clinical monitoring possibly with 100% data verification. However, there is growing consensus that a risk-based approach to monitoring focusing on the most critical data elements is more likely to ensure participant protection and protect trial quality. Central monitoring may improve the ability to ensure the quality of trial data while saving valuable resources. Data anomalies (e.g., fraud, including data fabrication) may be more readily detected by centralized monitoring than on-site monitoring (US FDA, 2013). Statistical and data management methodologies for the detection of fraud, missing and delinquent data, and safety signals, will be important considerations as monitoring approaches evolve.

References

Choi SC, Smith PJ, Becker DP. 1985. Early decision in clinical trials when treatment differences are small. *Controlled Clin Trials* 6:280–288.

DeMets DL, Furberg CD, Friedman LM. 2006. *Data Monitoring in Clinical Trials.* New York: Springer.

Ellenberg SS, Fleming TR, DeMets DL. 2003. *Data Monitoring Committees in Clinical Trials: A Practical Perspective.* England: Wiley.

Evans SR, Hamasaki T, Hayahi K. 2011. Design and data monitoring of clinical trials using prediction. *Joint Meeting of the* 2011 *Taipei International Statistical Symposium and the 7th Conference of the Asian Regional Section of the IASC*, Taipei, Taiwan.

Evans SR, Li L, Wei LJ. 2007a. Data monitoring in clinical trials using prediction. *Drug Inf J* 41:733–742.

Evans SR, Simpson D, Kitch DW, King A, Clifford DB, Cohen BA, McArthur JC. 2007b. A randomized trial evaluating prosaptide™ for HIV-associated sensory neuropathies: use of an electronic diary to record neuropathic pain. *PLoS One* 2(7):e551. doi:10.1371/journal.pone.0000551.

Gould AL, Pecore VJ. 1982. Group sequential methods for clinical trials allowing early acceptance of *H0* and incorporating costs. *Biometrika* 69(1):75–80.

Grant RM et al. 2010. Preexposure chemoprophylaxis for HIV prevention in men who have sex with men. *N Engl J Med* 363:2587–2599.

Harrington DP, Fleming TR, Green SJ. 1982. Procedures for serial testing in censored survival data. In: Crowley J, Johnson RA (Eds.), *Survival Analysis.* . Hayward, CA: Institute of Mathematical Statistics. pp. 269–286.

Herson J. 2009. *Data and Safety Monitoring Committees in Clinical Trials*. Boca Raton, FL: Chapman & Hall.

Jennison C, Turnbull BW. 1984. Repeated confidence intervals for group sequential clinical trials. Controlled Clin Trials 5:33–45.

Kim KKG, DeMets DL. 1987. Design and analysis of group sequential tests based on the type I error spending rate function. *Biometrika* 74:149–154.

Lan KKG, DeMets DL. 1983. Discrete sequential boundaries for clinical trials. *Biometrika* 70(3):659–663.

Lan KKG, Simon R, Halperin M. 1982. Stochastically curtained tests in long-term clinical trials. *Commun Stat C* 1:207–219.

Li L, Evans SR, Uno H, Wei LJ. 2009. Predicted interval plots: a graphical tool for data monitoring in clinical trials. *Stat Biopharm Res* 1(4):348–355.

O'Brien PC, Fleming TR. 1979. A multiple testing procedure for clinical trials. *Biometrics* 35:549–556.

Pampallona S, Tsiatis AA. 1994. Group sequential designs for one-sided and two-sided hypothesis testing with provision for early stopping in favor of the null hypothesis. *J Stat Plann Inferace* 42:19–35.

Pampallona S, Tsiatis AA, Kim K. 2001. Interim monitoring of group sequential trials using spending functions for the type I and type II error probabilities. *Drug Inf J* 35:1113–1121.

Pocock SJ. 1977. Group sequential methods in the design and analysis of clinical trials. *Biometrika* 64:191–199.

Slud EV, Wei LJ. 1982. Two-sample repeated significance tests based on the modified Wilcoxon statistic. *J Am Stat Assoc* 77:862–868.

Tsiatis AA. 1982. Group sequential methods for survival analysis with staggered entry. In: Crowley J, Johnson RA (Eds.), *Survival Analysis.* Hayward, CA: Institute of Mathematical Statistics. pp. 257–268.

US FDA. 2013. Guidance for industry oversight of clinical investigations—A risk-based approach to monitoring. U.S. Department of Health and Human Services Food and Drug Administration Center for Drug Evaluation and Research (CDER) Center for Biologics Evaluation and Research (CBER) Center for Devices and Radiological Health (CDRH) Office of Good Clinical Practice (OGCP) Office of Regulatory Affairs (ORA) August 2013 Procedural.

Wang SK, Tsiatis AA. 1987. Approximately optimal one- parameter boundaries for group sequential trials. *Biometrics* 43:193–199.

Ware JH, Muller JE, Braunwald E. 1985. The futility index: An approach to the cost-effective termination of randomized clinical trials. *Am J Med* 78:635–643.

Whitehead J. 1997. *The Design and Analysis of Sequential Clinical Trials*, 2nd revised edn. New York: Halstead Press.

Whitehead J, Stratton U. 1983. Group sequential clinical trials with triangular continuation regions. *Biometrics* 39:227–236.

8

Analysis Considerations

We now discuss considerations for the analyses of a clinical trial. We first discuss preparations for analyses including the construction of an SAP and other analysis preparation activities. We then discuss issues during statistical analyses. We conclude with a discussion of issues in writing the report summarizing the analyses.

8.1 SAP

Before the 1990s, typically, after an initial analysis was completed and the data were presented, questions about the analyses would arise often resulting in further ad hoc analyses being conducted. However, it was recognized that this practice is subject to multiplicity concerns and selective reporting (i.e., conducting an analysis using several approaches and then reporting the one with the most desirable result). Using this approach also strained resources as if, often seemed like, analyses were never-ending. To avoid these issues, in the early 1990s, statisticians proposed the SAP (called a data analysis plan [DAP] in some organizations), a concept that was rapidly and widely accepted by the clinical trial community.

The SAP is a document containing detailed specifications regarding how the data from a clinical trial will be analyzed. It often contains detailed descriptions of the statistical tests, models to be fit, tables, figures, and listings that will be produced and serves as a template for the clinical study report (CSR) that summarizes the results of the trial. The SAP is finalized after the trial protocol is finalized and before the blind is broken.

A primary rationale for the SAP is to provide transparency. The SAP clearly prespecifies the planned analyses (endpoints, populations, time points, and analytical strategy) thus providing multiplicity context and assisting with control of the Type I error rate. The SAP further clarifies that any unspecified analyses are post hoc. This helps to limit issues of revisiting the data and iterative analyses that can create multiplicity concerns (although some post hoc analyses inevitably occur due to questions that arise from the data). The SAP also serves as a pseudo-contract between statisticians and other members of the project team helping to preserve statistical resources by preventing numerous iterations of analyses. It is very important to avoid or minimize post hoc analysis requests. The optimal strategy for preventing ad hoc requests is careful planning and review of the SAP. The SAP now also serves as a communication mechanism with regulatory agencies.

All the analyses required from the clinical trial are predefined in the SAP. This includes the definitions of the primary endpoint, primary comparison, primary analysis set, primary statistical model, and primary time point so that the prespecified α is well protected. Details regarding subgroup analyses, how missing data are to be handled, and how data will be interpreted are clearly described.

Deviating from the SAP can be problematic. The immunotherapy for prostate adenocarcinoma treatment (IMPACT) trial was a double-blind, placebo-controlled, multicenter phase III trial, that randomly assigned 512 patients in a 2:1 ratio to receive either Provenge (341 patients) or placebo (171 patients) administered intravenously every 2 weeks, for a total of three infusions for the treatment of prostate cancer. The primary endpoint was overall survival (OS). According to the Dendron's SAP, analyses would evaluate if Provenge increased survival compared to placebo in the overall population. It further stated that subgroup analyses would be performed for those above 65 years versus below 65 years of age. However, when the study results were published in the *New England Journal of Medicine* in 2010, the cutpoint for the subgroup analysis based on age was 71 years. Deviation from the prespecified analyses raised suspicion of data dredging and selective reporting of desirable results and raised questions about the appropriate interpretation of the trial results.

There are three main components to an SAP: the text, the list of tables (LoT), and the mock tables. The text part of an SAP typically follows an outline. One example of an SAP outline is given in Table 8.1.

Typically, the text of the SAP is followed by the LoT and the mock tables. The LoT presents the tables, figures, and listings to be used in the CSR. This list is often prepared according to the recommendations from International Conference on Harmonization (ICH) E3. The order of the tables is often something like patient disposition, demographics, baseline summaries, efficacy summaries, adverse events, laboratory data, and other safety tables. Data listings often follow the tables.

The mock tables are a skeleton of the tables that will be provided in the CSR but without real data. Mock tables are useful in communicating the ideas of data presentation and data analysis among the project team. In many companies, the demographic, baseline, and safety tables are standardized since these data are relatively similar regardless of disease area.

The SAP is carefully reviewed and scrutinized by study team. Project team members are forced to think carefully about what analyses are to be done and how it will be presented. Two key team members that the statistician needs to consult when writing the SAP are the statistical programmer and the clinician(s). The statistician works with the clinician to obtain a deep understanding of the relevant research questions and obtains agreement on

TABLE 8.1

Example SAP Outline

Introduction
  Study design
  Study objectives
  Sample size calculation
Interim analyses, final analyses and unblinding
Hypotheses and decision rules
  Statistical hypotheses
  Statistical decision rules (e.g., multiple comparison adjustment)
Analysis sets
  Full analysis set
  PP analysis set
  Safety analysis set
  Other analysis sets
  Treatment misallocations
  Protocol deviations
Endpoints and covariates
  Efficacy endpoint(s)
  Safety endpoints
  Other endpoints
  Outcome research endpoints
  Covariates
Handling of missing values
Statistical methodology and statistical analyses
  Statistical methods
  Statistical analyses
  Primary analysis
  Secondary analyses
  Safety analyses
References
Appendices
  Data derivation details
  Statistical methodology details

the data analysis and presentation to address those questions. It is also critical for the statistician to work closely with the programmer when drafting the SAP. For any planned table, the data structure must be ready to allow such a table to be produced.

A common practice prior to data analysis is to have a blinded data review (BDR). The BDR typically takes place after most of the data are entered into the database, but not too late to make changes, that is, there is still time to allow last minute revision of the statistical methods or data presentations prior to unblinding. During the BDR, the programmer runs the entire set of tables on the blinded data (e.g., by applying a dummy treatment code). The tables are reviewed, and issues such as data distributions, outliers, and so on are evaluated. The SAP may be revised based on these evaluations noting that this is still prior to unblinding.

8.2 Other Preparations for Analyses

8.2.1 Data Management Preparations for Analyses

Quality data is imperative for reliable analyses. Prior to analyses, the data are thoroughly cleaned through iterations of data checks, queries, and corrections. Data checks can take many forms, for example, range checks, logical checks, and missing data checks. Data managers, statisticians, and clinical people construct lists of checks that are conducted during the course of the trial often in an automated fashion. Errors and discrepancies are resolved prior to finalizing the database (i.e., database "lock" or "freeze"). An excellent discussion of data management issues can be found in McFadden (2007).

8.2.2 Clinical Data Interchange Standards Consortium

On July 21, 2004, the SDTM, defining a standard structure for clinical trial data tabulations was selected as the standard specification for submitting tabulation data for clinical trials to the FDA. Eventually, all data submissions will be expected to conform to this format. The Submission Data Standards (SDS) team called the Clinical Data Interchange Standards Consortium (CDISC) defines the SDTM. It is advisable to plan for CDISC SDTM before data are collected (protocol language, case report form [CRF] design, and output data format).

SDTM is built around the concept of observations collected about trial participants. Each observation can be described by a series of variables, corresponding to a row in a dataset or table. Each variable can be classified according to its role, determining the type of information conveyed by the variable and how it can be used. Variables can be classified into four major roles: (1) identifier variables, which identify the study, trial participant, the domain, and the sequence number of the record; (2) topic variables, which specify the focus of the observation (such as the name of a lab test); (3) timing variables, which describe the timing of the observation (such as start date and end date); and (4) qualifier variables, which include additional illustrative text, or numeric values that describe the results or additional traits of the observation (such as units or descriptive adjectives). A fifth type of variable role, Rule, can express an algorithm or executable method to define the start, end, or looping conditions in the Trial Design model. The set of Qqualifier variables can be further categorized into five subclasses.

8.2.3 Statistical Programming

Statistical programming develops during the course of the trial but is applied after database lock to create analysis ready datasets (appropriately formatted and labeled). Statistical programmers work with statisticians to ensure that the data necessary for analyses is created and is in usable form. Statistical analysis software (SAS) is the most commonly used software in clinical trials due to its data management and statistical capabilities.

Good programming practice is important for statistical programming. Quality programming in clinical trials generally refers to (1) fitness (whether the programs meet functional requirements), (2) robustness (accurate and reliable under different conditions; reduce risk of logical errors), (3) remodeling friendliness (easy to read, understand, and maintain; code clarity and easy facilitation of code review; ease of code transfer to other programmers), (4) extendibility (reusable), (5) economy (minimize the development effort by development and reuse of standard code and by use of dynamic and easily adaptable code; efficient and timely; maintenance and resource efficiency, and (6) compliance with regulatory requirements regarding validation and 21CFR Part 11.

Programming validation has become a very important issue in clinical trials. The FDA defines process validation as "establishing documented evidence that provides a high degree of assurance that a specific process will consistently produce a product meeting is predetermined specifications and quality attributes." Each clinical trial programming group develops a validation strategy that must be followed to ensure that the programs do what they purport to do and that it is reliable. Suggestions for producing bug-free code are (1) design programs to include debugging aids during the development phase and to exclude debugging aids during the production phase, (2) test every step of programming logic, (3) try various data scenarios, (4) falsify data errors to see if code breaks down, (5) devise a group review and change control process, (6) review initial program development and changes thoroughly, and (7) conduct independent testing (Chow, 2010). Many organizations institute double-programming for validation purposes.

Two major objectives of statistical programming are (1) preparation of analysis-ready datasets and (2) table/figure generation. Much of the data in clinical trials are collected using CRFs (many electronic these days) or laboratories. Data from these sources are organized into, for example, demographic, efficacy, and safety datasets, and prepared to use in statistical analyses. Statistical programming starts with the extraction of raw data from the clinical trial database (CRFs and labs). These data are stored and organized similar to the manner in which they were collected (e.g., a distinct dataset for each CRF). A major goal is to convert the clinical data into analysis-ready datasets (i.e., well formatted and labeled). The analysis datasets serve multiple purposes including for the production of summary tables, statistical analyses, data tabulations, and submission to regulators (in a development setting) or the public domain (in the government funded setting). When constructing analysis-ready datasets, it is critical to understand how the data will be used.

Broadly speaking, tables or figures that are included in a CSR can be classified as baseline/demographic, efficacy, safety, and supporting tables.

The efficacy tables are most challenging as baseline/demographic tables and safety tables can often be standardized from study to study. However, the efficacy data often varies across studies and are typically intervention-specific.

8.3 General Issues

In the analysis of clinical data, there are many challenging issues. These concerns are discussed in detail below.

8.3.1 Describe the Data

One important principle to remember is to describe your data thoroughly before moving on to inference. A common mistake is to rush through data description in order to get to the more exciting inference (with p-values and confidence intervals [CIs]). Researchers can be liberal with descriptive statistics. There is no statistical penalty for description.

Describing the population being studied and their characteristics can be accomplished through descriptive statistics (e.g., Ns, percentages, means, medians, modes, standard deviations, variance, percentiles) as well as via data displays (e.g., boxplots, scatterplots, Kaplan–Meier plots). This is useful for becoming familiar with the data and for identifying data anomalies. Graphical displays are particularly appealing when sample sizes become large.

8.3.2 Analysis Sets (ITT versus Per Protocol [PP])

It is not uncommon for participants to be excluded from analyses in randomized controlled trials (RCTs). For example, Svensson et al. (2010) reported the results of a RCT comparing compression-only cardiopulmonary resuscitation (CPR) versus standard CPR including ventilation with respect to a primary endpoint of a 30-day survival. 3809 study participants were randomized, however, the primary analyses were conducted using data from 1276 (33.5%) participants. In the same issue, Rea et al. (2010) reported the results of a RCT comparing CPR via chest compression alone verus chest compression with rescue breathing with respect to a primary endpoint of survival to hospital discharge. 5524 participants were randomized, however, the primary analyses were conducted using data from 1941 (35.1%) participants. Nuesch et al. (2009) reviewed 14 meta-analyses including 167 trials; 29 (23%) trials included all patients in the analysis. Given that exclusions are so frequent, it is important to evaluate when such exclusions are appropriate, the consequences of excluding participants, and how they should be reported.

8.3.2.1 The ITT Principle

Analysis of an RCT typically associated with an ITT analysis guided by the ITT principle. The ITT principle is a fundamental concept in clinical trials but is frequently misunderstood and misapplied. The ITT principle essentially states to "analyze as randomized." This means that once a study participant has been randomized then they will be included in the analyses (i.e., included in the "ITT analysis set") as part of the randomized regimen, regardless of adherence to/deviation from protocol, the actual treatment taken, early study withdrawal, or anything that happens after randomization. Requiring analysis of study participants who did not take their assigned therapy is counterintuitive to many, and thus a difficult concept to understand. The ITT analysis can be considered an evaluation of the treatment "strategy" rather than an evaluation of treatment under ideal conditions (e.g., perfect adherence). In order to conduct an ITT analysis, all randomized study participants must be followed to study completion, death, or until the outcome event in event-time trials. If data are missing, then imputation is generally conducted. The ICH E9 states:

> The intention-to-treat principle implies that the primary analysis should include all randomized subjects. Compliance with this principle would necessitate complete follow-up of all randomized subjects for study outcomes. In practice, this ideal may be difficult to achieve, ... In this document, the term full analysis set is used to describe the analysis set which is as complete as possible and as close as possible to the intention-to-treat ideal of including all randomized subjects. Preservation of the initial randomization in analysis is important in preventing bias and in providing a secure foundation for statistical tests. In many clinical trials, the use of the full analysis set provides a conservative strategy. Under many circumstances, it may also provide estimates of treatment effects that are more likely to mirror those observed in subsequent practice.

Occasionally a modified-ITT (mITT) will be defined which can exclude randomized study participants who did not meet entry criteria or that never received study treatment. Note, however, that never receiving study treatment is not necessarily sufficient for exclusion. For example, a study participant who learns of his or her treatment assignment in an open-label trial and then elects not to be treated should not be excluded since the reason for the exclusion may not be independent of treatment assignment. As an example for which an exclusion may be acceptable, consider a trial that randomizes study participants to one of the two treatments for a bacterial infection. A presumed (pending a laboratory screening test) eligible study participant might require immediate initiation of therapy once the infection was suspected, and thus would be randomized. However, laboratory tests may later confirm that the study participant did not have the suspected infection. This study participant may be excluded from the mITT analysis set since the reason for the exclusion is clearly independent of treatment assignment.

Although the strict definition of the ITT analysis set is all randomized participants, many published reports of RCTs have used alternative definitions (i.e., published reports claim that an ITT analysis has been conducted, but review of the analyses revealed that some participants were excluded from analyses). Kruse et al. (2002) reviewed 100 randomly selected RCTs that reported analysis using ITT with only 42 trials analyzing all randomized participants. Reasons for the exclusion of participants included (1) received no follow-up in 16 trials, (2) received no treatment in 14 trials, and (3) were found not to meet entry criteria in 12 trials. The number of participants utilized for the ITT analyses could be deduced in 92 trials with 10 of these trials excluding >10% of participants. Gravel et al. (2007) conducted a cross-sectional literature review of 403 RCTs reported in 10 medical journals in 2002; 249 (62%) reported the use of ITT among which 192 (77%) analyzed available patients as randomized. Hollis and Campbell (1999) conducted a survey of all randomized trials published in 1997 in the *BMJ*, *Lancet*, *JAMA*, and *NEJM* with 199 (48%) mentioning ITT analyses. They concluded that the ITT principle is often inadequately applied and described.

One alternative to an ITT analysis is a PP analyses. Study participants who do not take the assigned therapy are excluded in the analyses and may also be censored from the analyses when treatment is discontinued. PP analyses can also be defined where study participants who do not adhere to protocol (e.g., schedule of assigned therapy) or have protocol violations (e.g., took prohibited medications or the wrong therapy) might be excluded from the analyses. A PP analyses may exclude study participants who do not finish the trial or participants who do not have appropriate data for analyses (sometimes referred to as a "completers analyses") and can be viewed as a subgroup analyses (i.e., those that adhered to protocol). The exact definition of the PP analysis set is not unique and frequently varies depending on the goals of the analyses. The PP analyses may be viewed as an evaluation of the treatment under near ideal use of the therapy (e.g., good adherence), and thus may be informative when adherence can be influenced by clinician intervention. A PP analyses can intuitively be an attractive alternative as it is more consistent with the theory that "treatment not taken cannot affect the outcome" than ITT. However, the PP analyses can be biased as the decision to comply with therapy may be influenced by therapy itself (e.g., trial participants who are dissatisfied with therapy might fail to comply, often resulting in an overestimate of the treatment effect). Furthermore, nonadherers may not have the same characteristics as adherers, and thus results may only be generalizable to a subset of study participants.

There are important distinctions between an ITT analyses and a PP analyses. Firstly, they address different questions. The ITT analyses address a strategy question, comparing the strategy of randomized interventions. Thus, ITT analyses are important from a clinical application perspective.

Participants who do not adhere or experience toxicity of the intervention are part of the strategy and are included in the analyses. The PP analyses addresses a question of biological effects, comparing intervention when patients adhere to protocol (e.g., take assigned medications as prescribed). Thus, PP analyses may be attractive when evaluating mechanisms of action or causal pathways.

A second key difference is that the ITT analyses are strictly consistent with randomization (e.g., treatment comparisons are indeed randomized comparisons) (Lachin, 2000). Randomization is the foundation for statistical inference, ensuring the expectation of balance between all factors (known or unknown; measured or unmeasured) except for treatment assignment. It is this expectation of balance that distinguishes RCTs from nonrandomized epidemiological studies. Thus, inferences drawn from ITT analyses are strongly rooted in statistical theory. Only ITT analyses fully preserve the statistical validity of treatment comparisons established by randomization. By contrast, PP analyses do not analyze study participants as they are randomized (e.g., some patients consciously or unconsciously self-select themselves out of the analyses by not adhering to protocol) but instead analyze study participants as they adhered. The expectation of balance of all other factors has been lost, and thus, a PP analysis is subject to the same potential biases in nonrandomized studies (e.g., bias from self-selection or nonrandom dropout). Hence, a PP analysis is not rooted in the same foundation for statistical inference. In a randomized trial, if one observes treatment differences in an ITT analyses, then these differences may be due to differences in treatment or from *random* imbalances (and statistical inference techniques can discriminate between these two). However, if one observes treatment differences in a PP analyses, then these differences may be due to differences in treatment, random imbalances, or from a factor that is causing people not to adhere to protocol (and statistical techniques cannot necessarily isolate the treatment effect). An ITT analysis retains the integrity of a RCT whereas a PP analysis does not retain all of the integrity of an RCT.

Another key difference is generalizability. ITT analyses apply to patients sitting in a clinician's office waiting to be treated but a PP analysis may not. The PP analysis only compares adherent patients, but the future adherence of a patient sitting in a clinician's office is unknown. However, the ITT analyses of treatment strategy will apply to the patient. This is attractive since the possibility that the patient will not adhere should be considered when making treatment decisions. Another difference between ITT and PP analyses arises with initiation time with event-time endpoints. The time and date of randomization is generally used as the starting of the clock when measuring event time in ITT analyses whereas the time and date of treatment initiation might be used in a PP analysis.

Clinical trials should be conducted with the ITT principle in mind. Study participants should be followed regardless of adherence or treatment status. Occasionally, treatment may need to be withdrawn when there are concerns about patient safety. However, the patient should still be followed on-study and data should continue to be gathered. It is important to realize the distinction between "off-treatment" and "off-study." Researchers should try to keep study participants on-study regardless of treatment status in accordance with ITT principles. Scientifically, "off-treatment but on-study" is generally preferable to "off-study." Participants may go off-study when they withdraw consent, for safety concerns (e.g., are at risk of failing to comply with protocol so as to cause harm to oneself and for which taking the study participant off-treatment cannot alleviate the safety concerns), or for ethical concerns (e.g., the study is no longer in the best interest of the study participant). Guidance for adhering to the ITT principle is provided in Table 8.2.

In superiority studies, an ITT-based analysis tends to be conservative (i.e., there is a tendency to underestimate true treatment differences). As a result, ITT analyses are generally considered the primary analyses in superiority trials as this helps to protect the Type I error rate (i.e., making a false superiority claim). However, when conducting a noninferiority (NI) trial where the goal is to show NI or similarity, an underestimate of the true treatment difference can make the treatments appear more similar—thus, inflating the "false positive" (i.e., incorrectly claiming NI) error rate. Thus, ITT is not necessarily conservative in NI trials. For these reasons, an ITT analysis and a PP analysis are often considered as coprimary analyses in NI trials.

The analysis sets and analyses (ITT vs. PP) should clearly be specified and defined in the protocol and in the SAP plan with distinctions between primary and secondary analyses. Methods for handling missing data and censoring when conducting an ITT analysis should also be clearly stated.

Conducting both an ITT and a PP analyses should be considered to assess the robustness of the trial results along with the assessment of the differences between the ITT and PP analysis sets. PP analyses often result in a larger effect size since ITT often dilutes the estimate of the effect, but PP analyses frequently result in wider CIs since it is based on fewer study participants than ITT. Interpretation is easier with the ITT and PP analyses as they produce similar qualitative results. A between-treatment arm comparison of the eligibility for the PP analysis set might reveal hidden treatment effects. A comparison of baseline characteristics of the PP analysis set versus those randomized participants who are not in the PP analysis set may also reveal if there are subgroups of participants who cannot adhere and illustrate the nonrandomness of exclusion from the PP analysis set.

Are there circumstances where participants can be excluded from the analyses while still maintaining complete integrity of a RCT? Yes, but appropriate exclusions must be determined carefully. Participants can be excluded without compromising the integrity of a RCT when the reason for exclusion can clearly be documented as independent of treatment assignment. For example, in trials evaluated treatments for bacterial infections, participants are randomized based on suspected infection when confirmatory lab results are not immediately available and when timely treatment is warranted. However,

TABLE 8.2

Guidelines for Adherence to the ITT Principle

Guiding principle
- Participants can be excluded without compromising the integrity of a RCT when the reason for exclusion can clearly be documented as independent of treatment assignment

Design
- Design so that it is easy to retain and track participants (e.g., minimize dropout and maximize adherence)
- Refrain from discontinuing participants for toxicity or other reasons except for withdrawal of consent. Toxicity issues can usually be addressed by taking the participant off-treatment
- Schedule regular visits or phone calls to participants to avoid loss-to-follow-up (LFU)

Monitoring
- Continue to follow participants regardless of adherence
- Be aware of the important distinction between off-study versus off-treatment. Toxicity issues can usually be addressed by taking the participant off-treatment but try to keep the participant on-study

Analyses
- Consider the use of a blinded adjudication committee to assist in determining whether participants should be excluded
- Possible reasons for valid exclusions
  - Patient did not satisfy entry criteria (and assessment of criteria is standardized) … ideally entry criteria would be assessed prior to randomization but this is not always possible (e.g., mITT wrong infection)
  - Reiecved no therapy in a double-blind trial
  - LFU due to natural disaster
- Reasons for exclusion that could compromise the integrity of a RCT
  - Received no therapy in an open-label trial
  - Participant received some but not all of the treatment
  - Participant did not adhere to therapy or visit schedule
  - Participant had sparse follow-up or has missing data
  - Participant was lost-to-follow-up or dropped-out of trial
  - Participant had protocol violations
  - Participant experienced toxicity and was taken off study intervention or opted not to take study intervention

Reporting
- Clearly state the number randomized, and the number analyzed
- Clearly list/describe the reasons for any exclusions and describe either (1) justification of why these exclusions do not compromise the integrity of the RCT (i.e., independent of treatment assignment) or (2) the potential bias induced by the exclusions and the strategy for addressing the bias
- When reporting results that exclude participants:
  - Compare exclusions versus nonexclusions with respect to baseline factors to assess generalizability
  - Compare the exclusion rate and reasons for exclusions between arms
  - Conduct sensitivity analyses to assess the potential impact of exclusions

lab results may later confirm that the participant did not have the specific suspected infection. These participants may be excluded since they did not meet entry criteria and were inadvertently enrolled as long as an evaluation of these entry criteria is uniform. Excluding patients due to LFU due to a natural disaster or for not taking the intervention in a double-blinded trial also may not jeopardize the integrity of the RCT. However, many cited exclusions cannot be documented as independent of treatment assignment, and thus, resulting analyses are subject to the same biases of RCTs. Such reasons include excluding participants for poor adherence, LFU, not taking the intervention in an unblinded trial, toxicity, or for protocol violations. Discussions regarding postrandomized exclusions are available (Fergusson et al., 2002).

Another analyses set is often defined when focused on safety analyses, the "as-treated" (AT) analyses or the safety analysis set SAS. An AT analyses analyzes study participants as they were actually treated. For example, if subject A was randomized to treatment X, but was actually treated with Y, then in ITT, subject A is analyzed in treatment group X, but in the AT or safety set, subject A will be evaluated as part of treatment group Y.

8.3.2.2 Intent-to-Diagnose

ITD is an analog to ITT in diagnostic studies, albeit considerably less well-known and applied. The idea of ITD is that it evaluates the strategy of diagnostic use in practice. This means incorporating indeterminate results into the analyses (indeterminate results are typically discarded and ignored in the analysis of most diagnostic studies). An indeterminate result is a failure of the strategy of diagnostic use in practice, and thus should be included when evaluating the strategy of the diagnostic application.

8.3.3 Baseline Comparisons and Baseline as a Covariate

Baseline measurements of important variables are made prior to randomization. These variables may include demographics; background information such as medical history, previous medication, smoking status, alcohol consumption; medical/physical conditions such as physical examination, vital signs, electrocardiogram (ECG), disease diagnosis, and important trial endpoints such as efficacy measurements (e.g., pain, blood pressure, number of painful joints, Hamilton Depression Rating Scale [HAM-D-17] score). Measurement of baseline helps to describe the population being studied and provides a reference to interpret postbaseline measurements after treatment initiative, for example, via change-from-baseline summaries.

Baseline for important variables should be clearly defined in the SAP. Occasionally, more than one measurement of an important variable is collected prior to randomization. A question arises as to which measurement should be used as a baseline if the baseline is being utilized during analyses. Strategies may include using the measurement closest to the randomization or averaging the measurements. Measurements that are utilized in screening to determine entry criteria should be avoided when defining baseline. If screening measurements are used, then changes from baseline measurements can be biased due to a regression to the mean phenomenon. In these cases, a second independent measurement prior to randomization should be conducted and used as the baseline measurement.

During the analyses of a trial, summaries of baseline information are generally provided by randomized intervention. Examples include demographics and baseline disease severity. Only descriptive statistics is used in these summaries. Inferential comparisons of randomized treatment arms with respect to baseline variables are generally not appropriate as any imbalance between treatment arms is completely random (i.e., 5% of comparisons would have between-arm comparison p-values of <0.05, and thus are meaningless). This is not to say that baseline differences between randomized interventions do not exist or are not important. However, comparisons should be based on clinical importance rather than statistical significance.

Baseline measurements of primary outcome variables or variables that are considered to be potential confounders or effect modifiers are frequently used during the final analyses, for example, as terms in a regression model.

In an example of a study for the treatment of patients with rheumatoid arthritis, the number of painful joints is one of the key efficacy outcomes. There are 68 joints in the human body to assess pain. In a six-month study for the treatment of rheumatoid arthritis, the baseline and change from baseline in number of painful joints of the two treatment groups (test intervention and placebo) are displayed in Figure 8.1. The distribution of the baseline number of painful joints runs from very few to 66 (the x-axis). Most trial participants experienced a reduction in the number of painful joints (i.e., patient improvement). The two regression lines also indicate a larger reduction in the number of painful joints when the number of painful joints at baseline was higher (i.e., a common negative slope [a model assuming that the difference between the two treatments are the same regardless of how many painful joints there were at baseline]). Under this model, the number of painful joints at baseline is prognostic (i.e., effects outcome) but is not an effect modifier (regression lines for each treatment are parallel). Since the regression line for the test intervention is below the placebo, there may be benefits of the intervention but statistical significance would have to be assessed.

Since the minimum value of the number of painful joints at six months is zero, the largest change from baseline for a particular trial participant is the baseline measurement (e.g., for a participant with 32 painful joints at baseline, the maximum change from baseline is −32). Thus, all the points in Figure 8.2 are bounded below by the negative value of the baseline. If many patients reach this limit then an analysis of the proportion of patients that reduce the number of painful joints to zero might be considered.

If the slopes of the two regression lines were different indicating treatment by baseline interaction (or equivalently effect modification) then, in

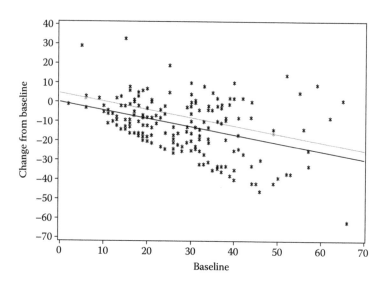

FIGURE 8.1
Baseline versus change from baseline to six months in the number of painful joints.

this case, the interpretation of the intervention is more complex. The intervention appears superior to placebo when the number of painful joints is >14 at baseline but inferior when it is <14. A careful evaluation of this interaction is required, and the impact of uncertainty in the estimation would be important.

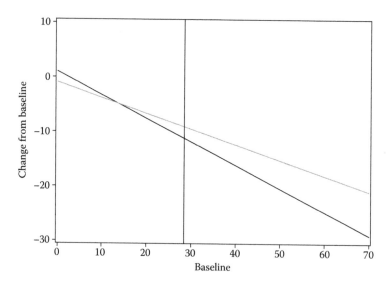

FIGURE 8.2
Baseline versus change from baseline (heterogeneous slopes).

8.3.4 p-Values versus Confidence Intervals

8.3.4.1 Poor p-Value Interpretation

The use of statistics in medical journals has increased dramatically over the past few decades. One unfortunate consequence has been a shift in emphasis away from the basic results toward a concentration on hypothesis testing (Gardner and Altman, 1986).

One of the biggest flaws in medical research is the over reliance on and misinterpretation of the p-value. The p-value is interpreted within the context of a hypothesis test where complimentary hypotheses, a null hypothesis (assumed to be true) and an alternative hypothesis (the claim that researchers wish to prove) are developed.

The p-value is defined as the probability of observing data as or more extreme than the observed data if the null hypothesis was true (note that the p-value is not the probability of a hypothesis being true given the data). If this probability is low (e.g., <0.05) then either (1) the observation of these data is a rare event or (2) the null hypothesis is not true. The standard practice is to reject the null hypothesis (in favor of the alternative hypothesis) when the p-value is acceptable low. If the p-value is not acceptably low then there is a failure to reject the null hypothesis.

Since the p-value is defined as the probability of observing data as or more extreme than the observed data if the null hypothesis was true, in order to appropriately interpret the p-value, a clear understanding of the null hypothesis is needed. For example, the null hypothesis for a two-sample t-test is the means of two groups are equal. Thus, if the null hypothesis is rejected, then one concludes that the means are unequal (the alternative hypothesis). The validity of the t-test is based on an assumption of normality, however, this assumption does not always hold. In such a case, statisticians often opt for a Mann–Whitney U test whose validity does not depend on the normality assumption. However, the null hypothesis for the Mann–Whitney U test is not the equivalence of means of the two groups but that the ranks of the two groups do not differ (i.e., if one were to rank the outcomes in the two groups combined that the ranks in one group are not higher or lower than the other group). Since the two-sample t-test and the Mann–Whitney U test have different hypotheses, the p-values from these two tests should be interpreted differently.

One may interpret the results of a hypothesis test similarly to the result of a court trial where the null hypothesis is the assumption of innocence, and the alternative hypothesis is that the person is guilty. If there is enough evidence to reject the null hypothesis of innocence (i.e., verdict of "guilty"), then one may conclude that there was sufficient evidence found to conclude guilt. However, if the null hypothesis of innocence is not rejected (i.e., verdict of "not guilty"), then it cannot be said that innocence was proven, only that there was a lack of evidence to conclude guilt. Thus, one does not prove the null hypothesis; you only fail to reject it. "Absence of evidence is not evidence of absence" (Altman and Bland, 1995).

For example, consider a trial comparing a new therapy versus placebo. Researchers would like to show that the new therapy is different (better) than placebo (and thus, this becomes the alternative hypothesis while its compliment, that the new therapy is not different from placebo is the null hypothesis). Suppose that researchers decide that a 5% false positive error rate is acceptable. Thus, if the p-value is <0.05 then the null hypothesis is rejected in favor of the alternative and the conclusion is that the new therapy is different from placebo. If the p-value is not <0.05 then the null hypothesis is not rejected and researchers cannot conclude that the new therapy is different from placebo. Note that "no difference" has not been proven; you are only unable to reject the possibility of "no difference."

The traditional cut-point of 0.05 is arbitrary, and p-values are not binary statistics. There is very little difference in the evidence of effect when a p-value is 0.052 versus 0.048. The 0.05 cut-off is used to control the "false positive" error rate (i.e., to ensure that it is not >5% when the null hypothesis is true). However, researchers can decide for themselves if a 5% false positive error rate is appropriate. If a false positive error is very costly (e.g., would result in a very expensive and invasive therapy being used when effective and safer alternatives are available) then researchers may opt for a 1% false positive rate (i.e., use 0.01 as a cutoff). If a false positive error is not costly, then a 10% (i.e., 0.10 cutoffs) could be used. For example, if a trial was conducted to evaluate a potential treatment for Ebola (a viral infection) or for progressive multifocal leukoencephalopathy (PML, a rare brain disease), both deadly diseases without a known treatment, then tolerance of a larger false positive error may be prudent. The standard 5% false positive error rate is often used since this is the regulatory hurdle for approval of a new intervention.

The p-value is a function of effect size, sample size, variability, and, of course, the null hypothesis. Larger effect sizes, larger sample sizes, and smaller variation all contribute to smaller p-values. Researchers often incorrectly interpret the p-value as providing direct information about the effect size. p-Values do not provide direct information about the magnitude or clinical relevance of the effect. Low p-values (e.g., <0.05) do not imply clinical or practical relevance and high p-values do not imply "no effect." Information about the effect size (or what effect sizes can be ruled out) can only be obtained by constructing CIs.

8.3.4.2 Need for CIs

p-Values are only one tool for assessing evidence. When reporting the results of a clinical trial, CIs should always be reported to identify effect sizes that can be "ruled out" (i.e., effect sizes that are inconsistent with the data). If a p-value is significant, implying an effect, then the next natural question is "what is the effect?" CIs directly address this question. If a p-value is not significant, implying that you were not able to rule out the possibility of "no effect," then the next natural question is "what effects could be ruled out?"

CIs again directly address this question. The under-reporting of CIs is a serious flaw in the medical literature.

CIs are not a *replacement* for p-values but instead should be provided *with* p-values. p-values are still very useful tools particularly when assessing trends and interactions.

CIs can also be misinterpreted. A 95% CI can be thought of as an interval that has 95% probability of covering the parameter of interest (note that this is distinct from a value having 95% probability of falling into an interval). This is to say that if a trial was repeated for a very large number of times and each time a 95% CI estimate of the treatment effect was obtained, then 95% of the CIs would cover the true treatment effect. A common misinterpretation is that the true treatment effect is more likely to be close to values in the center of the CI. However, this is not the case as the center of the interval is no more likely to cover the true treatment effect than values within the interval but away from the center. CIs are also subject to multiplicity concerns. If many 95% CIs are constructed, then it is expected that 5% of those CIs will fail to cover the true parameter.

8.3.5 Time Windows, Visit Windows

In outpatient clinical trials, trial participants are scheduled to visit the clinic according to a prespecified schedule defined in the protocol. Visits are often numbered or referred to by a time, for example, day 0 is usually defined as the day of randomization. Screening visits (prior to randomization) are frequently denoted with negative days while visits after treatment are positive days. This timeline can be used in constructing a schedule of enrollment, interventions, and assessments. The SPIRIT statement has a template for the schedule of events (Figure 8.3).

Day 0 is often referred to as "baseline" and should be clearly defined. Day 0 may also be the first day of treatment, but this can occur after day 0 in some instances. Baseline assessments of efficacy and other patient characteristics are often taken on day 0 prior to the treatment initiation.

The schedule may reflect that a trial participant must return for evaluations after 7 seven days and then again after 14 days. However, the first trial participant returns on day 6 or day 17 while the next trial participant returns on day 9 and day 13. Which data are to be used for day 7 and day 14 evaluations during statistical analyses? Typically, visit windows are defined within the SAP. The visit windows should be mutually exclusive and exhaustive.

Visit window issues become more complex in long-term trials, for example, when trial participants are followed for several years. In these trials, the earlier visits are often more frequent while the later visits are less frequent, for example, a trial that follows a participant for 3 years may have weekly visits for the first month, monthly visits from 1 month to 6 months, and then quarterly visits thereafter.

	Enrollment	Allocation	Postallocation					Closeout
TIMEPOINT**	$-t_1$	0	t_1	t_2	t_3	t_4	etc.	t_x
ENROLLMENT:								
Eligibility screen	×							
Informed consent	×							
[List other procedures]	×							
Allocation		×						
INTERVENTIONS:								
[Intervention A]			◄─────►					
[Intervention B]			×		×			
[List other study groups]			◄─────►					
ASSESSMENTS:								
[List baseline variables]	×	×						
[List outcome variables]				×		×	etc.	×
[List other data variables]			×	×	×	×	etc.	×

**List specific time points in this row.

FIGURE 8.3
Example template of recommended content for the schedule of enrollment, interventions, and assessments. (From the SPIRIT statement, www.spirit-statement.org.)

After all the visit windows have been defined in the SAP, it is important to define how to select the observations for a given visit to be used in analyses, when there is more than one observation within the visit window. One option is to use the last nonmissing observation within a given visit window. An advantage of this approach is its simplicity. Another option may be to average the values within the visit window.

When analyzing safety, it is often necessary to follow a patient for potential adverse events even after treatment discontinuation. Thus, it is common to define a "lag period" to follow trial participants after treatment discontinuation. The lag period is defined in the protocol and in the informed consent.

8.3.6 Multiplicity

Researchers are often interested in testing several hypotheses. Consider a clinical trial designed to compare a new therapy versus placebo. Researchers may wish to test the effect of the intervention versus placebo on several outcomes (e.g., a primary outcome and several secondary outcomes). Similarly, researchers may wish to test these endpoints in several subgroups of patients (e.g., defined by gender, race, age, or baseline disease status) or at several time points during therapy.

Each time a hypothesis test is conducted (e.g., each time a p-value is calculated), there is a chance to make an error (e.g., a false-positive error). For any single test, researchers can control the false positive error rate by deciding the "significance level." For example, it is common to claim that p-values below 0.05 are "significant." This decision rule sets the false positive error rate at 5% for a single test. However, if a researcher conducts several tests, then the probability of making at least one false positive error is >5%. The probability of at least one false positive finding increases as the number of tests that are conducted increases. If a researcher conducts 14 hypothesis tests when null hypothesis is true, then the probability of at least one false-positive finding is $[1 - (1 - 0.05)^{14}] = 0.512$ or 51.2%.

Thus, it is important for researchers to consider testing only important hypotheses to reduce the possibility of false conclusions. Significant results that are obtained when many tests were conducted without control of the trial-wise false positive error need to be validated with independent data.

It is important to report the results of nonsignificant tests so that significant results can be interpreted within the context of the number of hypothesis tests that were conducted. Researchers can either (1) clearly report the total number of hypothesis tests, the significance level of each test, and the number of expected false-positive tests by chance or (2) control the false positive error rate with an adjustment to the significance level of each individual test so that the probability of making a trial-wise false positive error is controlled. For example, if a researcher plans to conduct two hypothesis tests, then the trial wise error rate could be controlled at 5% by conducting each individual test at 0.025 (rather than 0.05), although this is a conservative approach.

For confirmatory trials, statistical procedures need to be well defined and documented before the blind is broken in order to ensure control of the type I error rate. Consider the example of a dose–response study with high dose, low dose, and placebo. There are at least two pairwise comparisons of interest: (1) high dose versus placebo and (2) low dose versus placebo. In order to control the trial-wise type I error, a multiple comparison adjustment is necessary.

There are many multiple comparison procedures (MCP). In the analysis of dose–response studies, popular MCPs include Bonferroni, Dunnett, Holm, Hochberg, and Gatekeeping. We briefly introduce some of these methods using the dose–response example. Suppose the type I error rate of α is prespecified, that is, the trial-wise error rate is α. There is also a comparison-wise error rate associated with each comparison. The concept of MCPs is to adjust the comparison-wise error rate so that the trial-wise error rate of α is protected.

The simplest adjustment is the Bonferroni method. Suppose there are k comparisons of interest, the Bonferroni adjustment divides the trial-wise error α by k and applies this to each comparison. In the dose–response example with two comparisons (high-dose vs. placebo and low-dose vs. placebo), if the desired trial-wise error rate is 0.05, then each pairwise comparison is performed using 0.05/2, and the trial-wise error rate is thus protected. The Bonferroni adjustment is very conservative, and hence, one of the least powerful methods. However, it is easy to understand and implement. The Bonferroni adjustment can also be applied without the need to consider the distributional properties of the data. It has historically been accepted by regulatory agencies.

Another simple strategy is to assume that the k comparisons are independent, and then each pairwise comparison is tested at $1 - (1 - \alpha)^{(1/k)}$ level. In the dose–response example, each pairwise comparison is performed at the $1 - (1 - \alpha)^{(1/2)}$ level. Although in practice, the pairwise comparisons can be correlated, they often are positively correlated, resulting again in a conservative procedure. This strategy is slightly more powerful than the Bonferroni adjustment and is easy to use and understand.

Another popular option for MCP adjustment is the Dunnett's procedure. If the data are normally distributed, then Dunnett suggests taking advantage of the positive correlations among the pairwise comparisons, and calculating the necessary critical points for the t-statistics. Suppose there are k pairwise comparisons, for example, k doses versus placebo. This MCP can be applied either in a step-down or a step-up fashion.

The step-down application starts with the largest observed absolute t-statistic. This observed t is compared with the critical value associated with the jointly distributed t calculated based on the Dunnett MCP from all k comparisons. If the observed t is less than the critical t, then the procedure stops (i.e., no further testing is conducted) and a conclusion that no dose has been shown to be different from placebo is reached. If the observed t is greater than the critical value, then there is evidence that the particular dose associated with this t is significantly different from placebo, and the next largest t is then evaluated. This second largest observed t is compared with the critical value calculated from the jointly distributed t based on the set of k − 1 comparisons. If the observed t is less than the critical value, then the procedure stops (i.e., no further testing is conducted) and a conclusion that no additional doses have been shown to be different from placebo is reached. If this the second largest observed t is greater than the critical value, then there is evidence that the particular dose associated with this t is significantly different from placebo, and the next largest t is then evaluated. The process continues until no further doses are shown to be different from placebo or until all comparisons have been made.

The step-up application of the Dunnett procedure starts with the smallest observed absolute t. This is compared this with the critical value obtained from a t-table (with a single comparison). If this observed t is greater than the ordinary critical value from the t-table, then there is evidence that *all* k doses are significantly different from placebo, and the procedure stops. Otherwise, the dose associated with the smallest t has not been shown to be different from placebo, and the second smallest observed absolute t is evaluated with the Dunnett critical value calculated from the jointly distributed t based on two comparisons. If the observed t is greater than the critical value, then there is evidence that the remaining k − 1 doses are significantly different from placebo (the dose associated with the smallest t is still not different), and the procedure stops. Otherwise, the dose associated with the second smallest t has not been shown to be different from placebo, and the third smallest observed absolute t is evaluated with the Dunnett critical value calculated from the jointly distributed t based on three comparisons. The process continues until either all pairwise comparisons have been performed, or when one of the t-statistics is greater than the corresponding critical value, indicating that the dose associated with this t-statistic and all doses with t-statistics greater than the current t-statistic are different from placebo.

The Holm's MCP is a step-down procedure using Bonferroni adjustment. The procedure is similar to the Dunnett step-down approach, but instead of assuming normality and using t-distributions, the Holm's procedure uses the observed p-values. It starts with the smallest observed p-value and compares it with α/k. If p is greater than α/k, then there is insufficient evidence to conclude that any dose is different from placebo, and the procedure stops. Otherwise, the dose is declared different from placebo and the second smallest observed p-value is compared with $\alpha/(k-1)$. If p is greater than $\alpha/k-1$, then there is insufficient evidence to conclude that any remaining doses are different from placebo, and the procedure stops. Otherwise, the respective dose is declared different from placebo, and the third smallest observed p-value is compared with $\alpha/(k-2)$. The process continues until there is a p-value that is not significant or until all doses have been compared. An advantage of the Holm's MCP is not limited to normally distributed data. A disadvantage is; it is less powerful than Dunnett's MCP because it is based on Bonferroni adjustment.

The Hochberg method is a step-up procedure. It is similar to the Holm's procedure because it also uses the Bonferroni adjustment. The first step of the Hochberg MCP is to test the dose with the largest p-value. If the p-value is less than α, then there is evidence to suggest that *all* k doses are different from placebo. If the p-value is not less than α, then the second largest p-value is compared with $\alpha/2$. If the second largest p-value is less than $\alpha/2$, then there is evidence to suggest that the remaining k − 1 doses are different from placebo. If the second largest p-value is not less than $\alpha/2$, then the third largest p-value is compared with $\alpha/3$. This process continues until a p-value is significant or until all the doses have been tested.

One very popular method for dose-finding studies is the *gatekeeping procedure* that can be expressed as a closed testing procedure based on the closed testing principle. To understand the closed testing principle, suppose there are k hypotheses $H_1, ..., H_k$ to be tested and the trial-wise type I error rate is α. The closed testing principle allows the rejection of any one of these elementary hypotheses, say H_i, if all the possible intersection hypotheses involving H_i can be rejected by using valid α level tests. It controls the trial-wise error rate for all the k hypotheses at α level in the strong sense. For example, suppose there are three hypotheses H_1, H_2, and H_3 to be tested and the overall type I error rate is 0.05. Then, H_1 can be rejected at level α if $H_1 \cap H_2 \cap H_3$, $H_1 \cap H_2$, $H_1 \cap H_3$ and H_1 can all be rejected using valid tests with level 0.05.

The gatekeeping strategy prespecifies an order of the testing and uses all α at each step. An intuitive approach to gatekeeping is to assume a monotonic dose–response relationship, and implement a comparison of the highest dose versus placebo as a first step, using the entire α. If the null hypothesis that, "there is no difference between the highest dose and placebo" fails to be rejected, then no further testing is employed and a conclusion that none of the doses is effective is reached. If the null hypothesis is rejected, then the next highest dose is tested against placebo again using the entire α. This process is repeated until either a null hypothesis is not rejected, or all of the doses are tested. The name gatekeeping comes from the fact that the highest dose serves as a gate, and if the gate is closed, then none of the following doses can be tested. The advantage of gatekeeping is that it is the most powerful MCP under a monotonic dose–response relationship. The disadvantage is; if the null hypothesis for any dose is not rejected, then no further testing is performed, that is, some doses are never evaluated.

Multiplicity concerns are not limited to multiple comparisons (e.g., from comparing multiple arms). Multiplicity can arise from multiple endpoints, analysis sets, time points, and analytical models. In all cases, it is important to be aware of the concerns with multiplicity. Methods for controlling errors associated with multiplicity should be clearly documented in the SAP.

8.3.7 Confounding ≠ Effect Modification

During analyses (and design) it is important to understand the distinction between confounding and effect modification. Within the context of trials with an objective estimating an intervention effect, confounding is a distortion of the intervention effect that occurs when the intervention groups differ with respect to other factors that influence the outcome. Effect modification is distinct from confounding, occurring when the magnitude of the intervention effect differs depending on the level of a third variable. In other words, the intervention effect is not unique but varies depending upon patient characteristics.

Consider the following models describing a patient response:

$$\text{Model 1}: F(\text{outcome}) = \beta_0 + \beta_1 C + \beta_2 T + \beta_3 T * C, \text{ and}$$

$$\text{Model 2}: F(\text{outcome}) = \beta_0 + \beta_2 T$$

where
  C is an indicator of whether a factor (potential confounder) is present
    (C = 1) or absent (C = 0) and
  T is an indicator of treatment (T = 1 for the experimental treatment and
    T = 0 for the control).

C is a confounder if the estimate of β_2 from Model 1 is different from Model 2 (i.e., the effect estimate of treatment differs depending on whether you control for C). C is an effect modifier if $\beta_3 \neq 0$ from Model 1 since the treatment effect depends on the factor (i.e., if the patient does not have the factor then the treatment effect is β_2, but if the patient has the factor then the treatment effect is $\beta_2 + \beta_3$).

Confounding and effect modification issues are addressed in different ways. Randomization and particularly stratified randomization (if the potential confounder can be measured prior to randomization) helps to control confounding in randomized trials since it provides the expectation of balance of the distribution of the potential confounder across the intervention group. For this reason, control for confounding in randomized trials is often not necessary unless analyses are restricted to subgroups (e.g., participants who complete the trial) for which selection can create confounding.

Control of confounding during analyses is particularly important in nonrandomized trials (e.g., historical or external controls) and for PP analyses or complete-case analyses. Epidemiological methods including restriction, matching, stratification, multivariable regression, propensity scores, and instrumental variables can be utilized to control for confounding during analyses. A good discussion of these methods can be found in Stafford et al. (2014).

Although randomization can prevent confounding, it cannot prevent effect modification. The only way to prevent effect modification is to limit the trial to a homogenous population. Otherwise, effect modification is usually addressed through subgroup analyses.

Stratified analyses are often confused with subgroup analyses, but these are distinct. The estimate of β_2 from Model 1 controls for C, and thus is considered stratified analyses. Subgroup analyses would imply separate analyses of intervention within each subgroup defined by C.

8.3.8 Stratification

Stratification is a strategy to control for potential confounders (baseline characteristics) that could distort estimates for intervention effects if such factors were not controlled.

In many clinical trials, there is concern that center is a potential confounder. Thus, many multicenter studies stratify randomization and analyses by center (analyses should be consistent with design). Stratified analyses are often conducted via regression modeling that could take several forms (e.g., a linear, logistic, or proportional hazards regression model) depending on the characteristics of the outcome and other conditions that affect the appropriateness of such modeling. A typical model includes an indicator for the intervention as well as effects for center and baseline measurement of the outcome. The intent of this model is to control for the potential confounding effects of center and baseline condition.

Simpler models could be considered. For example, a model utilizing only an intervention term could be considered. However, power to detect intervention effects can be low using such models since some of the variation associated with the outcome may be due to other factors. Appropriate model evaluation (e.g., via evaluation of residuals) will help to ensure that the results are robust.

8.3.9 Subgroup Analyses

A subgroup analysis is an evaluation of a treatment effect for a specific subset of patients defined by baseline characteristics (ideally, these characteristics are used as stratification factors in randomization). Analyses of subgroups that are defined by postrandomization characteristics are generally not advised in clinical trials as such analyses are subject to many biases and are difficult to interpret. A PP analyses is an example of such a subgroup analyses. It is often tempting to evaluate intervention effects in adherers (i.e., adherence is based on postbaseline information) or to compare treatment outcomes in nonadherers. If adheres have better outcomes than nonadherers, then this may (erroneously) be claimed as an intervention effect. However, the comparison of adherers versus nonadherers is not protected by randomization (expectation of balance is no longer guaranteed), that is, adherers and nonadherers may be different in ways that are independent of the intervention. Characteristics that influence adherence may also influence outcome, confounding the effect of the intervention.

Subgroup analyses are subject to multiplicity issues, and thus should be conducted carefully (Assman et al., 2000; Wang et al., 2007). Many suggest that an evaluation of whether a treatment effect varies across subgroups (i.e., treatment effect heterogeneity) should be conducted prior to conducting subgroup analyses to avoid subgroup analyses that can produce spurious results. This evaluation is typically conducted via statistical tests for interaction. Only if the treatment effect varies across subgroups should specific subgroup analyses be undertaken. For example, there may be interest in evaluating whether a treatment effect is similar for men versus women. If the treatment effect varies by gender then subgroup analyses may be undertaken. However, if the treatment effect is not dissimilar then there is no reason to conduct subgroup analyses within each gender. When evaluating whether a treatment effect varies by subgroup using an evaluation of the statistical interaction, it is important to consider how this will be assessed (e.g., the α level that will be used for the interaction test and the magnitude of the interaction considered to be clinically important). Most studies are not sized to detect statistical interactions, and thus interpretation of tests for interaction can be challenging.

There are two distinct types of interactions. Quantitative interaction refers to when one intervention is always better than the other, but by varying degrees depending on the subgroup. Qualitative interaction refers to when one treatment is better than the other for one subgroup (e.g., males), and worse for another (females). It is especially important to identify qualitative interactions since patients could be harmed with a poor intervention selection.

When evaluating whether the intervention effect varies across subgroups, it is important to clarify the metric upon which comparisons are being made. Consider the data in Table 8.3 displaying the response rate for a new therapy and a control for three age subgroups. For each treatment group, the response rate increases with age. However, there is no interaction (heterogeneity of treatment effects) on the relative risk scale but there is on the absolute scale.

When conducting subgroup analyses, one cannot determine whether the treatment effect varies by subgroup or by evaluating subgroup-specific p-values. Consider a trial that compares a new therapy versus a control where the primary outcome is a clinical response (vs. no response). Suppose the results for men were 32 of 40 in the new therapy arm responded versus 16 of 40 in the control arm. This yields a p-value of <0.01. Suppose the results for women were 4 of 10 in the new therapy arm responded versus 2 of 10 in the control arm. This yields a p-value of 0.49. Does this imply that the treatment is effective in males but not in females? No. Note that the relative risk in each gender is two. It is only the smaller sample size that leads to the nonsignificant result in females. Note that most clinical trials are powered to detect overall treatment effects and not necessarily for effects within particular subgroups (where sample sizes are clearly smaller).

When subgroup analyses are conducted, then they should be presented regardless of significance. A forest plot is an effective method for presenting reporting of the results of subgroup analyses. The number of subgroup analyses conducted should be transparent so that results can be interpreted within the appropriate multiplicity context. However, both prespecified and post hoc subgroup analyses are subject to multiplicity concerns. However, prespecification of subgroup analyses in the SAP provides transparency and context for multiplicity evaluation.

TABLE 8.3

Response Rate for a New Therapy with Control for Three Age Subgroups

Age Subgroup	New Therapy Rate	Control Therapy Rate	Relative Risk	Risk Difference
Young	0.1	0.05	2	0.05
Medium	0.4	0.2	2	0.2
Old	0.7	0.35	2	0.35

8.3.9.1 Subpopulation Treatment Effect Pattern Plot

The discussion of subgroups thus far has focused on subgroups defined by categorical characteristics. If there is a desire to evaluate how the intervention effect may vary as a function of a single continuous covariate, then a STEPP can be constructed (Bonetti and Gelber, 2000, 2004). STEPP is a moving average approach to examine intervention effect heterogeneity as a continuous characteristic varies. Confidence bands for the intervention effect can be constructed for subgroup-specific or overall inference. STEPP was utilized in the international breast cancer study group (IBCSG) IX trial of 1715 postmenopausal, node negative women who were randomized to receive Tamoxifen for five years versus CMF (a drug combination consisting of cyclophosphamide methotrexate, and fluorouracil) and Tamoxifen for five years (IBCSG, 2002). Evidence suggested that differences between the treatments depend on estrogen receptor (ER) status (often measured on the log scale), with CMF resulting in a higher disease-free survival (DFS) for women with low ER status but not for women with higher scores. STEPPs were constructed displaying the intervention differences in five-year DFS as a function of ER status (Figure 8.4a). Evans (2012) extended this idea to describe the heterogeneity of the sensitivity/specificity of a diagnostic in diagnostic trials using sensitivity of sensitivity/specificity (SOS plots) and sensitivity of predictive value (SOP plots).

In many situations, it may be desirable to tailor medical decision-making on the basis of more than one covariate. To address this issue, the strategy of two-stage modeling may be useful. In this case, an initial "working model" is fit in order to reduce all the patient-level information into a single stratification score. The resulting score can then be used in the second stage to estimate patient risks or responses to treatment among all patients with a particular value of the stratification score. These treatment responses may represent a single benefit or risk event or a combination thereof. Cai et al. (2011) estimated the effect of three-drug versus two-drug treatment combinations on the change in CD4 count after 24 weeks of treatment among HIV patients in the AIDS Clinical Trials Group (ACTG) 320 trial using CD4, viral load and age at baseline (Figure 8.4b). Li et al. (2011) was able to estimate subject-specific 5 year risks of death due to prostate cancer, cardiovascular disease, or other causes, among patients receiving treatment for prostate cancer. The index scoring system was constructed as a combination of a patient's age, weight, performance rating, history of cardiovascular disease, serum hemoglobin, size of the primary lesion, and the Gleason score. In both cases, the subgroup-specific outcomes are estimated using nonparametric smoothing and 95% confidence bands are constructed for subgroup-specific and global inference.

8.3.10 Multicenter Trials

Clinical trials are commonly conducted at a number of different investigator sites or centers. One main reason for this practice is to ensure timely

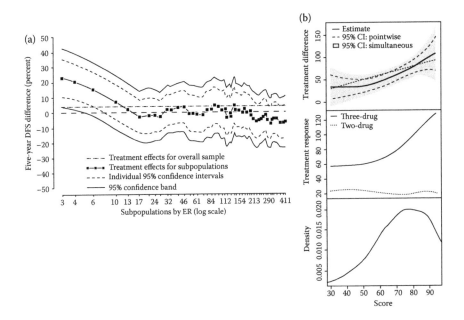

FIGURE 8.4
Examples where treatment effectiveness depends on patient characteristics expressed through (a) an individual covariate (ER status) or (b) a combination of patient covariates (index score based on seven covariates).

enrollment of sufficient number of patients. Another benefit of multiple center studies is that results obtained from these studies can represent a wider variety of patient background, and thus may be more generalizable and less subject to influence from a single center.

Centers can be a significant source of variation in clinical trials. This is particularly true for device trials where the effectiveness of the intervention may not only be a function of the intervention but also a function of how the intervention is applied (e.g., the skill and experience of a surgeon in trials evaluating surgically implanted devices). This variation can be limited by selecting surgeons with appropriate experience, implementing a training program as a part of the clinical trial, or by having the surgeon complete a specific number of operations prior to participating in the trial.

Since center may affect outcome, stratification by center in design and analyses is worth considering. This will help to control the confounding effect of center. If center is an effect modifier, however, then stratification is a first step but may not be sufficient. The interaction between treatment and center can be evaluated to determine if center is an effect modifier. If center is an effect modifier, then separate treatment effects must be made for each center. If not, then a single treatment effect can be made controlling for the confounding effect of center. It is important to note, however, that most trials are not sized to detect such interactions with high power.

Different centers may have different recruitment rates creating an imbalance in the number of patients recruited from various centers. Some centers may fail to provide enough patients for randomization into each treatment group, and the treatment-by-center interaction may become nonestimable. For example, in a dose–response study, there is often a need to include many doses in one study. A typical dose–response study may include a placebo and three test doses—a high dose, a medium dose, a low dose and placebo (a total of four treatment groups). In a multiple center study, it is desirable to include at least four patients (one in each treatment group) from each center. However, in some cases, a center may fail to recruit up to four patients. There can also be situations where one particular dose is over (or under) represented in many centers. When the treatment by center interaction is nonestimable due to a paucity of trial participants at a particular site, a common practice is to pool in those centers with low enrollment. Pooling by geographic region or type of center are also options that may combine trial participants into more homogeneous groups.

In the rare case, when a treatment by center interaction is detected and determined to be real, it is important to investigate the type of interaction. Is the interaction quantitative or qualitative? The difference between a qualitative interaction and a quantitative interaction can be explained using a simple example. In a study with two treatment groups and two centers, the qualitative interaction means that treatment A is better than B in one center but is worse than B in another center. Thus, a qualitative interaction is extremely important to detect. A quantitative interaction means that although treatment A is better than B in both centers, the treatment difference in the first center is much larger than the treatment difference in the other center. A forest plot of 95% CI estimates of the treatment effect for each center will help to illustrate treatment effect heterogeneity across centers.

It is also important to ensure an appropriate infrastructure to retain patients and ensure protocol compliance particularly when recruiting patients at sites with limited resources. Study dropout and noncompliance can threaten the integrity of study results.

8.3.11 Multinational Trials

Several design and statistical issues arise when planning and conducting multinational trials. First, there is a potential for considerable site variation of responses. For example, baseline results of neuropsychological tests from ACTG 5199 suggest considerable country variation in neuropsychological function (Robertson et al., 2012). Possible reasons for these differences include site differences in HIV viral genetics and associated neurotoxicity, the distribution of HIV-associated complications and opportunistic infections, cultural or socioeconomic factors, or a variation in test administration (lack of standardization and training). Generally, increased variation results in a loss of precision and power and decreases the ability to detect intervention effects. As a result, stratified or subgroup analyses may be required when conducting such multinational studies. Researchers can try to minimize variation by using standardized methods and definitions, objective and simple endpoints, central scoring and reading of data whenever possible, and design options such as matching or cross-over studies.

Many issues associated with multinational studies are similar to those associated with multicenter studies. However, some issues may be unique to multinational studies. For example, variation in languages creates many challenges. If the participating nations are non-English speaking, then all the protocol documents need to be translated into the local languages. CRFs, also need to be translated and then back-translated for validation. If the study is designed to use an active control, then this needs to be selected very carefully. Some interventions may be approved for use in some nations but not in others. Even if an intervention is approved in all participating nations, the label, the use, and its perception may vary across countries. Furthermore, medical practices may differ from nation to nation.

It is critical to develop standardization in the application of evaluations to have excellent translation of tests and associated instructions. For example, the Semantic Verbal Fluency test is based upon asking questions of comparable difficulty over time. However, due to translation complexities, questions that are of comparable difficulty in one language/culture may not have comparable difficulty in another. This will affect the interpretation of changes over time. Randomized treatment comparisons will still be valid; however, site comparison of responses may not be interpretable if translation is not consistent. Appropriate standardization can be accomplished through appropriate quality assurance and site training.

Since assessment of interventions is often less well studied in international settings than in the United States or Europe, assumptions made in the design of trials are questionable when using data from U.S. trials to design international trials. Studies may require interim assessment of these assumptions for validity. Adaptive designs ("internal pilot studies") may be warranted. For example, sample size calculations may be conducted utilizing estimates of variability or response rates obtained from studies conducted in the United States. These assumptions may not be valid in international settings, and thus sample-size reestimation at an interim time point may be prudent.

Another consideration is that clinical trial subjects may not be representative of the general population as hoped. Consider HIV clinical trials where sites in the United States have difficulty identifying eligible patients; African sites have large pools of potentially eligible subjects. In addition to entry criteria restrictions for particular trials, site investigators may select patients with more favorable prognoses for enrollment into trials that have site enrollment limits. This can result in a selection bias that can affect the generalizability of the trial results. For these reasons, clinical trials may not be the most appropriate mechanism for estimating prevalence and incidence in these settings.

It is also important to be aware of the challenge of interpreting differences in cultural and regional norms. For example, a particular challenge exists for studying neuropsychogical function. Appropriate normative data to improve interpretation is wanting in international settings. Without such normative data, there are difficulties in assessing the clinical relevance of changes and identifying individual patient impairment. Statistical significance of changes can still be obtained and valid treatment comparisons can still be made in randomized studies and confounding can be controlled using statistical methods, but there will be difficulty interpreting the clinical relevance of any identified changes or treatment differences.

8.3.12 Missing Data

Missing data is the lack of observation of all data from all randomized participants. Missing data is one of the biggest threats to the integrity of a clinical trial. It undermines randomization, the lynchpin of inference, and thus, can be the cause of biased treatment comparisons or estimates of treatment effects. An excellent discussion of missing data methods can be found in Little and Rubin (2002) and the National Research Council's (NRC) report on the prevention and treatment of missing data in clinical trials (National Research Council, 2010).

When "missingness" is not random (e.g., is related to treatment, outcome, or both), it creates a considerable challenge during data analyses and resulting interpretation; consider a complete case analysis which only analyzes observed (nonmissing) data. However, if participants who perform poorly or drop out of the trial and only remaining participants are analyzed, then a distorted view of the effectiveness of the intervention will be obtained. In order to comply with the ITT principle, the general approach is to impute data that are missing, that is, use a strategic "guess" (or multiple guesses) at what the data might have been if the observation had been made. For example, if the primary endpoint was binary (i.e., treatment success vs. treatment failure) then an assumption of treatment failure may be imputed and then the data would be analyzed. However, the validity of any imputation strategy relies on unverifiable assumptions. Sensitivity analyses can be conducted to see how robust the results are to varying assumptions about the missing data. We discuss these issues in more detail in the sections that follow.

8.3.12.1 Preliminary Analyses for Missing Data

Prior to implementation of any imputation strategy, a few preliminary analyses of missing data can be performed to help understand the nature and impact of the missing data. These include (1) documenting and examining the reasons provided for the missing data (e.g., reasons for early study discontinuation) as this can inform imputation strategies; (2) examining the amount of missing data as this can inform regarding the magnitude of the impact; (3) comparing the rate of missing data between treatment arms noting that imbalances may represent treatment effects; and (4) comparing trial participants with missing data to those without missing data with respect to important baseline characteristics noting that differences may indicate a selection bias and a limitation to generalizability. These analyses can help evaluate whether missingness is random and whether participants with non-missing data are representative of all the enrolled participants, and thus, whether analyses based on observed data is biased. Researchers may also use missing data (i.e., missing vs. nonmissing) as an outcome in a regression model to examine baseline factors that are potentially associated with missing data. These analyses may reveal a selective subgroup of participants who are not able to tolerate therapy.

8.3.12.2 Definitions

The properties of missing data methodologies depend on the mechanisms leading to missing data. It can be useful to distinguish three scenarios.

1. Data are *missing completely at random* (MCAR), implying that the missing data are unrelated to the trial variables. If the data are MCAR, then the complete cases are representative of all the randomized cases (i.e., outcomes for those with missing data would be similar to those without missing data). In this case, missing data can be ignored without bias although a loss of precision will still be incurred by discarding data.

2. Data are *missing at random* (MAR) implying that the recorded characteristics can account for differences in the distribution of missing variables for observed and missing cases. MAR implies that outcomes for participants who, for example, dropped out are expected to be similar to outcomes for participants who did not drop out and that had similar baseline characteristics and similar intermediate measures up to that time. Thus, missing outcomes can be modeled on the basis of outcomes of similar participants who did not have missing data.

3. Data are *missing not at random* (MNAR) implying that recorded characteristics do not account for differences in the distribution of the missing variables for observed and missing cases.

Only the assumption that the data are MNAR allows for the possibility that events that were not observed (e.g., severe toxicity or disease progression occurring after the last visit) may have influenced the decision to drop out, resulting in outcomes that are likely to be different from those of similar participants who did not drop out. Models that are based on the assumption that the data are MNAR must make further assumptions about the effect of such possibilities.

If the data are MAR or MCAR, then missing data is often termed *ignorable*. If the data are MNAR then the missing data is term *nonignorable*.

8.3.12.3 Analyses Methodologies for Missing Data

The ICH E9 guidance provides recommendations regarding missing values:

> Missing values represent a potential source of bias in a clinical trial. Hence, every effort should be undertaken to fulfill all the requirements of the protocol concerning the collection and management of data. In reality, however, there will almost always be some missing data. A trial may be regarded as valid, nonetheless, provided the methods of dealing with missing values are sensible, particularly if those methods are predefined in the protocol. Definition of methods may be refined by updating this aspect in the statistical analysis plan during the blind review. Unfortunately, no universally applicable methods of handling missing values can be recommended. An investigation should be made concerning the sensitivity of the results of analysis to the method of handling missing values, especially if the number of missing values is substantial.

There are no universal methods and no easy fixes for handling missing data in clinical trials. There are many methods that have been proposed including complete case analysis, single imputation methods (e.g., last observation carried forward [LOCF] and baseline observation carried forward [BOCF]), methods based on a statistical model (e.g., maximum likelihood, Bayes, multiple imputation [MI]), and estimating equation methods (e.g., inverse probability-weighted methods). Each method relies on unverifiable assumptions. Evaluating which assumptions are defensible will help inform which analyses may be appropriate. We describe the primary ideas behind these methods here.

8.3.12.3.1 Complete Case Analysis

Perhaps the simplest approach to dealing with missing data is the "complete case analysis." In the complete case analysis, participants who have missing data are deleted from the analysis. This approach is valid when data are MCAR. For example, if data are missing because of a natural disaster, then this can be deemed independent of intervention assignment, and complete case analyses may be valid. However, this assumption is usually unrealistic. There is also a loss of efficiency with the deletion of information. Thus, complete case analyses are often discouraged unless missing data are extremely sparse. Complete case analyses can be useful as part of sensitivity analyses.

8.3.12.3.2 Single Imputation Methods

Many single imputation methods (meaning a single imputation is made for each missing value) are available. Two of the most popular imputation methods are the LOCF and BOCF.

8.3.12.3.3 LOCF

We illustrate the concept of LOCF with an example. Suppose in a four week clinical trial, participants are evaluated for efficacy at baseline (week 0) at weeks 1, 2, 3, and 4 after baseline. The primary comparison is with respect to changes in efficacy measurements from baseline to week four. Suppose N participants are randomized; further suppose that, there were n_1 participants who dropped out before week two, an additional n_2 participants who dropped out before week three, and finally an additional n_3 participants who dropped out before week four. If a complete case analysis was conducted then there would only be $n = N - n_1 - n_2 - n_3$ participants available for analyses. In LOCF, the n_1 participants who dropped out before week two would have the respective week one observations imputed; the n_2 participants who dropped out before week three would have the respective week two observations imputed; and the n_3 participants who dropped out before week 4 would have the respective week three observations imputed.

The LOCF imputation approach is very common but is viewed with mixed opinions. Advantages to LOCF include its simplicity, that is, ease of application and explanation. It is generally used for analyzing data at the last time point of the trial, instead of evaluating the longitudinal pattern of response. Critics of LOCF note that it is based on a strong assumption that the participant's status does not change after dropping out. Bias induced by the invalidity of this assumption can go in either direction (depending on whether the participant would improve or decline). Furthermore, LOCF fails to account for uncertainty properly, and thus does not provide valid tests and CIs. BOCF has developed as a variant of LOCF.

8.3.12.3.4 Other Single Imputation Methods

Other single imputation methods have been used, but they again rely on strong assumptions. As a simple example, suppose that a trial has a binary endpoint (e.g., success vs. failure of an intervention). If data are missing, then it could be assumed that the outcome was a "failure" (worst case) or a "success" (best case). In a randomized trial, a best case could be imputed for one intervention and a worst case for the other intervention to evaluate how extreme results could have been.

8.3.12.3.5 Multiple Imputations

MI has become a popular analysis method for missing data. MI is a three-step process. First, sets of plausible values for missing observations that reflect uncertainty about the missing data model are created. Each of these sets of plausible values is used to fill-in the missing values, creating a complete dataset. Second, each of these datasets is analyzed using appropriate complete-data methods. Lastly, the results are combined, allowing the uncertainty regarding the imputation to be incorporated.

Advantages of MI include (1) it propagates imputation uncertainty, (2) the imputation model and the analysis model can differ (e.g., auxiliary variables that are not included in the analysis model can be used in the imputation model), and (3) estimates of the treatment effect and associated standard errors are unbiased. Disadvantages of MI include its complexity relative to single-imputation methods and that the imputations rely on parametric assumptions.

8.3.12.3.6 *Inverse Probability Weighting*

Another popular alternative for the analysis of missing data is the use of inverse probability weighting (IPW) (Robins and Rotnitzky, 1995; Robins et al., 1995; Scharfstein et al., 1999). The idea behind IPW is to assign sampling weights to complete cases (i.e., participants with similar characteristics that do not have missing data), to recreate the population that would have been observed without missing data, thus, adjusting for the selection bias imposed by ignoring cases with missing data. Weights are estimated to represent the inverse probability of missingness given factors affecting the likelihood of missingness. IPW appropriately adjusts for bias when the data are MAR, and if the model for the probability of being observed is correct.

Disadvantages of IPW include that it can yield large variance estimates when some individual weights are high. It also requires two additional assumptions: (1) there are no covariate patterns in which the outcome cannot be observed and (2) the support of the missing data distribution is the same as that for the observed data distribution.

IPW can be used with generalized estimating equations (GEE) in repeated measures settings to yield consistent estimators when the response probability is correctly specified. However, IPW with GEE does not fully utilize information from incomplete cases.

Augmented IPW with GEE has been developed to remedy this. Augmented IPW with GEE also remedies the issue of a large variance when individual weights are large. It also offers efficiency improvements over standard IPW and provides bias protection when the model for the missingness probabilities is incorrectly specified. IPW estimators have a double-robustness property, that is, they yield valid inferences with either one of the outcome regression or the missingness model is correct.

8.3.12.3.7 *Sensitivity Analyses*

Sensitivity analysis is an extremely important component of the analyses of missing data. The main idea of sensitivity analyses is to evaluate the robustness of the conclusions given the uncertainty introduced by missing data. Thus, several (ideally prespecified) imputation methods (that utilize a range of plausible assumptions) are conducted, and the qualitative results of the different analyses are compared. If results are similar across a reasonable range of assumptions and imputation strategies, then there is confidence in the overall trial interpretation. However, if results are dissimilar, then the missing data are affecting the qualitative interpretation of the trial data. In this case, the results of the trial may be inconclusive. If the amount of the missing data is small, then results are likely to be robust. However, if there are lots of missing data, then disturbing discrepancies between the various analyses can occur.

One attractive sensitivity analyses methodology is "tipping point analyses." A typical tipping point analyses consists of a series of analyses based on cycling through all (or many) reasonable imputations for missing data and summarizing the results. Consider the following simple example of a randomized trial comparing two interventions A and B with respect to a binary (success vs. failure) primary endpoint. At the end of the trial, arms A and B are missing data from two and three participants, respectively. A 3×4 matrix is constructed. The first dimension consists of the imputation of success for 0, 1, and 2 participants from the two missing participants in arm A. The second dimension consists of the imputation of success for 0. 1, 2, and 3 participants from the three missing participants in arm B. Twelve analyses are conducted, one for each imputation combination. An "X" is placed in the matrix entry if the result is significant for the corresponding imputation while an "O" is placed if not, creating a two-dimensional view of the robustness of the results. Effect size contours or p-value contours could also be implemented. The tipping point analyses have been extended to continuous and event-time outcomes.

8.3.12.3.8 Summary

Many of the methods (single imputation) do not propagate imputation uncertainty, and thus, lead to underestimates of standard errors and p-values, resulting in inflated type I errors. Thus, these methods are not generally recommended as the primary approach to the treatment of missing data.

The panel that developed the NRC report (2010) favored estimating-equation methods and methods that are based on a statistical model for the data. Weighted estimating equations and multiple-imputation models have an advantage as they can be used to incorporate auxiliary information about the missing data into the final analysis, and they give standard errors and p-values that incorporate missing-data uncertainty. Analyses that are performed with such methods often assume that missing data are MAR. This assumption appears reasonable in many cases, however, the observed data can never verify whether this assumption is correct. Therefore, to assess the robustness, sensitivity analyses are recommended (Little et al., 2012).

Little et al. (2012) described a useful set of six principles for drawing inferences in the presence of missing data (Table 8.4).

8.3.13 Competing Risks

Standard survival data in clinical trials usually measure the time from randomization to an event of interest. For example, HIV trials may measure the

TABLE 8.4

Six Principles for Drawing Inferences in the Presence of Missing Data

1. Determine if the values that are missing are meaningful for analysis.
2. Formulate an appropriate and well-defined primary measure of the treatment effect in terms of the data that were intended to be collected. Distinguish what is being estimated from the method of estimation, which may vary according to assumptions.
3. Document the reasons why data are missing (e.g., the patient moved vs. severe toxicity). Knowing the reasons for missing data can help formulate sensible assumptions about missing observations. Collect auxiliary variables that may be predictive of the outcome and of dropping out, since analysis methods such as MI and weighted estimating equations can exploit these data to reduce bias from missing data and improve the precision of estimates.
4. Decide on a primary set of assumptions about the missing-data mechanism. In some cases, MAR may suffice. Assumptions about the missing-data mechanism must be transparent and accessible to clinicians.
5. Conduct a statistically valid analysis under the primary missing-data assumptions.
6. Assess the robustness of inferences about treatment effects to various missing-data assumptions by conducting a sensitivity analysis. Sensitivity analysis is a relatively new area, and further research on the best methods is needed for the clinical endpoints implies the null hypothesis for the surrogate.

Source: Adapted from Little, R.J. et al., 2012, *NEJM*, 367(14):1355–1360.

time to virologic failure or oncology trials may measure the time to disease progression. However, other events can prevent the event of interest from being observed. For example, a patient may die before observing virologic failure in an HIV trial or prior to disease progression in an oncology trial. Caution is needed when estimating effects of interest when these *competing risks* are present. One may be tempted to treat the competing risk events as censored observations. However, if the true event time is associated with censoring, then Kaplan–Meier estimates will be biased when using this strategy. In addition, the risk of the event of interest may change after a competing risk is observed. Patient dropout can be viewed as a particularly important competing risk.

One approach to the challenge of competing risks is to define a composite event-time endpoint, specifically one that includes competing events, particularly, if they are important events. This essentially makes the competing risk part of the endpoint. For example, in cardiovascular trials, the endpoint may be the time from randomization to the first of any of the following events: stroke, myocardial infarction, hospitalization, or death. The Antibacterial Resistance Leadership Group (ARLG) developed response adjusted duration of antibiotic use (RADAR) for use in stewardship trials. RADAR brings competing risks such as mortality into the outcome producing a desirability of outcome ranking (DOOR) for each trial participant. When applying these approaches, researchers need to acknowledge the differing levels of importance of the events. Competing risk and multistate models can be used to conduct more complex analyses (Putter et al., 2007).

8.3.14 Censoring in Survival Data

Censoring of survival data is common in clinical trials, for example, due to a competing risk. Censoring is informative (or dependent) if individuals censored at a time t are not representative of all individuals at risk (with a similar covariate pattern) at time t. Equivalently, this implies that there is a correlation or systematic relationship between the censoring time and the true survival time. In this case, the hazards for censored individuals are systematically different from individuals who are not censored.

Methods to analyze survival data when censoring is noninformative such as Kaplan–Meier methods and proportional hazards regression are well-developed. However, if censoring is informative and the prevalence of censoring is not trivial, then analyses that ignore the censoring mechanism such as most standard survival methods, are not appropriate.

When censoring is informative, standard survival analysis methodology is subject to bias. The magnitude of the bias depends upon (1) the degree of informativeness and (2) the proportion of censored values. As the dependence between the censoring time and true survival time increases, the bias also increases. The bias also increases as the proportion of censored values increases.

Marginal survival will be overestimated using methods that ignore the censoring mechanism, if a positive correlation exists between censoring times and true survival times; and will be underestimated, if a negative correlation exists between censoring times and true survival times. Estimates of treatment differences may also be biased if the effect of informative censoring differs between treatment groups. As an illustration, consider a controlled clinical trial in which two treatment groups (a new investigational drug A and a control group B) are compared with respect to a time to cure. If subjects likely to have a longer time to cure are more likely to drop out (a competing risk) when in treatment A than in treatment B, then treatment B will artifactually display a longer time to cure relative to treatment A, with all other things being equal. This will inflate the type I error rate in a superiority study with a placebo control. If this was an NI trial with an active control, then one may spuriously claim NI.

8.3.14.1 Design and Monitoring

Ideally, censoring should be avoided. Censoring can be minimized using sound study designs and good monitoring. Studies should be designed such that (1) endpoint data are easy to collect, (2) subjects and investigators can easily comply with the study directives, (3) the study is long enough to collect event data on a high proportion of subjects, and (4) diligence in subject follow-up. Because the impact of informative censoring is decided by the degree of informativeness and the proportion of the censoring when the proportion of censoring is very small, we can ignore the censoring mechanism even if it is highly informative. For example, If censoring is primarily due to the administrative end of a study and it is anticipated that there will be a high proportion of censoring, then researchers may consider increasing the length of study follow-up, to reduce the proportion of censoring and the potential bias due to informative censoring.

It is also helpful to keep a record of everything related to the censoring. For example, one can collect all the covariates that can potentially affect censoring. Note that, if the true survival time is closely related to these covariates, controlling the observed the covariates can avoid modeling the censoring mechanism even if it is informative. If it is hard to collect many covariates, it is very important to record the detailed reasons for censoring. Censoring reasons can help evaluate if the censoring is informative. Moreover, they can help us to build a model accounting for the censoring mechanism, if the proportion of informative censoring is high and inferences from models ignoring the censoring mechanism are not reliable.

8.3.14.2 *Preliminary Investigation*

Once censoring has occurred, then the following preliminary steps may help determine if censoring is informative:

- Examine and report the reasons for censoring. If the reason for censoring is related to latent survival time, then censoring may be informative for estimating marginal survival. If the reason for censoring is related to treatment assignment, then censoring may be informative for estimating treatment differences. If censoring is administrative, then it may not be informative unless time is a confounder.

- Consider censoring itself as an endpoint and compare the censoring rates between treatment groups. Different censoring rates may imply informative censoring with respect to estimating treatment differences. At the very least, a difference in censoring rates may indicate a safety or compliance issue that is related to treatment.

- Compare censored individuals to completers with respect to important variables. Differences may imply informative censoring.

- Examine the relationship between explanatory variables and censoring. An association between an explanatory variable and the probability of being censored suggests informative censoring. This can easily be done by examining contingency tables, plotting, or fitting a simple logistic regression model using the censoring indicator and the explanatory variables.

8.3.14.3 *Sensitivity Analyses*

Similar to sensitivity analyses for missing data, a sensitivity analysis for informative censoring can assess the impact of informative censoring. The principle of a sensitivity analysis is to examine if the inference ignoring the censoring mechanism is different from the inference accounting for the informative censoring mechanism. Therefore, an informative censoring assumption is necessary to perform a sensitivity analysis. One can build an informative censoring assumption according to the recorded censoring reasons or any clinical plausible censoring reasons. If the impact is modest, we can trust the inference from the methods ignoring the censoring mechanism. Note that, sensitivity analysis is not used to judge whether the censoring is informative or not. In fact, without recorded censoring reasons, we cannot judge the nature of censoring based on the censored data. Even if it is clear that censoring is informative, it is still worthwhile to perform a formal sensitivity analysis to exam the impact of informative censoring.

8.3.14.4 *Extreme Sensitivity Analyses*

One may easily perform sensitivity analyses without using sophisticated nonignorable models by using "extreme" imputation for the censored values (Allison, 1995). Consider estimating marginal survival in a one-sample case. One may first estimate marginal survival using standard survival methods. Then consider two extreme analyses: (1) assume using imputation that all censored cases were events immediately after the censoring time (i.e., censored subjects are at high risk of the event) and (2) assume using imputation that all censored cases have a longer survival time than anyone in the sample (i.e., censored subjects are at low risk of the event). These two analyses define a range (or upper and lower bounds) for the marginal survival. If this range is tight, then censoring is not very sensitive, and one may ignore the censoring mechanism. However, if the range is wide and the inferences vary greatly from a practical perspective, then one needs to further examine the censoring process.

One may also use a similar extreme in estimating treatment differences. Consider a study investigating a new treatment versus a placebo control where the time to failure will be compared between the two groups. If one wishes to protect the type I error rate (i.e., incorrectly concluding that the new treatment is superior to placebo), then the most conservative approach is to treat all censored cases for the treatment group as events at the time of censoring, and treat all censored cases from the control group as having a longer time to failure than anyone in the sample. If the superiority of the treatment is then declared, then it is clear that it was not due to bias induced by censoring.

8.3.15 Adherence

Adherence, sometimes referred to as compliance although this term is losing favor, it can be defined as the extent to which participants' behavior coincides with the protocol, for example, whether they are taking their assigned intervention. There are many potential reasons for poor adherence (e.g., forgetfulness; a lack of understanding the instructions; a change of mind regarding participation; inconvenience, unpleasantness, toxicity, or ineffectiveness of the intervention). Even surgical trials can have adherence issues since some surgeries can be reversed.

Monitoring adherence in trials is important since it can help to identify problems in the trial, creating an opportunity to potentially intervene and improve adherence. This also allows evaluation of the impact of adherence on the trial results.

Many adherence issues can be prevented/minimized using sound design strategies such as shorter trials, adequate supervision, using simple interventions and schedules, and selective participant enrollment. Entry criteria may exclude participants who have characteristics of poor adherers such as drug/alcohol users or people commuting long distances. Adherence may be improved by frequent reminders to trial participants via phone calls, texts, mail, and other prompts; lots of personal attention, providing friendly support, and keeping participants well-informed; self-recording medication logs; and involving family member of trial participants.

A trial may also include a "run-in" phase that can be used to improve adherence. In this design, participants enroll into a first stage in which everyone receives an intervention (possibly placebo) and adherence is documented. Participants who adhere well are allowed to go to the second stage that consists of the randomized part of the trial. Poor adherers are not randomized. Disadvantages of using a run-in phase are a longer trial, generalizability challenges due to the selective nature of the participants, and that participants may notice a change in intervention after randomization potentially impacting the success of the blind.

Adherence can be measured in many ways. Perhaps the most common approach in drug trials is "pill count" methods, for example, using a medication event monitoring system (MEMS). Drug trials can also utilize PK methodology and measure the level of the assigned agent in the blood or urine. The half-life of the agents must be considered when utilizing this method. This method can also be complicated in placebo-controlled trials. Additive chemicals (e.g., riboflavin) can be considered as a marker to address this issue. More modern techniques include microprocessors installed in bottle caps or inhalers that record the date/time of bottle opening inhaler use.

The analyses and interpretation of adherence data is a challenging issue. An important question is whether adherence should be considered an outcome or a predictor of outcome (or both). Evaluation of adherence as an outcome should generally be conducted as it can provide information regarding hidden intervention effects. Analyses of adherence as an outcome can be conducted using similar methods to other outcomes if reliable methods for measuring adherence are implemented.

It is considerably more difficult to reach reliable conclusions when evaluating adherence as a predictor of outcome. It is not uncommon to compare the outcomes of good adherers versus poor adherers, often resulting in good adherers perform better. This is often misinterpreted as implying the effectiveness of the intervention. Research suggests that adherers are frequently very different than nonadherers in ways that are independent of intervention. For example, higher levels of education have been correlated with higher adherence. This may mean that adherers have a better diet or other healthier lifestyles resulting in better outcomes. Documenting the reasons for poor adherence can be informative. This may help to determine whether poor adherence is associated with the interventions and inform methods for analyses. Comparing poor adherers to good adherers (or modeling adherence as a dependent variable) may help to define who the poor versus good adherers are.

An important distinction should be made as to whether there is an interest in comparing the *strategies* of the interventions or whether there is interest in comparing the intervention conditional upon acceptable adherence of the participants. If the interest is in comparing the strategies of interventions, then adherence is part of the strategy. In this case, descriptively summarizing the extent and time pattern of adherence by intervention arm may be helpful, but the primary interest is comparing patient outcomes between the two strategies *regardless of adherence*.

Alternatively, consider a trial designed to compare two interventions A and B *controlling for adherence*. This is a very challenging task since adherence is a postrandomization factor and may be associated with the intervention. One strategy that has been employed is to compare only the adherers from each intervention group. However, nonadherence cannot be assumed to be unrelated to the intervention. Excluding nonadherent participants from analyses can create bias even when the percentage of excluded participants is similar between arms. Adherers in arm A may not be comparable with respect to adherers in arm B since the comparison is no longer protected by randomization. Adherers are frequently very different from nonadherers, suggesting that generalizability may be a concern. Another strategy is to use regression modeling with adherence as a covariate when estimating the intervention effect. These analyses can be severely biased since adherent participants can be a highly select group. It is also important to prespecify how adherence is defined so as to avoid data dredging when conducting these analyses.

8.3.16 Rescue Medications

The use of rescue medication (a nontrial treatment) is common in long-term clinical trials evaluating interventions for chronic diseases. For example, trials evaluating interventions to treat pain may include rescue medication use, should primary interventions fail. Asthma trials may include oral steroid use for participants who suffer exacerbations. If rescue medication use is random (i.e., not associated with assigned intervention) then the effect will be to reduce the observed intervention difference. However, if not random, then the effect could theoretically go in any direction. Any trial with rescue medication should report the extent and time pattern of use by intervention arm.

The analyses and interpretation of rescue medication use are particularly challenging issue. An important distinction should be made as to whether the use of rescue medication is part of the intervention strategies being evaluated or whether use of rescue medication is an endpoint. If the research objective is to compare the strategy of intervention A (that includes use of rescue medication should intervention A fail) versus the strategy of intervention B (again possibly with use of rescue medication should intervention B fail), then rescue medication is part of the strategy. In this case, descriptively summarizing the frequency of rescue medication use may be helpful but the primary interest is comparing patient outcomes between the two strategies regardless of rescue medication use.

Alternatively, consider a trial designed with the objective to compare intervention A versus intervention B. A rescue medication may be offered to participants who fail one or both of these interventions to ensure the optimal health and comfort of the participants. In this case, the primary interest is in comparing intervention A versus B but there is less scientific interest in evaluating the rescue medication as part of the strategy. Here, rescue medication is viewed as in indication of the failure of intervention, and thus could be viewed as an endpoint or part of a composite endpoint indicating failure of the intervention. Analyses of rescue medication use could then be conducted, as with any other endpoint, but need to be integrated with other endpoints that measure intervention success.

The analyses and interpretation of rescue medication data are a challenging issue. An important question is whether rescue medication is considered an outcome or a predictor of outcome (or both). Evaluation of rescue medication as an outcome should generally be conducted as it can provide information regarding intervention effects.

Some researchers try to compare two interventions A and B *controlling for rescue medication use.* This is a very challenging task since rescue medication use is a postrandomization factor and may be associated with intervention. One strategy that has been employed is to ignore data captured after initiation of rescue medications. These analyses are subject to bias since the data that are ignored are often associated with participants with worse outcomes. Another strategy is to use regression modeling with rescue medication as a covariate when estimating the intervention effect. These analyses can be severely biased since rescued participants can be a highly select group. Regression models that include rescue medication as a predictor often display results indicating that rescue medication is detrimental. This is often because of a confused direction of causality with the poor outcomes creating rescue medication rather than rescue medication affecting outcomes.

8.3.17 Treatment Crossover

In some disease areas (e.g., oncology), it is common for patients to switch from the treatment to which they were randomized to the alternative trial treatment ("treatment crossover"). Crossover can occur for both ethical and practical reasons: (1) it may be unethical not to switch treatment for a patient who is not responding well to a trial evaluating interventions to treat serious diseases and (2) the possibility of treatment switching within a trial protocol may improve participant and site enrollment, as this provides them with comfort regarding appropriate treatment during the course of the trial. In addition, the motivation to maintain a current treatment may be lessened by incorporating crossover itself or a surrogate into the trial endpoint. For example, oncology trials are often designed to estimate differences in progression-free survival (PFS) as a primary endpoint, a surrogate for OS that has been accepted by the regulatory agencies (FDA and EMA). Once the progression has occurred, participants may crossover to the other treatment. Since PFS is the endpoint (and not OS specifically), there may be less motivation for continued follow-up to evaluate OS (although, careful thought is warranted here).

The reason for a crossover is often related to an individual's prognosis. Commonly patients switch based on a poor participant response and a responsibility to do what is best for the health of the participant. Crossover may also occur due to adverse side effects. If a trial participant crosses-over to other trial intervention in these cases, then the crossover is selective, creating challenges during analyses and interpretation. The intervention received also becomes a mixture of the trial interventions.

An important distinction should be made as to whether the intervention received after the crossover is part of the intervention *strategy* being evaluated, whether crossover is an endpoint, or whether there is a desire to evaluate what the intervention effect would have been had crossover not occurred (a result that is not observed). If the research objective is to compare the *strategy* of intervention A that includes potential crossover should intervention A fail, versus the strategy of intervention B (again possibly with crossover should intervention B fail), then the crossover (and resulting treatment) is a part of the strategy and not adjustment for crossover is needed. In this case, descriptively summarizing the frequency of crossover may be helpful but the primary interest is comparing patient outcomes between the two strategies regardless of crossover.

An ITT approach is often used to evaluate the strategy question. Trial participants are analyzed as randomized regardless of crossover. This is a pragmatic approach that reflects the overall effectiveness of an intervention policy, if it were introduced in practice (when treatment crossover is an option in practice). The results from an ITT analysis should always be given, regardless, as it is free from selection bias and reflects the design consistent with randomization. While analysis of this type is valid, it may underestimate the effect of the intervention alone (independent of crossover) when this is of primary interest. For example, if the experimental treatment truly is superior to the control treatment, and some patients have switched from control to experimental and are, therefore, receiving the benefits of this, using an ITT analysis will make the treatments appear more similar than they would have been, had crossover not occurred. This is particularly problematic in NI trials.

It is often of interest to estimate the effectiveness of the intervention alone. For example, what would have been the OS effect had no participant's crossed over? Some have approached this question using a PP analyses excluding participants who crossover, or censoring them at the time of crossover. This approach can lead to severe selection bias if those excluded differ in prognosis from those retained in the analysis, which may very well be the case since participants often switch treatments because their condition has deteriorated. Consider the oncology example above with PFS and OS being evaluated as endpoints. A PP-based estimate of the effect on OS, that is, analyses that censors participants at the time of the crossover (progression), is likely biased, since participants who are censored differ in prognosis from those who are not censored.

The ITT and PP approaches can also produce important different results. The National Institute for Health and Clinical Excellence (NICE, 2002) evaluated a key trial of trastuzumab for the treatment of metastatic breast cancer (MBC). Crossover was a feature of the trial in which 75% of patients randomized to the control treatment eventually switched to the trastuzumab. When these participants were excluded from the analysis (PP analyses), a median survival gain of 17.9 months was found. However, when all participants were included (ITT analyses), the median survival gain was reduced to 7 months. If one were estimating the survival gain independent of crossover, then the true median survival gain is likely to be between 7 and 17.9 months.

A relatively simple alternative is to use treatment as a time-dependent covariate in a model. However, this approach is also subject to selection bias if crossover is related to prognosis.

A more complex popular alternative is the inverse probability of censoring weight (IPCW) method, an approach whereby participants are censored at the point of crossover. The IPCW is then used to adjust for the selection bias associated with time-dependent confounders affected by treatment. Weights are estimated to represent the inverse probability of (informative) crossover censoring given factors affecting the likelihood of crossover. For participants with similar characteristics that did not crossover, IPCW assigns weights to recreate the population that would have been observed without crossover. However, the bias associated with IPCW is substantial when proportion crossing over is large.

Other complex statistical methods have been developed to estimate the effectiveness of the intervention alone (i.e., had there been no treatment crossover) including observational data-based methods (e.g., structural nested models), and randomization-based efficacy estimators (e.g., rank preserving structural failure time models [RPSFTM], an iterative parameter estimation algorithm, two-stage estimation). An excellent discussion and evaluation of the methods for survival data can be found in Morden et al. (2011).

None of the methodologies for addressing treatment crossover is perfect. Many rely on unverifiable assumptions. Thus, sensitivity analyses are important to evaluate the robustness of the results to varying reasonable assumptions.

During analyses it is important to (1) report and compare the number of crossovers in each arm, (2) report and compare the distribution of the timing of crossovers, (3) report the reasons for crossover, and (4) evaluate the relationship between crossover and important prognostic variables by comparing the participants who crossed over versus the participants who did not crossover with respect to important baseline characteristics.

8.3.18 Association ≠ Causation

A common mistake of clinical researchers is to interpret significant statistical tests of association as causation. Causation is a much stronger concept than association. There are no formal statistical tests for causation (only for association). Although criteria for determining causation are not universal, a conclusion of causation often requires ruling out other possible causes, temporality (demonstrating that the cause precedes the effect), strong association, consistency (repeatability), specificity (causes result in a single effect), biological gradient (monotone dose–response), biological plausibility, coherence (consistency with other knowledge), and experimental evidence. Clinical trials try to address the causation issue through the use of randomization and the ITT principle. However, even in randomized controlled trials, replication of trial results via other randomized trials is usually needed. This is particularly true for evaluating causes other than randomized treatment. A more common concern is to conclude causation between a nonrandomized factor and a trial outcome. Researchers should be very careful about concluding causation without randomization.

When reporting the results of clinical trials, it is important to report measures of variation along with point estimates of the treatment effect and CIs. Reporting both relative-risk and absolute risk-measures, of adverse events, for example, are helpful for interpreting the impact of the events. Creative and interpretable data presentation helps to convey the overall message from the trial data. Reporting both benefits and risks (categorized by severity) provides a more complete picture of the effect of a therapy. Providing reference rates (e.g., of no therapy or an alternative therapy) can further help put the results into perspective and aid other clinicians in making treatment decisions.

8.3.19 Causation ≠ Determination

It is also important to remember that causation is not deterministic. Suppose that a randomized trial compared a new intervention to placebo and the response rates were 60% and 30%, respectively, and the 95% CI for the difference in response rates was (20%, 40%). Given the randomized nature of the trial, one can conclude a significant causal effect of the intervention assuming that the trial was well-conducted. However, this does not imply that the intervention is deterministic of outcome. The response rate was 60% meaning that 40% of the patients that receive the intervention will not respond to the intervention. Thus, although the intervention is clearly superior to placebo, it does not help everyone, and the response of any particular individual given the intervention has a level of uncertainty.

8.3.20 Diagnostic Trials

The clinical trial community has a growing appreciation for diagnostic trials due to the importance of accurate diagnostics for the optimal treatment of patients. Diagnostic tests, applied to patients with suspected disease often triggered from preliminary patient data, are distinct from screening tests, usually applied to assumed healthy patients without suggestive preliminary data.

Analyses of diagnostic trials are different from intervention trials. Research questions of interest in diagnostic trials include (1) How well does the diagnostic identify patients with disease, (2) How well does the diagnostic identify patients without disease, and (3) Once the diagnostic result is obtained, what is the probability that it is correct? This evaluation requires a comparison to a reference ("gold standard") test that is, a diagnosis that can be regarded as the "truth" or nearly so. Many reference standards are imperfect raising challenges during analyses and interpretation. If a perfect reference standard is available or can be measured at least on a subgroup of participants, then such data will be useful for evaluating the imperfect reference test, as well as the new diagnostic test. It is important that what is known about the performance characteristics of the reference test be made clear during analyses.

Table 8.5 summarizes important analyses to be considered when evaluating diagnostic trial data. A thorough evaluation usually begins with estimating the sensitivity (the probability of a positive test when truly positive) and the specificity (the probability of a negative test when truly negative). These are quantities of interest as you walk into a hospital or a doctor's office, for example, if I am truly diseased/nondiseased, then what is the probability that the test will indicate as such?

However, once a diagnostic test is obtained, and you are walking out of the hospital or the doctor's office, then the question is what is the probability that my diagnosis is correct? Interest then shifts to estimating the positive predictive value (PPV, the probability of a true positive given a positive test) and the negative predictive value (NPV, the probability of a true negative given a negative test). PPV and NPV are a function of the prevalence of the disease. A common mistake is to report the PPV and NPV based on the prevalence observed in the study. However, since the prevalence may geographically

TABLE 8.5

Important Analyses in Diagnostic Trials

- Sensitivity: The probability that a diagnostic test will yield a positive test result when truly positive (via a reference test). Sensitivity is estimated by the proportion of truly positive participants who yield positive diagnostic results. Confidence intervals for sensitivity should be displayed.
- Specificity: The probability that a diagnostic test will yield a negative test result when truly negative (via a reference test). Specificity is estimated by the proportion of truly negative participants who yield negative diagnostic results. Confidence intervals for specificity should be displayed.
- False positive rate (FPR): 1 − specificity, i.e., the probability that a diagnostic test will yield a positive test result when truly negative. The FPR is estimated by the proportion of truly negative participants who yield positive diagnostic results. Confidence intervals for the FPR should be displayed.
- False negative rate (FNR): 1 − sensitivity, i.e., the probability that a diagnostic test will yield a negative test result when truly positive. The FNR is estimated by the proportion of truly positive participants who yield negative diagnostic results. Confidence intervals for the FNR should be displayed.
- Receiver operating characteristic (ROC) curve: A plot of sensitivity versus (1 − specificity) when the diagnostic result is measured on a continuum and multiple "cutpoints" for discriminating a positive versus a negative test are being considered. A single cutpoint results in a single point (sensitivity and specificity pair) on the plot. The ROC curve provides a visual display of the trade-off between sensitivity and specificity as the cutpoint varies since sensitivity and specificity are inversely related. An example may be found in Ellis et al. (2005). Area under the curve (AUC) can be calculated and used for diagnostic evaluation and comparison.
- Positive predictive value (PPV) plot: The probability that a person is truly diseased given a positive diagnostic test. This probability can only be calculated from knowledge of the prevalence of the disease. Given the prevalence, the PPV can be calculated using Bayes theorem. A common mistake is to report the PPV based on the prevalence observed in the study. However, since the prevalence may geographically and temporally vary, it is advisable to plot the PPV as a function of prevalence (over the plausible range of prevalence).
  - PPV = (sensitivity × prevalence)/[(sensitivity × prevalence) + ((1 − prevalence) × (FPR))]
- Negative predictive value (NPV) plot: The probability that a person is truly nondiseased given a negative diagnostic test. This probability can only be calculated from knowledge of the prevalence of the disease. Given the prevalence, the NPV can be calculated using Bayes theorem. A common mistake is to report the NPV based on the prevalence observed in the study. However, since the prevalence may geographically and temporally vary, it is advisable to plot the NPV as a function of prevalence (over the plausible range of prevalence).
  - PPV = [specificity × (1 − prevalence)]/
    {[specificity × (1 − prevalence)] + [(prevalence) × (FNR)]}
- Positive likelihood ratio (PLR): [sensitivity/(1 − specificity)]. PLR is the ratio of the probability of a positive result in people with disease to the probability in people who do not have the disease. Interpretation: how many more (or less) times likely patients with the disease are to have a positive result than patients without the disease. Display with confidence intervals.
- Negative likelihood ratio (NLR): [(1 − sensitivity)/specificity]. NLR is the ratio of the probability of a negative result in people with disease to the probability in people who do not have the disease. Interpretation: how many more (or less) times likely patients with the disease are to have a negative result than patients without the disease. Display with confidence intervals.

(Continued)

TABLE 8.5 (*Continued*)

Important Analyses in Diagnostic Trials

- Likelihood ratio graph (when comparing the results of two or more diagnostics): A plot of the true positive rate versus the false positive rate (Biggerstaff, 2000). Consider a new test (A) being compared with a standard test (X). Plot the true positive versus false positive rate for X (see below). Then draw a line through X and (0, 0), i.e., the positive likelihood ratio line. Then draw another line through X and (1, 1), i.e., the negative likelihood ratio line. This divides the space into four regions. The two tests X and A can be compared by noting which region the true positive versus false positive rate for A falls.
  - If A falls within region I: Then A is superior overall to X since it has a larger positive likelihood ratio and a smaller negative likelihood ratio.
  - If A falls within region II: Then A is superior for confirming the absence of disease since it has a smaller positive likelihood ratio and a smaller negative likelihood ratio.
  - If A falls within region III: Then A is superior for confirming the presence of disease since it has a larger positive likelihood ratio and a larger negative likelihood ratio.
  - If A falls within region IV: Then A is inferior overall since it has a smaller positive likelihood ratio and a larger negative likelihood ratio.

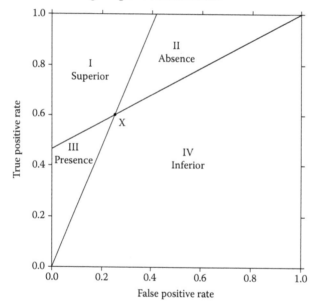

- Sensitivity of sensitivity or sensitivity of specificity (SOS) plots: Sensitivity (or specificity) plotted as a moving average of a continuous variable. For example, if it is desirable to see if the sensitivity of a test varies with age, then the sensitivity may be plotted as a function of age using a moving average strategy.

and temporally vary, it is advisable to plot the PPV and NPV as a function of prevalence (over the plausible range of prevalence). Note that it is possible when comparing two tests to have a test that has a lower sensitivity or specificity but also has both a larger PPV and a larger NPV.

The interpretation of these measures depends on the disease being studied, implications of therapy upon diagnoses, and alternative diagnostics. If a disease is very serious and requires immediate therapy, then a false negative is a very costly error, and thus high sensitivity is very important. However, if a disease is not life threatening but the therapy is costly and invasive, then a false positive error is very costly (i.e., high specificity is necessary).

Since false positive and false negative errors can have vastly different consequences, and thus levels of importance, care should be taken regarding methods that combine results into a single measure of "agreement." The Kappa statistic is one such agreement measure that is also affected by prevalence (in addition to the level of agreement). It also suffers from a lack of clarity regarding interpretation, that is, Kappa = 0.80 is often regarded as "good" but the quantity is not easily interpreted. Thus, Kappa is not generally recommended (Gwet, 2002).

8.3.20.1 ITD and the Impact of Interpretable Tests

When a diagnostic test result is uninterpretable or missing, it is important to evaluate the consequences and develop a plan for how this is to be handled during analyses. Here we distinguish uninterpretable results from indeterminate (or equivocal) results that are not missing per se but reflect an intermediate result. There may be value in repeating an uninterpretable result, that is, the repeat test may result in a positive or negative result. There is less value in repeating an indeterminate result. Uninterpretability is an important consideration for practical reasons (e.g., costs and utility), and thus, the magnitude of interpretability and resulting effect should be reported.

An important question is how to handle uninterpretable results during analyses. Uninterpretability can be considered analogous to missing data in an intervention trial. Consider estimating sensitivity and specificity when there are uninterpretable diagnostic tests. If the uninterpretable results are treated as negative then sensitivity will decrease; while if treated as positive then specificity would decrease. Alternatively, discarding the results for these cases may artificially inflate both sensitivity and specificity. Thus, it is important to evaluate the nature and cause of the interpretability as well as the consequence of different strategies for handling the uninterpretable results.

Consider estimating sensitivity. There are (at least) two distinct types of sensitivities that may be of interest depending on whether uninterpretable tests would be repeated in practice until interpretable results are obtained. If, in clinical practice, a test is to be measured only once (and thus, uninterpretable results will never be clarified via retesting), then it would be desirable to estimate the *pragmatic sensitivity*. The pragmatic sensitivity is directly estimable from the trial by including the uninterpretable result in the denominator of the calculation, but the numerator only includes interpretable positive tests.

However, if in clinical practice, uninterpretable tests would be continually repeated until an interpretable test was obtained, then we would be interested in estimating the *true sensitivity*. The true sensitivity is not directly estimable from the study if uninterpretable tests were not retested. Could the pragmatic sensitivity or a naïve sensitivity (where uninterpretable results are discarded) be used to estimate the true sensitivity? Note that the pragmatic sensitivity is always smaller than the naïve estimate in the presence of uninterpretable results.

To address this question, consider three assumptions regarding the nature of the uninterpretability: (A1) uninterpretability is random, (A2) uninterpretability is dependent on the disease alone, and (A3) uninterpretability is dependent on the disease and correlated with the test result. If A1 or A2 is true, then the naïve estimate, which can be directly calculated, is an unbiased estimate of the true sensitivity. However, the pragmatic estimate is biased (underestimating the true sensitivity). If A3 is true, then both the naïve and pragmatic are biased. Repeat testing would be needed to obtain an unbiased estimate of the true sensitivity (Begg et al., 1986).

8.3.20.2 Example

Susceptibility testing for antibiotics usually takes 48–72 h. During this time patients are treated empirically based upon presumptive antibiotic susceptibility since clinicians are often reluctant to withhold antibiotics until after susceptibility testing due to the presumed impact of early administration of antibiotics on mortality and morbidity in serious and life-threatening diseases. This often results in suboptimal patient care, inappropriate use of antibiotics, and the spread of antibiotic resistance, a growing public health concern. Rapid and accurate diagnostics are needed.

Rapid molecular diagnostic (RMD) platforms are a promising diagnostic tool for identifying antibiotic susceptibility and resistance. The Platforms for Rapid Identification of MDR-Gram-negative bacteria, an Evaluation of Resistance Study (PRIMERS) conducted by the Antibiotic Resistance Leadership Group (ARLG) evaluated four RMD platforms.

Each of the platforms, in a blinded fashion evaluated each of 196 *Escherichia coli* and *Klebsiella pneumonia* isolates (two organisms rated "urgent" hazard level by the CDC), for the presence or absence of the genetic targets that have been associated with antibiotic-specific resistance. The platform result was considered "resistant" when any of the targeted genes were found; the result was considered "susceptible" when none of the targets were found. The minimum inhibitory concentrations (MICs) for each strain were used as a reference for defining susceptibility versus resistance for each of 14 antibiotics using breakpoints defined by the Clinical and Laboratory Standards Institute (CLSI). The platform results were compared to MIC results for each antibiotic.

Discrimination summary (DIM-SUM) plots are used to display the 95% confidence interval (CI) estimates of (1) susceptibility sensitivity, defined as the probability that the platform result is susceptible when the MIC result is susceptible and (2) resistance sensitivity, defined as the probability that the platform result is resistant when the MIC result is resistant (Figure 8.5) for one of the platforms.

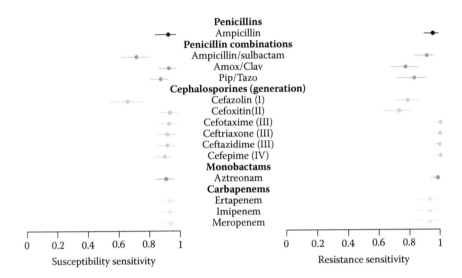

FIGURE 8.5
DIM-SUM plot for the PRIMERS study.

The susceptibility predictive value (SPV) was defined as the probability that a MIC result would indicate susceptibility when the platform result indicates susceptibility and the resistance predictive value (RPV) was defined as the probability that a MIC result would indicate resistance when the platform result indicates resistance. The SPV and RPV are functions of the prevalence of susceptibility. Since the prevalence of susceptibility varies geographically and temporally, the SPV and RPV were plotted as a function of the prevalence of susceptibility (with 95% confidence bands) to allow for broad interpretation. The SPV and RPV plot for one of the antibiotics (Imipenem) is displayed in Figure 8.6.

8.4 Report Writing

All data from a clinical trial is summarized and interpreted in a CSR. Statisticians work very closely with clinicians in preparing the CSR.

A CSR should (1) allow an independent statistician to replicate the analyses; (2) educate and inform the readers on how the trial was designed, conducted, and analyzed; (3) describe the logical argument that leads to the conclusion; and (4) enable a research team to construct a methods and results section of a manuscript reporting the results of the trial.

Good report writing is challenging and takes practice. Suggestions for report writing are summarized in Table 8.6. We discuss a few these suggestions.

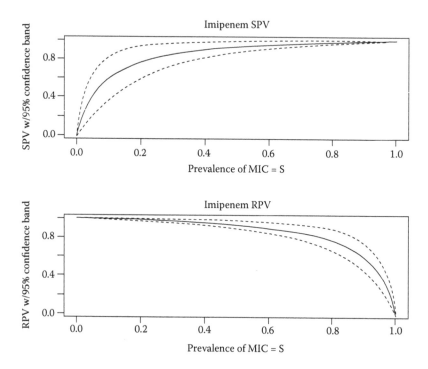

FIGURE 8.6
SPV and RPV plot for Imipenem from the PRIMERS study.

Tailor the technical level and language based on the background of the targeted audience. A final report does not have the same audience as a statistical methodology paper. Reports are drafted with the objective of communicating the messages to readers from diverse research backgrounds. Hence, it is critical to understand who the readers are, and what messages are to be communicated. Creating an outline based upon key points that need to be made can help organize the presentation so that the document has a natural flow, and when more or less detail is needed in various places.

A study report can be viewed as a story. There is background information describing why and how the study was designed, the research questions, and how the design addressed the research questions. Demographics and baseline observations are then described, helping the readers to understand the patient characteristics and the distributions of their important baseline data. Efficacy findings are then discussed along with the statistical methods used for the efficacy analyses and an evaluation of the robustness of the conclusions. Safety findings are then described. The report ends with a summary of the results and an interpretation leading to conclusions. The summary often consists in part of a benefit:risk discussion evaluating efficacy and safety in totality in the context of alternative medical options.

TABLE 8.6

Suggestions for Report Writing

- Strive to describe and illuminate rather than to convince.
- Know your audience as this helps determine when less versus more details are needed.
- Write a final report in past tense. Interim reports may be written in the present tense.
- Anticipate what questions people will have when reading the report. Be proactive in addressing them.
- Include a table of key definitions.
- Provide estimates of variation along with estimates of effect.
- Provide confidence intervals with p-values or point estimates of effect.
- Provide summaries for each important outcome (benefits AND harms).
- Provide accounting for all randomized participants.
- Include a disposition flow diagram of trial participants through the trial.
- Describe the duration of follow-up (consider a Kaplan–Meier plot) and the reasons for LFU discussing the potential bias from this loss.
- Describe the amount of missing data and its impact on analyses and interpretation. Compare the amount of missing data between arms (treatment effect). Compare baseline characteristics between those with missing data versus those without missing data (generalizability). Evaluate reasons for missing data as this can help with imputation strategies.
- Think carefully about the terms "lost-to-follow-up" and "off study treatment." There may be protocol mandated reasons or definitions. Death may not imply lost-to-follow-up or "off study treatment" in the context of describing completeness of follow-up.
- Consider a one-to-one correspondence between a protocol objective and the analysis section evaluating the objective.
- Provide transparency of the multiplicity context (e.g., how many tests were conducted or how many subgroups were examined).
- Summarize heterogeneity of effects across subgroups if appropriate.
- Consider using creative figures instead of tabular summaries as they are often easier to digest and display trends and signals than voluminous listings and tables (e.g., when displaying trends over time, use longitudinal plots rather than tables of summary statistics; use forest plots for displaying confidence interval estimates for various subgroups; use flow diagrams to display participant flow through the trial).
- If safety analyses are presented which summarize only events "related to treatment," be clear regarding who made the relatedness assessment and how it was made.
- Provide estimates of absolute AND relative risk as they are interpreted within the context of each other (e.g., 2 in 10 events vs. 1 in 10 events has a RR = 2; 2 in 1000 events versus 1 in 1000 events also has a RR = 2; but the first is more important than the second).
- Provide information for a reference comparison (i.e., risk without the intervention or risk for an alternative intervention); provide comparison data for risks of events that are well-understood.
- Ask others (particularly people who are not familiar with the trial details) to review a draft report and provide comments. Make revisions to address the comments.
- Use appendices for voluminous output.
- Interpret the results beyond reciting them.
- Be careful regarding the interpretation of a negative result (e.g., $p = 0.15$ does not imply "there is no effect"). Look at confidence intervals and rule out effects with reasonable confidence.
- Be careful claiming association is causal.
- Discuss assumptions of the design and analyses and evaluation of their validity.
- Discuss the limitations of the trial.

For inexperienced authors, it is common to find difficulty getting started on writing a report even if the conclusions are clear. The ICH E3 offers a standard table of contents for CSRs that offers a good template for building the CSR (Table 8.7). Beginning with a high-level bullet point outline of the major points to be made can help. Sub-bullets can be created in a next iteration, and appropriate background can be added to set the stage. When a general outline is obtained then a particular section can be constructed, building the story one section at a time.

Numerous reviews and revisions are not uncommon and generally help to improve the quality. It is often a good idea to have others review the report and provide comment as authors can sometimes be "too close" to the content and may be unable to recognize if something is unclear or needs additional detail.

8.4.1 Balanced Interpretation

It is human nature to root for a positive trial. A positive result may change medical practice, and be financially and professionally rewarding. Thus, it is not uncommon for researchers to over-interpret and overstate results of a trial. It is important for reporting and interpretation to be fair, cautious, and balanced.

To help provide context, the FDA's critical path document stated that ~8% of interventions reaching phase I are eventually approved for regulatory use. Let us assume that 10% of all interventions are truly effective and that one of these is selected to be evaluated in a trial (designed with 90% power and $\alpha = 0.05$). What is the probability that the intervention is truly effective, that is, what is the PPV of the trial? Using Bayes theorem, this can be easily calculated to be 67%.

Thus, a most critical issue is to evaluate the data appropriately to ensure that the conclusions are valid. The conclusions should be challenged (e.g., envision questions that critical reviewers might ask) to ensure that they are robust. Key message can then be formulated. It is often a good exercise to summarize the high-level results in one or two sentences. Sensitivity or secondary analysis can be prepared to support the key messages.

8.4.2 Drug Trials in a Regulatory Setting

In drug trials in a regulatory setting, one of the three situations occurs during analyses:

- The drug is a clear success: The primary (and in some cases, secondary efficacy) variables demonstrate clinical and statistical significance; there are few safety concerns; and the development decision is a convincing "Go."

TABLE 8.7

Suggested Table of Contents for a Clinical Study Report from ICH E3

1. TITLE PAGE
2. SYNOPSIS
3. TABLE OF CONTENTS FOR THE INDIVIDUAL CSR
4. LIST OF ABBREVIATIONS AND DEFINITIONS OF TERMS
5. ETHICS
    5.1 Independent Ethics Committee (IEC) or Institutional Review Board (IRB)
    5.2 Ethical Conduct of the Study
    5.3 Patient Information and Consent
6. INVESTIGATORS AND STUDY ADMINISTRATIVE STRUCTURE
7. INTRODUCTION
8. STUDY OBJECTIVES
9. INVESTIGATIONAL PLAN
    9.1 Overall Study Design and Plan: Description
    9.2 Discussion of Study Design, Including the Choice of Control Groups
    9.3 Selection of Study Population
        9.3.1 Inclusion Criteria
        9.3.2 Exclusion Criteria
        9.3.3 Removal of Patients from Therapy or Assessment
    9.4 Treatments
        9.4.1 Treatments Administered
        9.4.2 Identity of Investigational Products(s)
        9.4.3 Method of Assigning Patients to Treatment Groups
        9.4.4 Selection of Doses in the Study
        9.4.5 Selection and Timing of Dose for Each Patient
        9.4.6 Blinding
        9.4.7 Prior and Concomitant Therapy
        9.4.8 Treatment Compliance
    9.5 Efficacy and Safety Variables
        9.5.1 Efficacy and Safety Measurements Assessed and Flow Chart
        9.5.2 Appropriateness of Measurements
        9.5.3 Primary Efficacy Variable(s)
        9.5.4 Drug Concentration Measurements
    9.6 Data Quality Assurance
    9.7 Statistical Methods Planned in the Protocol and Determination of Sample Size
        9.7.1 Statistical and Analytical Plans
        9.7.2 Determination of Sample Size
    9.8 Changes in the Conduct of the Study or Planned Analyses
10. STUDY PATIENTS
    10.1 Disposition of Patients
    10.2 Protocol Deviations
11. EFFICACY EVALUATION
    11.1 Data Sets Analyzed
    11.2 Demographic and Other Baseline Characteristics
    11.3 Measurements of Treatment Compliance
    11.4 Efficacy Results and Tabulations of Individual Patient Data
        11.4.1 Analysis of Efficacy
        11.4.2 Statistical/Analytical Issues
            11.4.2.1 Adjustments for Covariates
            11.4.2.2 Handling of Dropouts or Missing Data

(Continued)

TABLE 8.7 (*Continued*)

Suggested Table of Contents for a Clinical Study Report from ICH E3

        11.4.2.3 Interim Analyses and Data Monitoring
        11.4.2.4 Multicenter Studies
        11.4.2.5 Multiple Comparisons/Multiplicity
        11.4.2.6 Use of an "Efficacy Subset" of Patients
        11.4.2.7 Active-Control Studies Intended to Show Equivalence
        11.4.2.8 Examination of Subgroups
    11.4.3 Tabulation of Individual Response Data
    11.4.4 Drug Dose, Drug Concentration, and Relationships to Response
    11.4.5 Drug–Drug and Drug–Disease Interactions
    11.4.6 By-Patient Displays
    11.4.7 Efficacy Conclusions
12. SAFETY EVALUATION
    12.1 Extent of Exposure
    12.2 Adverse Events
        12.2.1 Brief Summary of Adverse Events
        12.2.2 Display of Adverse Events
        12.2.3 Analysis of Adverse Events
        12.2.4 Listing of Adverse Events by Patient
    12.3 Deaths, Other Serious Adverse Events, and Other Significant Adverse Events
        12.3.1 Listing of Deaths, Other Serious Adverse Events, and Other Significant Adverse Events
            12.3.1.1 Deaths
            12.3.1.2 Other Serious Adverse Events
            12.3.1.3 Other Significant Adverse Events
        12.3.2 Narratives of Deaths, Other Serious Adverse Events, and Certain Other Significant Adverse Events
        12.3.3 Analysis and Discussion of Deaths, Other Serious Adverse Events, and Other Significant Adverse Events
    12.4 Clinical Laboratory Evaluation
        12.4.1 Listing of Individual Laboratory Measurements by Patient (Appendix 16.2.8) and Each Abnormal Laboratory Value (see Section 14.3.4)
        12.4.2 Evaluation of Each Laboratory Parameter
            12.4.2.1 Laboratory Values Over Time
            12.4.2.2 Individual Patient Changes
            12.4.2.3 Individual Clinically Significant Abnormalities
    12.5 Vital Signs, Physical Findings, and Other Observations Related to Safety
    12.6 Safety Conclusions
13. DISCUSSION AND OVERALL CONCLUSIONS
14. TABLES, FIGURES, AND GRAPHS REFERRED TO BUT NOT INCLUDED IN THE TEXT
    14.1 Demographic Data Summary, Figures, and Tables
    14.2 Efficacy Data, Summary Figures, and Tables
    14.3 Safety Data, Summary Figures, and Tables
        14.3.1 Displays of Adverse Events
        14.3.2 Listings of Deaths, Other Serious and Significant Adverse Events
        14.3.3 Narratives of Deaths, Other Serious and Certain Other Significant Adverse Events
        14.3.4 Abnormal Laboratory Value Listing (Each Patient)
15. REFERENCE LIST
16. APPENDICES

- The drug is a clear failure: There was a clear lack of efficacy; the development decision is a convincing "Go."
- Neither of the above, there is uncertainty regarding the result.

The statistician and the clinician carefully evaluate the results to determine which conclusion has been reached. A key is to provide a balanced evaluation in the study report so that an objective decision can be made.

In order for the drug to be a clear success, there must be no safety "signals of concern." If the primary objective of the trial is efficacy, then the primary endpoint must demonstrate both clinical and statistical significance. Secondary analyses as well as sensitivity analyses of the primary endpoint should further support the result. Results for secondary endpoints would also have to meet the prespecified criteria for a "Go" decision.

If the study results indicate a "No Go" decision, then the key message from this study is to stop further development of the drug. The primary reason for this decision can be based on PK.

The third situation (uncertain) is a difficult case. This often occurs when there is a weak efficacy signal and an accept safety profile, or when the efficacy is strong, but there are safety concerns.

References

Allison PD. 1995. *Survival Analysis Using SAS*. Cary, NC: SAS Institute Inc.

Altman DG, Bland JM. 1995. Absence of evidence is not evidence of absence. *BMJ* 311:485.

Altman DG, Bland JM. 1998. Improving doctors' understanding of statistics. *J R Stat Soc A* 154:223–267.

Assman SF et al. 2000. Subgroup analysis and other (mis)uses of baseline data in clinical trials. *Lancet* 355:1064–1069.

Begg CB, Geenes RA, Iglewics B. 1986. The influence of interpretability on the assessment of diagnostic tests. *J Chronic Dis* 39(8):575–584.

Biggerstaff BJ. 2000. Comparing diagnostic tests: A simple graphic using likelihood ratios. *Stat Med* 19:649–663.

Bonetti M, Gelber RD. 2000. A graphical method to assess treatment–covariate interactions using the Cox model on subsets of the data. *Stat Med* 19:2595–2609.

Bonetti M, Gelber RD. 2004. Patterns of treatment effects in subsets of patients in clinical trials. *Biostatistics* 5(3):465–81.

Cai T et al. 2011. Analysis of randomized comparative clinical trial data for personalized treatment selections. *Biostatistics* 12(2):270–282.

Chow SC. 2010. *Encyclopedia of Biopharmaceutical Statistics*. CRC, London: Marcel Dekker Inc.

Ellis RJ et al. 2005. Clinical validation of the NeuroScreen. *J NeuroVirol* 11:503–511.

Evans SR. 2012. SOS Plots. A presentation at the annual meeting of the AIDS Clinical Trials Group.

Fergusson D, Aaron SD, Guyatt G, Hebert P. 2002. Post-randomization exclusions: The intention to treat principle and excluding patients from analysis. *BMJ* 325:652–654.

Gardner MJ, Altman DG. 1986. Confidence intervals rather than *p*-values: Estimation rather than hypothesis testing. *BMJ* 292:746–750.

Gravel J, Opatrny L, Shapiro S. 2007. The intention-to-treat approach in randomized controlled trials: Are authors saying what they do and doing what they say? *Clin Trials* 4(4):350–356.

Gwet K. 2002. Kappa statistic is not satisfactory for assessing the extent of agreement between raters. *Statist Methods Inter-Rater Reliability Assessment*, No. 1, April 2002, pp. 1–5.

Hollis S and Campbell F. 1999. What is meant by intention to treat analysis? Survey of published randomised controlled trials. *BMJ* 319:670–674.

International Breast Cancer Study Group (IBCG) (Castiglione et al.). 2002. Endocrine responsiveness and tailoring adjuvant therapy for postmenopausal lymph node-negative breast cancer: A randomized trial. *J Natl Cancer Inst* 94:1054–1065.

Kruse RL, Alper BS, Reust C, Stevermer JJ, Shannon S, Williams RH. 2002. Intention-to-treat analysis: Who is in? Who is out? *J Fam Pract* 51(11):969–971.

Lachin JM. 2000. Statistical considerations in the intent-to-treat principle. *Controlled Clinical Trials* 21:167–189.

Li Y et al. 2011. Estimating subject-specific dependent competing risk profile with censored event time observations. *Biometrics* 67(2):427–435.

Little RJ et al. 2012. The prevention and treatment of missing data. *NEJM* 367(14):1355–1360.

Little RJA, Rubin DB. 2002. *Statistical Analysis with Missing Data*, 2nd edn. New York: Wiley.

McFadden E. 2007. *Management of Data in Clinical Trials*. New York: Wiley.

Morden JP et al. 2011. *BMC Med Res Methodol* 11:4. http://www.biomedcentral.com/1471-2288/11/4.

National Research Council. 2010. *The Prevention and Treatment of Missing Data in Clinical Trials*. Washington, DC: National Academies Press (http://www.nap.edu/catalog.php?record_id-12955).

NICE. 2002. The clinical effectiveness and cost effectiveness of trastuzumab for breast cancer. TA 34 (http://guidance.nice.org.uk/TA34).

Nuesch E, Trelle S, Reichenbach S, Rutjes AWS, Burgi E, Scherer M, Altman DG, Juni P. 2009. The effects of excluding patients from the analysis in randomized controlled trials: Meta-epidemiological study. *BMJ* 339:b3244 doi:10.1136/bmj.b3244.

Putter H, Fiocco M, Geskus RB. 2007. Tutorial in biostatistics: Competing risks and multi-state models. *Stat Med* 26:2389–2430.

Rea TD, Fahrenbruch C, Culley L, Donohoe RT, Hambly C, Innes J, Bloomingdale M, Subido C, Romines S, Eisenberg MS. 2010. CRP with chest compression alone or with rescue breathing. *NEJM* 363:423–433.

Robertson K et al. 2012. Improved neuropsychological and neurological functioning across three antiretroviral regimens in diverse resource-limited settings: AIDS Clinical Trials Group Study A5199, the International Neurological Study. *CID* 55(6):868–876.

Robins JM, Rotnitzky A. 1995. Semiparametric e_ciency in multivariate regression models with missing data. *J Am Stat Assoc* 90:122–129.

Robins JM, Rotnitzky A, Zhao LP. 1995. Analysis of semiparametric regression models for repeated outcomes in the presence of missing data. *J Am Stat Assoc* 90:106–129.

Scharfstein DO, Rotnizky A, Robins JM. 1999. Adjusting for nonignorable drop-out using semi-parametric nonresponse models (with comments). *J Am Stat Assoc* 94:1096–1146.

Stafford KA, Boutin M, Evans SR, Harris AD. 2014. Difficulties in demonstrating superiority of an antibiotic for multi-drug resistant bacteria in non-randomized studies. *CID* 59:1142–1147.

Svensson L, Bohm K, Castren M, Pettersson H, Engerstrom L, Herlitz J, Rosenqvist M. 2010. Compression-only CPR or standard CPR in out-of-hospital cardiac arrest. *NEJM* 363:434–442.

Wang R et al. 2007. Statistics in medicine—Reporting of subgroup analyses in clinical trials. *NEJM* 357(21):2189–2194.

9

Analysis of Safety, Benefit:Risk, and Quality of Life

Although the efforts of most clinical trialists focus on the evaluiation of efficacy, the analyses of safety, benefit:risk, risk and quality-of-life (QoL) are also critically important. We discuss these issues in this chapter.

9.1 Safety

Concern for safety should always be considered with paramount importance. However, most clinical trials are designed to evaluate efficacy rather than safety. For example, most clinical trials are sized to detect efficacious effects but not potentially harmful effects. This creates many challenges in the evaluation of safety. These include the following:

- Harmful events may be rare and thus difficult to detect given that trials are usually sized to evaluate efficacy. Thus, the absence of harm in a trial does not imply that the intervention is safe.
- Harms may be time dependent. The trial duration may be too short to detect harmful outcomes (possibly the effect of cumulative exposure). Trial stopping rules may further shorten the trials.
- Harms can vary by subpopulations and patient factors. Enrollment criteria may be restrictive to susceptible subgroups. Awareness of safety issues in special populations (e.g., children and elderly) is important.
- Harm can vary by dose, route, surgeon, and other intervention administration factors.
- Efficacy outcomes are targeted and expected. Safety issues may be unexpected and thus difficult to plan for.
- Instruments for safety monitoring and evaluation may not be sensitive or specific enough to detect important events.
- There is usually no explicit safety hypothesis to be tested. Some suggest testing a null hypothesis of "safe," but failure to reject would not conclude that the intervention was safe. Others suggest testing a

null hypothesis of "not safe" but most trials are not powered to test this hypothesis. If tests are conducted, a decision needs to be made regarding the number of tests and how error rates will be controlled in the context of multiplicity. Many suggest avoiding formal testing.

- No intervention has zero risk. When are risks acceptable? Some interventions are applied to healthy people (e.g., vaccines with live biological agents). Can we ignore "first do no harm"?

Given these challenges and many recent safety concerns associated with widely used interventions raises the question as to whether we should be designing more trails to evaluate safety more comprehensively. Although a minority, some clinical trials are designed to specifically evaluate safety. For example, the long-acting beta-antagonists (LABA) trials are a series of trials evaluating approved medications for asthma to rule out a relative risk of 2 with respect to serious asthma outcomes (a composite of asthma-related hospitalization, intubation, and death) that have been reported with use of these medications. Another example can be seen with a popular class of drugs for the treatment of arthritis, the nonsteroidal anti-inflammatory drugs (NSAIDs), and an associated adverse event (AE) of ulcer or gastro-intestinal (GI) bleeding. By the turn of the century, a new class of anti-inflammatory drugs (Cox-2 inhibitors) was available and had decreased GI toxicity. In most of the Cox-2 development programs, clinical trials were designed to study GI-related AEs. In 2004, a Cox-2 inhibitor, Vioxx®, was removed from the market by Merck because of cardiovascular AEs. These results may in turn force trials evaluating similar drugs to include evaluation of cardiovascular and GI safety.

Formal statistical analyses may be useful for common events, but uncommon events can be more challenging. A solid strategy is to comprehensively describe events that occurred and estimate the incidence rates (and between-group differences in rates) using confidence intervals (CIs) (avoid p-values). The CIs can then be used to rule out rates (or rate differences) with reasonable confidence. Creative graphical summaries help to visualize signals and are often easier to digest for interested parties. It is important to utilize biological insight to make data displays more informative.

In general, all participants that receive any amount of intervention (often termed the safety population) should be included in the safety analyses. Hopefully, this safety population is close to the intention-to-treat (ITT) (all randomized) population. If there is a substantial difference between the safety and ITT populations (e.g., more than a few participants, then it is important to document the reasons for this). Large differences could impact the generalizability or the validity of intervention comparisons.

As part of the evaluation of safety, it may be important to summarize (ICH E1, 1994), the

- Extent of exposure: For example, number of patients exposed; the duration of the exposure; associated doses, and so on

- Reasons for study and intervention discontinuation as the discontinuation may be a signal of harm and can informatively censor future events
- AEs
- SAEs
- Laboratory measurements
- Vital signs
- Targeted events of special interest

We discuss issues involved with the evaluation of these measures.

9.1.1 Adverse Events

Analyses of AEs are a primary component of most safety evaluations. We begin with a few important definitions.

9.1.1.1 Definitions

An *AE* is any untoward medical occurrence in a study participant who has received the study intervention that may or may not have a causal relationship with the intervention (ICH E2A). The event could be a sign, symptom, disease, or laboratory abnormality. *Signs* can be objectively measured, usually by a clinician (e.g., heart rate, respiration, blood pressure, temperature, and reflexes). *Symptoms* are subjective and noticeable by the patient (e.g., pain).

AEs can be expected or unexpected. An *expected AE* is any adverse reaction whose nature and intensity are consistent with that documented in the investigational brochure or the general investigational plan. An *unexpected AE* does not have the specificity and intensity that is consistent with the investigational brochure or other risk observations whether or not it has been anticipated (e.g., because of the pharmacologic properties).

AEs can often be classified based on relatedness to the study intervention. A commonly used scale is definitely related versus probably related versus possibly related versus not related versus unknown. This relationship is typically determined by an experienced clinician or monitor at the study site. In some cases, the study sponsor (e.g., pharmaceutical company or the NIH) may reclassify this relationship based on an independent review. The temporal relationship between intervention administration and the event, the medical history of the participant, and concomitant therapy use is carefully considered when making this determination.

AEs are also frequently classified by severity. Severity is an intensity rating (e.g., mild, moderate, severe, life-threatening, and death). A *severe AE*, in this case, is an AE with at least a sever rating on the severity scale. Severity ratings are not standardized, that is, each clinical research team or organization can construct and utilize their own severity rating system.

A *serious adverse event* (SAE) is defined as any adverse drug experience occurring at any dose that results in any of the following outcomes:

- Death
- A life-threatening adverse experience
- Inpatient hospitalization or prolongation of existing hospitalization
- A persistent or significant disability/incapacity
- A congenital anomaly/birth defect (ICH E2A, 1994)

An important medical event that may not result in death, be life-threatening, or require hospitalization may be considered an SAE when, based upon appropriate medical judgment, they may jeopardize the participant and may require medical or surgical intervention to prevent one of the outcomes listed above.

Seriousness is distinct from severity. An AE can be severe (e.g., headache with severe pain) but may not be serious because it is not life-threatening or result in hospitalization. Conversely, chest pain may not be severe but may result in hospitalization and is thus qualifies as an SAE.

Seriousness (not severity) serves as a marker for regulatory reporting obligations. SAEs are subject to expedited reporting, that is, rapid notification to the regulatory agencies is required. Given the speed at which SAEs must be reported, SAE data are often stored in a special database distinct from the rest of the clinical database.

Some participants may have preexisting conditions prior to receiving the study intervention. These conditions may also be observed after intervention initiation. To distinguish events that may be associated with a preexisting condition from those that emerged only after the intervention, *treatment-emergent signs and symptoms* (TESS) were defined. An AE is a TESS if, and only if, it was (1) not observed prior to intervention initiation or (2) severity of the event had worsened from baseline.

ICH E2A provides a summary of important definitions.

9.1.1.2 Coding

AEs are coded using a medical dictionary. Older dictionaries include the World Health Organization (WHO), the International Classification of Diseases (ICD), and the coding symbols for a thesaurus of adverse reaction terms (COSTART). However, the currently used AE dictionary for regulatory reporting is the medical dictionary for regulatory activities (MedDRA).

MedDRA is a clinically validated international medical terminology used by regulatory authorities for data entry, retrieval, evaluation, and presentation.

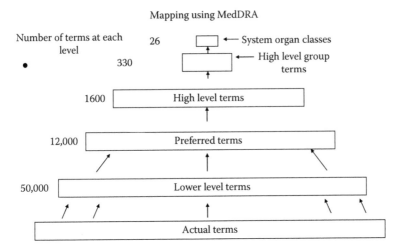

FIGURE 9.1
MedDRA.

MedDRA is also the AE classification dictionary endorsed by the ICH. Terms in MedDRA are classified in the following categories:

- Symptoms
- Signs
- Diseases
- Diagnosis
- Therapeutic indications
- Names and qualitative results of investigations, including (PK)
- Surgical and medical procedures
- Medical/social/family history

MedDRA is organized into a five-level hierarchy: the highest level is system organ classes, then high level group terms, high level term, preferred term, and lowest level term (Figure 9.1). An example is displayed in Figure 9.2. There are 26 system organ classes, and they comprise grouping by anatomical or physiological system, body organ, etiology, and purpose.

9.1.1.3 Spontaneous versus Active Collection

AEs can be collected using active probing (e.g., creating a specific CRF asking for specific event data of interest), or via spontaneous reporting (i.e., report the event if observed; otherwise no reporting is done). The advantage of spontaneous reporting is that it focuses on the most important events and can detect unexpected events as they do not have to be prespecified. However, spontaneous reporting cannot distinguish missing data from a

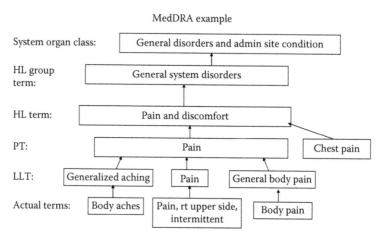

FIGURE 9.2
MedDRA example.

negative response (i.e., if no events are reported, it is not known whether this is because no events were observed or because they were not reported). Active reporting is standardized and makes data collection more complete.

9.1.1.4 Targeted AEs

Concerns regarding specific AEs can arise with particular interventions or disease areas. For example, an undesirable property of some nonantiarrhythmic drugs is their ability to delay cardiac repolarization, an effect that can be measured by the prolongation of the QT interval (ICH E14, 2005) on the surface electrocardiogram (ECG). A delay in cardiac repolarization creates an electrophysiological environment that favors the development of cardiac arrhythmias, most clearly torsade de pointes (TdP). People that take beta-agonists to treat asthma are often monitored for prolongation of the QT interval. People who take specific drugs to treat multiple sclerosis are now carefully monitored for progressive multifocal leukoencephalopathy (PML), a rare and deadly brain disease. People that take antiepileptics or antidepressants are carefully monitored for suicidality associated with these medications.

When concerns regarding specific AEs are identified prior to trial initiation, then trials can be designed, monitored, and analyzed with these specific AEs in mind. It may be prudent to consider adding the AE as a specific trial endpoint or even making it a part of a composite primary endpoint. These targeted AEs can then be analyzed with similar vigor to efficacy endpoints.

9.1.1.5 Analysis Issues

In the presentation of AEs, it is important to consider summaries of the proportions of participants in each group experiencing an event and the total incidence of AEs that may include multiple events per participant.

Fundamentally, it is desirable to use the participant as the unit of analyses. The advantage of counting participants rather than events is that participants are independent whereas events may not be, for example, if multiple events are observed on a single participant. Two participants each suffering one event is stronger evidence than a single participant suffering two events. Moreover, most conventional statistical methods require independent observations. An argument for also summarizing events is that the participant analyses do not take advantage of all of the data and that modern methods (e.g., random-effects modeling) can take into account the nonindependence of the data. A useful display when analyzing recurrent events is the mean cumulative function (MCF), a graphic summarizing the mean cumulative number of AEs up to time t (as a function of time) (Siddiqui, 2009).

A first step to the analysis of AEs is to estimate the incidence rates usually summarized by organ class. Only AEs with a particular severity may be selected for these analyses as well. Two primary types of (intervention-specific) incidence rates for an AE can be presented. The crude incidence rate is calculated as the number of participants that experience the AE, divided by the total number of participants. The exposure-adjusted incidence rate (also called person–time or person–time at risk) is calculated as the number of participants with AE divided by the total time of exposure (e.g., person years). The total exposure time is obtained by summing up all of the exposure time from all of the participants. In this calculation, if a participant did not develop the AE, exposure is calculated from the first intervention initiation (e.g., dosing date) to the last intervention dosing date. For a participant that develops the AE, the exposure is calculated from the first intervention initiation date of dosing to the first occurrence of the AE.

A very informative summary of incidence rate estimates is with the use of a modified forest plot that displays incidence rates for each intervention group and CIs for the between-group difference in incidence rates. AEs could be ordered by the point estimate of the differences (Figure 9.3) or clustering results by organ class.

Description of the temporal nature of AEs is very important. There are two dimensions to this issue: when the AE starts and the AE duration. In many drugs that treat chronic diseases, AEs tend to appear early in the study. For example, Clifford et al. (2005) reported that neurological symptoms occurred primarily in the first week of Efavirenz (EFV) therapy to treat HIV. However, most of the symptoms were gone one month after therapy initiation.

To illustrate the onset time of a particular AE of interest, the time-to-the-first occurrence of this event can be displayed using a Kaplan–Meier plot. For participants that developed an AE, the event time is the time of randomization to the time when the first AE was observed. Participants that did not experience that the AEs are considered as censored at the point of their last observation.

Creative graphical summaries are attractive tools for illustrating trends. Figure 9.4 is a bar chart of AEs incidence by dose and the time of the dose. Figure 9.5 is a bar chart of AE incidence over time by intervention group.

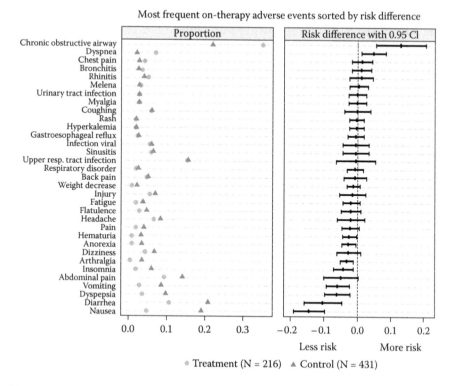

FIGURE 9.3

Forest plot summary of AEs by intervention and the risk difference between interventions.

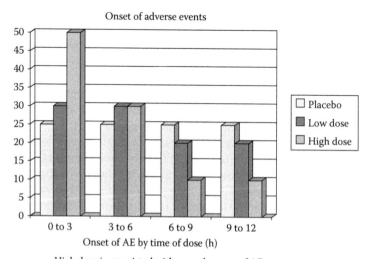

FIGURE 9.4

AEs by dose and time of dose.

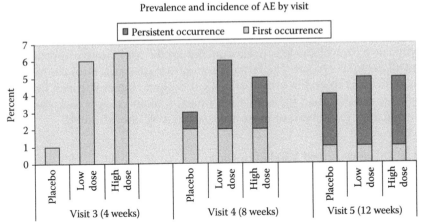

FIGURE 9.5
AEs by dose and time.

Individual participant profiles are useful for visualizing the temporal nature of the events. If one or several AEs are of particular interest, then the individual AE experiences may be reported graphically.

In many clinical trials, participant follow-up may not stop on the last day of treatment or even at the last day of participant visit at the clinic. Participants can experience AEs even after treatment discontinuation. For example, a drug may remain in the body for quite some time after drug discontinuation. Hence, it is important to consider the follow-up time (i.e., "lag period") after treatment discontinuation and evaluate events observed in this window to reduce informative censoring that would be created by concluding follow-up (Lagakos, 2006). It is important to consent participant with this lag period in mind.

A particularly important aspect of an AE is whether it is reversible, for example, with intervention withdrawal or with concomitant therapy. If an AE (and any damage created by the AE) is reversible, then the toxicity may be an acceptable risk. However, if the damage incurred by the AE is not reversible, then serious consideration must be given to whether the risks are acceptable. Thus, it is generally of interest to describe whether there were any actions taken to manage the AE (e.g., lower the dose, stop medications, other medical procedures) and what the associated outcome of this intervention was.

The assessment of causality can be a particular source of confusion when analyzing AEs. Many researchers base causality on *relatedness*, a partly subjective case-by-case assessment generally by an experienced clinician. However, statisticians in particular believe that the appropriate assessment of causality lies in the comparison of randomized intervention groups, for example, an increase in AE incidence of an active intervention versus placebo is the causal effect of the intervention (note that this is generally how the causal efficacy effect of an intervention is obtained). For the relatedness approach, a control group is irrelevant. But for the comparative approach, the control group is crucial. The latter comparative approach is, of course, more reliable as it is rooted in strong statistical design. However, given that most trials are not designed for evaluating safety, a balanced approach consisting of evaluation using both of these approaches may be warranted.

9.1.2 Laboratory and Vital Sign Data

Laboratory measurement (e.g., chemistries and hematologies for drug interventions) and vital signs also provide an opportunity to evaluate safety as many of these measures are surrogates for safety outcomes. "Labs" and "vitals" have a few advantages: (1) they may identify risks that may not be otherwise observable to the trial participant or upon clinical evaluation and (2) they tend to be less-biased as they can be objectively measured. Results are generally summarized for each lab test and vital sign individually.

Specimens for laboratory testing can be obtained from the blood, urine, saliva, or from other bodily tissues. The specimens are then sent to a laboratory for analyses. The measurement units and normal ranges are test-specific and laboratory-specific. Use of a central lab for all specimens can eliminate laboratory variation and, is thus, desirable under most situations. It is similarly important to standardize methods (e.g., with respect to rest, the time of day or time relationships to food, exercise, and sleep) for the collection of vital signs as measurements can vary by these factors.

9.1.2.1 Analysis Issues

An important consideration when analyzing lab data is whether it is more informative to evaluate general population trends, that is, via assessments of central tendency (e.g., means, medians, and modes) or extreme values (e.g., cases that may be concerningly abnormal). Measures of central tendency may be useful for trying to understand biological mechanisms of action but are often uninformative for evaluating safety associated with infrequent AEs. If measures of central tendency are utilized then order statistics (e.g., median and quartiles) may be more informative due to the potential for extreme values.

If priory interest is in the evaluation of infrequent events, then the outliers are of the greatest interest. In this case, labs can be classified as normal, below the lower limit of normal (LLN), or above the upper limit of normal (ULN) based on normal ranges (also called reference ranges) provided by the lab conducting the analyses. If the value for a specific lab measurement is between the LLN and ULN (Gilbert et al., 1991), then the test is considered normal. If the observed value is outside of this range, that is denoted as an abnormal lab value. Levels of abnormality can also be specified, for example, three times the ULN. The proportions of participants that reach various levels of abnormality can be summarized and compared. Changes in normality status can also be summarized using shift (or transition) tables, that is, a cross-classification of post-intervention normality status versus baseline normality status.

One informative presentation for visualizing lab changes with intervention initiation is a scatterplot, plot of the last observation (vertical axis) versus the baseline observation (horizontal axis). A line of equality could be superimposed representing no change from baseline.

In multiple-dose drug trials, it is important to evaluate the association of drug concentration with laboratory endpoints. The scatterplots described above can be useful to evaluate this potential association. The contrast between dosing groups could be visualized by having a separate plot for each dose or by using a different color for each dose group within a single plot. Consider the simulated data in Figures 9.6 (placebo), 9.7 (low dose), and 9.8 (high dose) for evaluating how dose may be associated with these laboratory changes. The placebo group does not have any recognizable trends in the change from baseline (Figure 9.6). The low-dose group tends to experience a slight decrease in this lab parameter as most of the points are slightly below the equivalence line (Figure 9.7). The trend is even stronger in the high-dose group (Figure 9.8) indicative of a dose–response relationship.

Such a scatter plot is not only useful in detecting a dose-related trend, but can also be helpful in spotting outliers and other patterns. These plots may be less informative in long-term trials. However, the plots could be modified to evaluate the change from baseline to the most extreme value or a particular time point of interest.

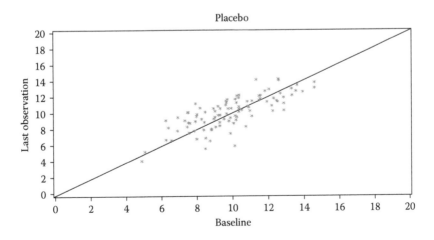

FIGURE 9.6
Change from baseline for placebo.

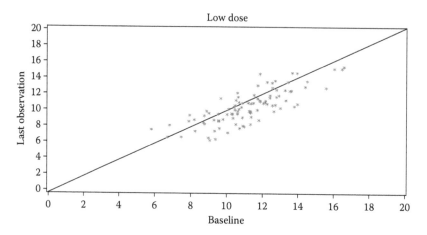

FIGURE 9.7
Change from baseline for the low dose.

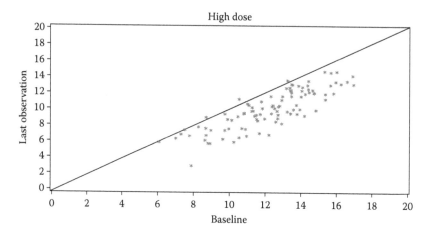

FIGURE 9.8
Change from baseline for the high dose.

Side-by-side box-plots (or jitter plots) can also be useful for visualizing trends over time and contrasting intervention groups (e.g., Figure 9.9).

9.1.3 Safety Analyses Using Observational Data

Since many clinical trials are not designed to detect safety signals, safety evaluation often involves large postapproval studies (i.e., phase IV) that evaluate safety and utilize meta-analyses using other epidemiological data, for example, from postapproval reporting. In 2011, the FDA received 179,855

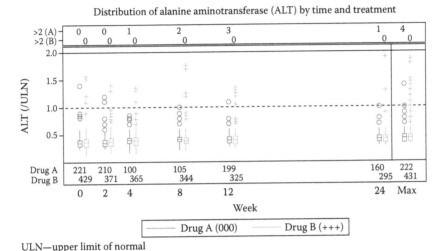

FIGURE 9.9
Side-by-side jitter plot of ALT over time.

reports of serious or fatal AEs. This may represent only ~1% of the total events as it is believed that ~99% of events go unreported. With such underreporting of observational data, confounding is a major concern. Propensity score matching can be applied, but confounding cannot be completely controlled unless information on the confounding factors (or surrogates) is uniformly collected. Regardless, however, the results of studies that utilize such data must be interpreted with caution.

In regulatory submissions, for example, new drug applications (NDAs), safety results are reported for each individual study but an integrated summary of safety (ISS) is also included. The ISS is a comprehensive analysis that combines the results of several studies together. This may come in the form of a meta-analysis or other aggregate summaries.

Periodic safety update reports (PSURs) are periodic updates of the worldwide safety experience of an approved intervention. PSURs create the opportunity for safety reevaluation and to indicate whether changes should be made to the available information in order to optimize the use of the intervention. Several ICH documents provide useful references including the following:

- *ICH E2C:* Clinical safety data management: PSUR for marketed drugs
- *ICH E2D (2003):* Postapproval safety data management: definitions and standards for expedited reporting
- *ICH E2E:* Pharmacovigilance planning

9.2 Benefit:Risk Evaluation

Medical interventions such as drugs, biologics, devices, or procedures have multidimensional effects. The beneficial and harmful effects are weighed by clinicians, patients, regulators and their advisory committees, sponsors, institutional review boards, and data monitoring committees (DMCs) to determine the overall utility of the intervention. Thus, benefit:risk evaluation is a fundamental element of most clinical trials and development programs (Evans, 2008).

Although the benefit:risk profiles of interventions have been assessed for many years, evaluation has often been conducted in an unstructured manner. There has been a call recently for a more systematic and transparent assessment based on a congressionally mandated study by the Institutes of Medicine (IOM) as well as the European Committee for Proprietary Medicinal Products (ECPMP) (EMA, 2005; IOM, 2007). More recently, EMA leaders (Eichler et al., 2009) called for a regulator refinement of methods to assess benefit:risk including (1) a shift from implicit to explicit descriptions of decision criteria and data interpretations with valuations (e.g., weighting factors for outcomes), (2) a shift from qualitative to quantitative descriptions of net health effects, (3) development of a consensus definition of "tolerable risk," acknowledging that the elimination of all risk is not feasible, and that some risks are acceptable if benefits are sufficient, and (4) incorporation of patient values and preferences, noting that patients may vary in their willingness to accept risk in exchange for a potential benefit. The proposal also involves a shift in the objectives from "ensuring intervention safety" and "communication of risk" to "ensuring a positive benefit:risk profile" and "communication of benefit:risk."

This leads to the motivating question of "how do we revise our traditional approaches to design, monitoring, analysis, and reporting of clinical trials to address the challenges of benefit:risk evaluation?"

9.2.1 Challenges

In order to develop better measures and an understanding of benefit:risk evaluation, we must first understand the challenges. The challenges may be classified as those relating to measurement, data summarization, and assessment.

9.2.2 Measurement

The term *benefit:risk analysis* is frequently used but not well defined. Although many researchers claim to have conducted a benefit:risk analysis, it is often difficult to discern what analyses have actually been conducted due to the lack of a consensus definition. Thus, an initial challenge is to clearly define *benefit:risk*, outline how it will be measured, and how data will be integrated for assessment. While facing these initial decisions, clinicians often defer to statisticians' quantitative advice, while statisticians defer to a clinicians' experience treating patients. One initial strategy is to define the two terms separately. For example, benefit may often be defined by the primary endpoint in a trial (i.e., the clinical effect that the intervention was intended to induce). Risks (or "harms") are generally more difficult to define. SAEs, signs and symptoms, abnormal chemistries and hematologies, quality of life, new diagnoses, and so on, may all contribute to the concept of "risk," but formalization of a definition integrating these outcomes is needed. In general, risks may be thought of as all negative consequences of a particular therapy. We may attempt to formalize this definition by considering all measurable patient-level quantities that are negatively impacted by the intervention, as assessment will be limited only to those quantities that are actually measured. Thus, a necessary condition for the proper assessment of risks is the foresight to anticipate and accurately measure such risks. In some special cases (e.g., interventions for which harms are recognizable to the patient through symptoms), patients could be asked to rate their perception of overall risks. However, this strategy can be problematic when symptoms do not equate with risk (e.g., silent risk associated with abnormal lab results).

One of the challenges to benefit:risk evaluation is censoring, an inability to observe important events. This may occur because of competing risks that prohibit future observations of interest such as premature trial discontinuation or the end of the trial. There is often a temporal difference between the times at which benefits and risks are observed. Risks often present later than benefits, perhaps as a result of cumulative exposure to the intervention, and thus may be censored by the end of the study. Thus, the absence of observed safety events or harms does not necessarily imply an absence of risk. Similarly, efficacy data can be censored by observed SAEs. Thus, diligent consideration of the duration of treatment and follow-up after treatment discontinuation (early discontinuation or the end of the study) is critical as relevant events may occur in this window (Lagakos, 2006). ITT principle requires strict enforcement to observe outcomes after treatment discontinuation (noting the important distinction between off-treatment and off-study) and to help protect against censoring.

The asymmetry of the attention and focus paid to benefits and risks during data collection and analyses also creates challenges. Most late stage trials are designed for efficacy endpoints and are not powered to assess safety outcomes, particularly when safety outcomes are rare. This can be problematic when these rare harms are serious. Many "safety noninferiority trials" (e.g., the LABA trials evaluating long-acting beta-antagonists for asthma) are being conducted with the goal of ruling out the minimum unacceptable harm within the context of the benefit provided to address this concern. Most statistical analyses also focus on efficacy rather than safety evaluations with AEs/SAEs often being summarized with simple counts and percentages. The passive collection of some safety data (e.g., report only if abnormal) affects data quality in that the lack of harms cannot be distinguished from missing data, potentially resulting in the under-reporting of harms. Primary efficacy outcomes may also be adjudicated in a blinded fashion whereas safety outcomes are not (i.e., reported only by investigators). A more balanced approach to the collection and analyses of harms and benefits can help to improve evaluation (O'Neill, 2007).

In order to conduct a thorough evaluation of benefit and risk, the appropriate data must be carefully collected in trials and through industry risk management strategies. It has been argued that the standard postapproval case reports are not sufficient to conduct quality benefit:risk analyses, as this process suffers from selective reporting with only specific harm events being reported when observed. Consider the diagnosis of PML, a deadly brain disease with no proven treatment options, which has been observed in patients taking Tysabri and rituximab. A few natural questions are (1) what is the incidence of disease, (2) what is the reference rate (e.g., without treatment or with alternative treatment), and (3) what is the benefit:risk profile of the intervention? To estimate incidence, the number of patients who are treated with the intervention (and possibly the duration of treatment) is needed, but this cannot be easily estimated from postapproval case reports of PML (but may potentially be estimated with moderate accuracy from sales data). The reference rate suffers from similar complications, and since benefit data is not uniformly obtained, a thorough benefit:risk evaluation is challenging. As a consequence, results of exploratory safety analyses should be interpreted with caution. The reliability of such analyses may be evaluated by considering the likelihood that the findings could occur by chance, the biological plausibility of the findings, and whether the findings can be confirmed by independent and prospective data (Fleming, 2008).

9.2.3 Summarization

When interpreting benefits and risks, it is important to consider both relative and absolute effects as they are best interpreted within the context of each other. Suppose an intervention increases the risk of a particular AE from 1 in 10 to 2 in 10 (a relative risk of 2). Now suppose instead that an intervention increases the risk from 1 in 100,000 to 2 in 100,000 (also a relative risk of 2). The two cases are interpreted very differently because of the difference in absolute risk (10% vs. 0.001%) increase. The first case is very concerning while the latter is far less so. This is important to remember, particularly in the context of estimating relative risk in event-driven trials when events are rare. For example, the Prospective Randomized Evaluation of Celecoxib Integrated Safety versus Ibuprofen or Naproxen (PRECISION) trial was designed to rule out the hypothesis that celecoxib is associated with an increase in cardiovascular death, myocardial infarction, or stroke relative to naproxen in patients with rheumatoid arthritis or osteoarthritis

(OA). The trial was sized based on relative risk with 508 events required to have 90% power to rule out a 33% increase under the null hypothesis of no differences in risk. It was anticipated that 7000 patients per group with 2–3 years of follow-up would be required to meet this event threshold. This sample size is large because the quantity being estimated is the relative risk, whose precision is determined by the number of events. Because events are rare, a large and/or long trial is required. However, the absolute risk can be measured with great precision, as its uncertainty decreases with overall sample size rather than events. Suppose intervention A has one event in 500 and intervention B has three events in 500. The relative risk is 0.6%/0.2% = 3 but is measured with limited precision, as evidenced by the wide 95% CI of (0.3, 158). The absolute risk difference is 0.6%–0.2% = 0.4% and is measured with high precision as evidenced by the narrow 95% CI of (−0.4%, 1.2%). Now suppose intervention A has one event in 5000, and intervention B has three events in 5000 (sample size increased by a factor of 10). The relative risk estimate and associated precision is unchanged at 0.06%/0.02% = 3 with a 95% CI of (0.3, 158) since the number of events has not changed. However, the absolute risk difference is 0.06%–0.02% = 0.04% and is measured with even greater precision as evidenced by the narrow 95% CI of (−0.04%, 0.12%).

9.2.4 Assessment

Once benefits and risks have been defined, measured, and summarized, the next step is to define a metric or to transform the benefits and risks so that they can be weighed on a common scale. There are many strategies and mathematical transformations that can be implemented. However, care must be taken to retain a sense of clinical relevance when transformations are made.

Once benefits and harms are transformed to a common scale, there is the challenge of weighing them. How should we weigh a rare but serious side effect against a real but nonlife-threatening benefit in a large proportion of users? There is no single correct answer. At an FDA Advisory Committee Meeting (ACM), the committee was asked to evaluate the NeuroStar™ device for the treatment of major depressive disorder. The committee was asked if the device demonstrated "substantial equivalence" to a predicate device, electroconvulsive therapy (ECT), and an effective device that is known to carry increased risk of seizures. To demonstrate substantial equivalence, the NeuroStar did not need to be as effective as ECT, provided that the clinical data demonstrated that any reduction in effectiveness was offset by an improvement in patient safety/risk. Although this is a reasonable concept, the methods for weighing the benefits and the harms were left unspecified, resulting in a lack of clarity regarding the assessment.

Weighing the benefits and risks is partly subjective, and the relative weights can vary from person to person. Analyses should, therefore, allow weights to vary. Sensitivity analyses can help to evaluate how inference (and resulting decision-making) changes as weights vary across reasonable ranges. Such analyses will help individuals to make their own choices.

Context plays a critical role in determining weights. The disease, the individual patient, the population, and the availability of alternative interventions all play a role in defining the context in which a treatment decision will ultimately be made. Some toxicities may be acceptable for some indications but not others (e.g., headaches may be acceptable when treating cancers but not stomachaches). Patient-specific characteristics such as age can affect the willingness to accept risk. Population characteristics (culture, socioeconomic status, geographical region, and ethical norms) may also affect benefit:risk valuations. Weights further depend on the availability of alternative interventions and their respective benefit:risk profiles. Many of these context-defining factors can and will change over time, affecting any weights that may be assigned. As a result, the benefit:risk profile of an intervention is dynamic. Consider the benefit:risk profile of antimicrobials with the development of resistance. Years ago, penicillin and similar drugs were effective against staph infections (e.g., methicillin-resistant Staphylococcus aureus [MRSA]). However, these bacterial infections are now frequently resistant to these drugs. Thus, the benefit:risk profile of these drugs has changed within the context of these bacterial diseases. Benefit:risk profiles also change due to evolving medical practice and the availability of new interventions.

9.2.5 Combining Separate Marginal Analyses

Several approaches to benefit:risk evaluation have been proposed that follow the strategy of combining separate marginal analyses of benefits and risks. Proposed methods can be classified by two characteristics: (1) within-intervention versus comparative and (2) whether they reduce benefits and risks into one dimension versus multidimensional (e.g., bivariate approaches). Chuang-Stein et al. (2008) describes several of these approaches.

9.2.5.1 One Dimension: Within-Intervention Measures

Although the term *benefit:risk ratio* is frequently used to generically refer to the benefit:risk profile of an intervention, benefits and risks do not have to be exclusively viewed in a ratio metric. However, there is a statistic that can accurately be described as a benefit:risk ratio when benefit and risk can each be characterized as dichotomous (i.e., benefit vs. no benefit; harm vs. no harm) based on occurrence of specific events (Payne and Loken, 1975). The benefit:risk ratio for a particular intervention is the incidence of benefit divided by the incidence of risk. For example, Mascalchi et al. (2006), estimate the ratio of fatal cancers prevented to those induced via lung cancer screening devices in smokers, ex-smokers, and nonsmokers. For a male current smoker, the rate of fatal cancers induced by multi-detector computerized tomography (MDCT) is 11.7 per 100,000, while the rate of prevention of fatal cancers due to screening is 523 per 100,000, resulting in a risk:benefit ratio of $11.7/523 = 0.02$, or alternatively, a benefit:risk ratio of $523/11.7 = 45$. The advantage of the benefit:risk ratio is that it is easy to understand and to communicate. Its disadvantages include that the ratio does not consider the relative timing of the events, severity of the events, censoring of events from competing risks, and the challenge of relative interpretation (e.g., what magnitude constitutes acceptable, unacceptable, or neutralized risk?).

9.2.5.2 One Dimension: Comparative Measures

To address the challenge of relative interpretation, comparative measures of benefit:risk can be constructed. The number-needed-to-treat (NNT) is the expected number of patients that must be treated with an intervention in order to experience one additional occurrence of an event relative to a control group (Bender, 2005; Cook and Sackett, 1995). The event under consideration could be classified as benefit (NNTB) or harm (NNTH). Let Π_A and Π_B be the incidence rates for a beneficial outcome in intervention groups A (control) and B (intervention of interest). Then the NNT for benefit (or harm) is $NNT = 1/(\Pi_B - \Pi_A)$, where it is generally taken that $\Pi_B > \Pi_A$. So that NNT is expressed as a positive value. Thus, NNT is a comparative measure, with a comparison to a specific control. Consider a recent meta-analysis of the effect of aspirin for the prevention of cardiovascular events which found a highly significant reduction in nonfatal myocardial infarction from 0.23% to 0.18% per year in low-risk patients, implying that one nonfatal multiple imputation (MI) is prevented for every $1/(0.23\%-0.18\%) = 2000$ patients taking aspirin regularly (ATT 2009). Although the NNT is primarily used for binary endpoints, there have been extensions to other scales of measurement (Sackett et al., 1991; Altman and Andersen, 1999; Savor, 2004; Froud et al., 2009; Haas et al., 2010) and to further include both intervention benefits and harms, resulting in the concept of "unqualified success" and "unmitigated failure" (Mancini and Schulzer, 1999). The probability of unqualified success refers to the probability that a patient will experience treatment benefits without harms while the probability of unmitigated failure refers to the probability that a patient will experience harms, but no benefits. The simple interpretation of NNT is attractive; however, there are concerns with the use of NNT for statistical inference. For instance, when no significant treatment effect is found, the resulting CI for NNT is often difficult to interpret, as it will contain positive values, negative values, and infinity; and if not reported correctly, may appear to exclude the estimated NNT (Altman, 1998; Lesaffre and Pledger, 1999; Hutton, 2000; Walter, 2001; Walter and Irwig, 2001; Grieve, 2003; Thabane, 2003; Duncan and Olkin, 2005; Alemayehu and Whalen, 2006; McAlister, 2008).

The benefit:risk index is a measure that is calculated using a strategy that utilizes weights to combine benefits and risks (Freedman et al., 1996). The approach was utilized in a randomized controlled trial evaluating the effects of hormone replacement therapy (HRT) as part of the women's health initiative (WHI). HRT can prevent menopausal symptoms but can also have harmful effects. The strategy was to estimate the effects of HRT on each of five endpoints of interest (heart disease, breast cancer, endometrial cancer, hip fracture, and mortality). The effects would then be combined using a weighted composite benefit:risk index:

$$W = w1d1 + w2d2 + \cdots + w5d5$$

where di is the estimated difference in proportions for outcome i. Using mortality as a reference outcome, the trial used weights of 0.50, 0.35, 0.15, 0.18, and 1.0 for heart disease, breast cancer, endometrial cancer, hip fracture, and mortality, respectively. However, sensitivity analyses to varying weights could be conducted to evaluate the robustness of the results or to evaluate results for people with differing values (weights).

A common form of weighting benefits and risks on the same scale is through the use of quality-adjusted life-years (QALYs) that is often used in the field of multi-criteria decision analysis (MCDA). QALYs are estimated using weights reflecting the quality or desirability of being in a particular health state (e.g., 0 = death, 1 = perfect health), and can be used as a common currency with which to compare intervention options. Weights or utilities are subjective, and a matter of preference and thus sensitivity analyses to varying weights is crucial. Minelli et al. (2004) utilized MCDA to evaluate the benefits (e.g., relief of menopausal symptoms, prevention of hip fractures, and decreased risk of colorectal and endometrial cancers) and harms (e.g., increased risk of breast cancer, CHD, stroke, and pulmonary embolism) of HRT. The net benefit:harm was calculated as the sum of the benefits minus the sum of the harms. The sums of the benefits and harms were defined as a function of the change in risk associated with treatment, the quality-of-life weighting associated with a particular outcome, and the probability of death associated with the outcome (see the schema in Figure 9.10). The net effect on QALYs, as well as the probability of a positive overall effect, was then plotted versus baseline risk, and it was shown that the results depended heavily on the value placed on the relief on menopausal symptoms.

Regulators, particularly at the Center for Devices and Radiological Health (CDRH) have considered the use of MCDA for analysis and review of regulatory submissions. Advantages of MCDA include an increased transparency in evaluation and a deeper understanding of the complexity of decision-making via dissection of the potential outcomes and discussion of the desirability of those outcomes (Hughes et al., 2007; Temple, 2007).

Other weighting methods include the *principle of threes*, which is based on multiplying together scores which assess the severity, duration, and incidence in order to evaluate the disease, the intervention benefits, and

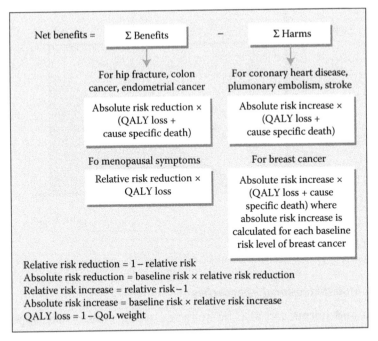

FIGURE 9.10
Comparison of net benefits and net harms of HRT through the usage of quality-adjusted life-years.

the AEs. In this approach, the disease severity and duration are subjectively categorized as high, medium, or low, with (arbitrary) scores assigned to each category. Similarly, the transparent uniform risk benefit overview (TURBO), a quantitative and graphical approach to benefit:risk assessment, proposes the calculation of both "B" and "R"-factors, each on a 7-point scale, which incorporates the relative likelihood and degree of intervention benefit and risk, respectively. A given intervention's final location on the 7×7 grid (Figure 9.11) is used to determine the intervention's overall "therapeutic score," which can then be compared to other interventions for the same disease (European Medicines Agency [EMEA] Committee for Medicinal Products for Human Use [CHMP], 2008).

Wittkowski (2004) proposed a rank-based approach that can be useful when patients can be ranked with respect to each of benefits and risks. Patients can then be ranked with respect to combined benefits and risks based on their benefit-specific and risk-specific ranks. Ranks can then be compared between interventions. Challenges with this approach include understanding the clinical relevance of the effects, handling ranking-ties, and identifying a decision rule for the relative ranking of two patients when each is ranked higher than the other on one of the two scales.

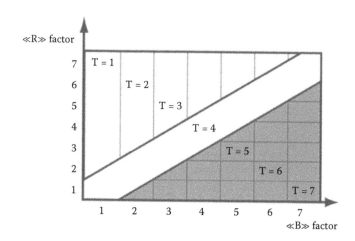

FIGURE 9.11
TURBO grid, used to assess an intervention's relative benefits and risks on a seven-point scale.

9.2.5.3 Multidimensional Approaches

Benefits and harms can also be analyzed using multivariate approaches (e.g., a bivariate approach). Consider a randomized trial comparing the benefits and harms of two interventions. If benefits and harms can each be dichotomized (i.e., each patient can be characterized as experiencing benefit or no benefit and harm or no harm), then a between-group difference in the proportion of patients with benefits and a between-group difference in the proportion of patients with harms can be calculated and plotted on a two-dimensional plot (one dimension for the between-group difference in proportions with benefit and the second dimension for the between-group difference in the proportions with harms). A two-dimensional "confidence region" (i.e., analogous to a standard one-dimensional CI) can then be constructed for this bivariate estimate using an assumption of multivariate normality or with other advanced statistical methods (e.g., bootstrapping). These analyses were conducted for the PROPHET trial; a randomized trial comparing hydrocortisone versus placebo for very-low-birth-weight babies to prevent chronic lung disease. "Benefit" was defined as survival without supplemental oxygen at 36-week postmenstrual age while harm was defined as GI perforation (Shaffer and Watterberg, 2006). Results are shown in Figure 9.12.

Such a two-dimensional approach can be used for coprimary hypotheses for benefit and risk outcomes. For example, if the goal of a trial is to claim "noninferior with respect to efficacy and superior with respect to safety," then the associated hypotheses could be formulated as follows:

H0: Not "noninferior with respect to efficacy and superior with respect to safety"

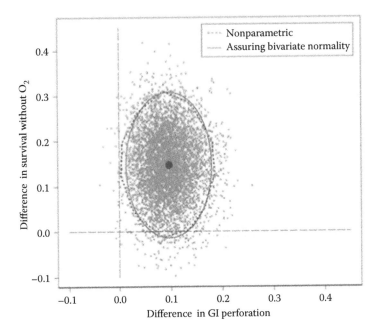

FIGURE 9.12
Confidence region for the joint effect of treatment on risks (x-axis) and benefits (y-axis). (From Shaffer, M.L., and Watterberg, K.L., 2006, *BMC Med Res Methodol*, 6:48.)

HA: "Noninferior with respect to efficacy and superior with respect to safety"

Test statistics can then be developed to test this joint hypothesis. Just as the rejection of a one-dimensional null hypothesis corresponds to a CI which excludes a particular value of interest, the rejection of this complex null hypothesis can be envisioned as a confidence region which excludes particular horizontal and vertical lines on a two-dimensional plot, as displayed in Figure 9.12 for the PROPHET trial.

Extending the concept of a two-dimensional confidence region, risk–benefit contours can also be constructed in two dimensions. Consider again a randomized trial comparing the benefits and harms of two interventions where benefits and harms are each dichotomous. A confidence level that the difference in proportions experiencing benefit (or harms) is at least X%, can be constructed for any X. For illustration, consider a two-sided 95% confidence interval (CI) for the between-group difference in the proportions of patients with benefits. Then, there is 97.5% confidence that the effect is greater than the lower-bound of the CI (and 2.5% confidence that the effect is greater than the upper-bound). Similar confidence levels could be constructed for the difference in harms. Then, by assuming a correlation between benefits and risks, a risk:benefit contour can be constructed. Shakespeare et al. (2001) utilized this approach to compare chemoradiotherapy versus radiotherapy

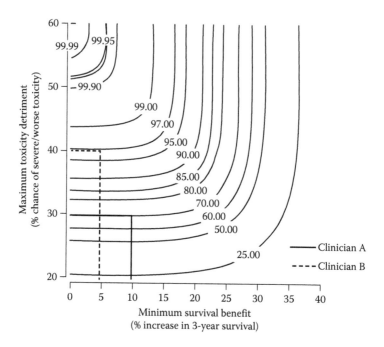

FIGURE 9.13
Two-dimensional confidence region expressed via significance contours.

alone with respect to survival (benefit) and acute toxicity (harm) for the treatment of locally advanced nasopharyngeal carcinoma. In Figure 9.13, we see that clinician A's desired treatment effect, a minimum 10% increase in survival with a maximum 30% increase in toxicity, cannot be confirmed with 95% confidence (i.e., there is only 70% confidence in this result), while clinician B's desired treatment effect (>5% survival increase, <40% toxicity increase) is plausible with a confidence level >95%.

9.2.6 Within-Patient Analyses

When an intervention is applied in clinical trials, some patients may benefit while some may experience toxicity. Are the patients who experience benefits also the patients who experience toxicity? The association between benefits and risks often goes without evaluation but should be considered an important part of benefit:risk evaluation. If the group of patients experiencing toxicity and the group of patients experiencing benefit are largely disjoint, then it is important to apply methods to distinguish these two groups from one another. The value of the intervention will depend on the ability to identify and avoid intervention use in the first group while targeting the intervention towards the second group. However, if these two groups are largely overlapping, then a thorough assessment of the relative risks and benefits must be conducted in order to determine the net effect of the intervention. In this case, the value of the intervention will depend upon whether the benefits outweigh the risks and upon identifying which patients have values or preferences that would result in an overall positive net benefit:risk profile.

Currently, the most common approach to the analysis of clinical trials is to conduct a marginal (population-averaged) analysis of efficacy and a separate marginal analysis of safety. The two marginal analyses are then combined but often in an unstructured manner. However, this common practice does not address the issue of the association between benefits and risks and thus does not distinguish between the two scenarios described above. Evaluating the association between benefits and harms is important for medical decision-making since individual patients are treated based on anticipated benefits and harms on an individual level. One approach to addressing these issues is to modify the order of operations during the statistical analyses. While the common practice is to construct separate summary measures of efficacy and safety before integrating the benefits and risks at the population level, a within-patient analysis instead integrates or combines the efficacy and safety outcomes at the patient level and then a summary of the combined benefit:risk outcome is created for each intervention. Interventions can then be compared with respect to the combined benefit:risk outcome. In other words, the traditional approach aggregates data for each intervention, compares the interventions, and then integrates the comparison of the beneficial and harmful outcomes. The within-patient approach first integrates beneficial and harmful outcomes, aggregates these data for each intervention, and then compares the interventions. Although this approach may not replace the traditional separate marginal analyses, it could be considered as supporting analyses.

Within-patient analysis essentially consists of creating a composite outcome for each trial participant consisting of benefit and risk components. Composite endpoints can be attractive since they can provide a more comprehensive characterization of the intervention effect. Strategies for developing a composite endpoint include linear combinations, composite event-time endpoints, use of two-way tables for ordinal data, and use of scatterplot methods for continuous data.

9.2.6.1 *Linear Combinations*

Chuang-Stein (1994) proposed a benefit-less-risk measure that discounts benefits for the presence of harms at a patient-level according to a prespecified algorithm: $BLR_j = B_j - f(R_j)$, that is, for any particular patient, the benefit-less-risk score is the benefit score minus a transformation of the harm score, where the transformation is meant to put benefits and risks on a comparable scale. The measure is simple and intuitive but reducing benefits and harms each to a single score is often very challenging as is identifying an appropriate transformation.

9.2.6.2 Composite Event-Time Endpoints

The commonly used techniques for event-time data may be applied to assess both benefits and risks of treatments. This is accomplished by redefining the event time of interest to be the time of "treatment failure," where failure may refer to any negative patient outcome. For example, Schneider et al. (2006) evaluated the effectiveness of atypical antipsychotic drugs in patients with Alzheimer's disease using an event-time endpoint of the minimum of (1) the time to discontinuation of treatment due to lack of efficacy or (2) the time to discontinuation of treatment due to toxicity or adverse effects. Many trials within the AIDS Clinical Trials Group (ACTG) evaluating the benefits and harms of antiretrovirals utilize the concept of "treatment failure," defined as the time from randomization to the first of any of the following events: death, disease progression, or virologic failure. In the evaluation of the cardiovascular drug Vorapaxar, Merck utilized the *net clinical outcome* defined as the time to the first of death, myocardial infarction (MI), stroke, or a serious bleeding event (a composite of benefits and harms). When using this approach, researchers must be careful with the interpretation when the components of the composite have differing levels of importance. One intervention may be associated with shorter event times, but the events could be less serious than those observed with a comparator intervention, creating interpretation challenges.

Follman (2002) used the idea of a composite outcome to rank all participants' clinical histories in a trial with respect to an overall benefit:risk outcome using a combination of event times. For example, participants were first ordered with respect to overall survival time. All participants surviving till the end of the trial would then be ranked according to their time of first hospitalization. Once such a ranking system has been established, it is possible to compare rankings between two interventions. A challenge of this rank-based method is evaluating the clinical relevance of effects.

9.2.6.3 Ordinal Data

Regulators in Japan have historically utilized a composite global utility rating of clinical usefulness. In this approach, overall safety for each patient was rated on an ordinal scale (e.g., safe vs. minor safety issues vs. moderate safety concerns vs. not safe) while overall effectiveness for each patient was rated on an ordinal scale (e.g., much improved, moderately improved, slightly improved, no change, or got worse). The safety and effectiveness ratings could be viewed in a two-dimensional cross-tabulation. Each cell of the table was then assigned an overall usefulness rating based on desirability (e.g., very useful, useful, slightly useful, not useful, or undesirable for use). For example, for a patient with much improvement and for whom the intervention was deemed safe, then the intervention was deemed very useful. Numerical scores could then be assigned to these ratings and interventions could then be compared. Boers et al. (2010) applied this strategy as part of the outcome measures in rheumatology (OMERACT) initiative.

Chuang-Stein et al. (1991) proposed the global benefit:risk score, a multinomial outcome based on efficacy and safety data. For example, the following ordered categories may be defined (in order of desirability): efficacy without serious side effects, efficacy with serious side effects, no efficacy with a lack of serious side effects, no efficacy with serious side effects, and toxicity leading to dropout. An algorithm could be created to classify each patient objectively or a blinded adjudication committee could be used. Weights could be assigned to indicate the relative importance of each category or a distance metric could be specified to indicate the relative desirability of the categories. A summary measure could then be computed for each intervention, and between-intervention comparisons could be made. Pritchett and Tamura (2008) applied this strategy to define the primary endpoint in a trial comparing two antidepressants by using remission status to categorize benefit and four possible AE outcomes to describe risk (no AEs, mild or moderate AEs, SAEs, or discontinued due to AEs). Norton (2011) extended these methods and developed a graphical display that summarizes aggregate treatment effects, within-patient changes longitudinally, and the temporal relationship and association between benefits and harms.

Ranks could be used as an alternative to assigning weights to the ordinal outcome. Evans (2014) and Evans et al. (2015) developed the desirability of outcome ranking (DOOR) as a composite benefit:risk measure utilizing the ordinal outcome proposed by Chuang-Stein et al. (1991) (i.e., efficacy without serious side effects, efficacy with serious side effects, no efficacy with a lack of serious side effects, no efficacy with serious side effects, and toxicity leading to dropout). A specific example was the response-adjusted for duration of antibiotic risk (RADAR), an application of DOOR to antibiotic stewardship trials when evaluating whether new stewardship strategies could reduce antibiotic use but not at the expense of clinical outcome. Ranks were based on two rules: (1) when comparing two patients with different clinical benefit:risk outcomes, the patient with the better clinical benefit:risk outcome receives a higher rank and (2) when comparing two patients with the same clinical benefit:risk outcome, the patient with a shorter duration of antibiotic use receives a higher rank.

9.2.6.4 Scatterplot Methods for Continuous Data

Assume that benefits and risks can each be measured on a bounded continuous scale (e.g., 0–10) or can be transformed as such. A scatterplot of patient scores could then be plotted. For any particular patient coordinate (i.e., bivariate score), the distance to the "ideal outcome" (i.e., B [benefit] = 10 and R [risk] = 0) could be constructed, effectively reducing the patient outcome to a single dimension (Figure 9.14a). The distances could then be summarized by intervention and compared. An extension to this approach could be made to incorporate the potential differential importance of one unit on the benefit versus risk scale by defining a tolerability curve through two or more points. For example, the curve could be drawn through (R = 0, minimum tolerable benefit when R = 0) and (maximum tolerable risk when B = 10, B = 10). Suppose we define R10 to be the maximum tolerable risk when B = 10 and B0 to be the minimum tolerable benefit when R = 0. Then a curve could be drawn, connecting the points (R10, 10) and (0, B0). The tolerability curve would divide the plane into two regions, one where benefits outweigh risks and the other where risks outweigh benefits. All points on the tolerability curve would have an equivalent benefit:risk profile. The tolerability curve could then be used to standardize the calculated distances. For example, if a is the distance between the observed bivariate benefit and risk outcome and the ideal outcome, and b represents the distance between the point of intersection between the tolerability curve and a straight line from the observed outcome and the ideal outcome (Figure 9.14b), then the standardized distance

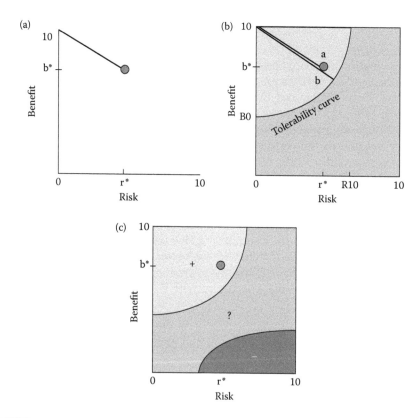

FIGURE 9.14
(a) Distance from actual patient outcome (r*, b*) to ideal patient outcome (0, 10); (b) tolerability curve representing range of responses (light gray) where benefits outweigh risks (i.e., a/b < 1); (c) ordered ranges of desirability, such that darker regions represent worse patient outcomes.

is a/b. Thall et al. (2006) used this approach for adaptive dose selection. Sensitivity analyses could be performed to investigate the robustness of results as the definition of the tolerability curve changes.

If benefits and risks can each be measured on a bounded continuous scale (e.g., 0–10) as above, then a plane of plotted scores could also be separated into ordinal regions of desirability, preferably according to prespecified definitions (Figure 9.14c). Between-intervention comparisons of the proportions of patients that fall into these regions could then be compared using methods that compare ordinal measures. Sensitivity analyses could be performed as definitions of the regions vary.

9.2.6.5 Adjudication Committee (AC) Approach

Another approach is to utilize blinded Adjudication Committees consisting of disease experts to provide an overall rating for the response for each patient using an appropriate disease-specific scale defined by benefits and harms. Important data for each patient is sent to the AC for review. Although it is important to clearly and objectively define patient outcomes, the AC approach can be an attractive approach when a clear algorithm for patient response cannot be clearly identified or when there are many unexpected events. However, this approach takes considerable planning and can be impractical for very large trials.

9.2.7 Tailored Medicine

The effectiveness of interventions with respect to risks and/or benefits often varies by patient characteristics. Although some patients benefit from interventions, others do not, and still others experience toxic side effects. Many drugs prescribed in the United States today are effective in <60% of treated patients, reflecting the variability of metabolism or other factors from person to person (Federal Coordinating Council for Comparative Effectiveness Research, HHS [FCC for CER], 2009). Twelve million women regularly take statins for high LDL cholesterol; however, statins carry a risk for serious side effects, for example, diabetes, while uncertainty remains as to whether statins prevent heart disease in women (O'Callaghan, 2010). Adverse drug reactions account for 100,000 patient deaths annually in the United States, represent 100 billion dollars in health care costs, account for 2 million hospitalizations, and are the fourth leading cause of mortality. In the United Kingdom, the average patient with blood pressure problems is prescribed four drugs before the optimal drug is identified. Estimates show that 20%–40% of all patients may either be on the wrong drug or the wrong dose (Clarke, 2010).

However, trial-and-error and one-size-fits-all approaches to intervention evaluation remains a common medical problem (FCC for CER, 2009). Traditionally, analyses of clinical trials have focused on the average effects across broad populations. Clinicians then frequently make patient-specific decisions based on population-averaged effects observed and reported in studies. In fact, generalizations that result from comparisons that fail to consider subgroups and individual differences may have limited applicability (FCC for CER, 2009). Making patient-specific decisions based on estimated population-averaged effects can lead to suboptimal patient care (Kent and Hayward, 2007) and contributes to inefficiencies in public health practice. At Personalized Medicine 2010, FDA Commissioner Dr. Margaret Hamburg stated the need for new, comprehensive, and integrated approaches (i.e., trial design, flexible regulatory path) for personalized therapies. NIH Director Dr. Francis Collins once stated "you wouldn't go into a shoe store and just pick shoes off the rack without noticing if they were your size and yet, oftentimes with medical recommendations, this is kind of what we do," concluding that it is necessary "to come up with recommendations that are more personalized." It is important to evaluate the effects of interventions at both the individual and population levels (IOM [Institute of Medicine], 2009).

Recent advancements in science and technology have led to the discovery of many biological and genetic markers associated with intervention responses. These new markers, combined with traditional clinical assessments, hold great potential for identifying subgroups of patients who are most likely to benefit or are at high risk for toxicity from a particular intervention. This information can be used to identify the most effective clinical options for individual patients by tailoring the patient-specific selection of treatment, diagnosis, and prevention strategies, based on an individual's own characteristics, so that the optimal benefit is realized and undesirable side effects and costs are minimized. In this realm of personalized or tailored medicine, interventions can then be targeted to well-defined groups that are likely to benefit and are at low risk for toxicity. For example, drug treatments for several major diseases are now recommended or required to have companion diagnostics to aid in patient-specific treatment decisions.

- *HIV:* CCR5 receptor status for Maraviroc and HLA-B*5701 for Abacavir
- *Cancer:* K-ras status for Vectibix and new biologics use in colorectal cancer; Philadelphia chromosome positivity for Dasatinib use in chronic myeloid leukemia; Her2/neu expression for Herceptin use in breast cancer
- *Cardiovascular disease:* CYP450 2C9 and VKORC1 tests for Warfarin use in blood clotting

The need to identify clinical heterogeneity and address the needs of emerging patient subgroups will grow as scientific knowledge of genomics, and molecular medicine improves and becomes an integral part of health care. Standard approaches to identifying clinical heterogeneity in clinical trials are fraught with challenges. The methodological challenges to tailored medicine are similar to those relating to subgroup analyses. Many issues are related to problems of "multiplicity," due to poor reporting of analyses (i.e., reporting only positive results without describing all analyses conducted). As a result, type I (false positive) error control is lost and p-values fail to retain their intended interpretation. For this reason, there is considerable need for clarity between hypothesis generation ("fishing for results") versus hypothesis confirmation, as unsurprisingly many subgroup results, have proven unable to be validated (Pfeffer and Jarcho, 2006). Additional concerns involve interaction testing, the most common method for assessing heterogeneity in treatment effects. Such interaction testing typically increases reliance on (often unverifiable) modeling assumptions, and also suffers from increased type II error (false negatives) due to small subgroup sizes or the increased variability (and thus decreased power) that is inherent to any assessment of a "difference of treatment differences." Even further complications occur when higher-order interactions (subgroups defined by combinations of multiple variables) must be introduced or when subgroups are defined based on factors measured after the initiation of the intervention.

9.3 Quality of Life

QoL is an important evaluation in many clinical trials in many therapeutic areas such as rheumatology, neuroscience, cardiology, immunology, respiratory diseases, urology, ophthalmology, and oncology. Broadly speaking, QoL concerns (1) how a patient feels (e.g., emotional well-being, pain, and energy) and (2) how a patient functions. It encompasses physical functioning (e.g., self-care activities such as bathing, social functioning, ability to work, ability to exercise, and ability to sleep) and neuropsychological functioning (e.g., depression, anxiety, memory, recognition, spacial skills, and motor coordination). QoL assessments are patient-centered focusing on the illness rather than the disease.

Patients and clinicians may not agree when assessing the outcomes of an intervention. In a study of hypotensive drugs, improvement was concluded when evaluation was judged by clinicians as they primarily base their evaluation on blood pressure control. However, patient evaluations revealed a reduction in QoL (Jachuck et al., 1982). Furthermore, when a physician makes an assessment, the physician compares the patient's current disease status with other patients with whom the physician is familiar or with published cases. However, when the patient makes an assessment, the patient is generally using his or her own personal history as a reference. These assessments are called *patient-reported outcomes* (PROs). Advocates of PROs often state that no one knows the patient's condition better than the patient herself/himself does, and thus such outcomes should be considered as extremely important when evaluating interventions.

9.3.1 QoL Instruments

Instruments that collect QoL data can be self-administered or interviewer-administered. Interviewer-administration tends to reduce missing data, but there can be reluctance on behalf of the participant to answer questions of a personal nature (e.g., regarding sexual function or depression). Illiteracy may force the use of interviewer-administered instruments in some trials. To reduce interviewer bias and interviewer variation, interviewer training is very important but can also increase costs. Both methods can be used to understand the varying perceptions of clinicians and trial participants.

QoL instruments that collect QoL data may be generic (i.e., have applicability to broad populations) or may be a condition/population-specific. Examples of generic instruments include the McGill pain questionnaire (MPQ), the profile of mood states (POMS), the health utilities index (HUI), the health survey (SF-36), the sickness impact profile (SIP), the assessment of motor and process skills (AMPS), Centers for Epidemiological Studies-Depression (CES-D), the Pittsburgh sleep quality index (PSQI), the state-trait anxiety index (STAI), and the functional activity questionnaire (FAQ). An example of a condition-specific questionnaire was developed to measure EFV-related symptoms as a part of A5097s, a trial conducted by the ACTG. The questionnaire focused on toxicities that were reportedly related to EFV use (e.g., sleep disturbance and neurological symptoms such as dizziness).

Many QoL instruments consist of several questions. Responses are often measured on (1) a Likert scale (e.g., 0 = "not at all," 1 = "slightly," 2 = "moderately," 3 = "quite a bit," and 4 = "greatly"), (2) a visual analog scale (VAS) (i.e., a measured distance on a line segment with "not at all" on one end and "greatly" on the other), or (3) a dichotomized scale (i.e., yes vs. no). It is important to obtain measurements at baseline and the end of the trial to fully understand participant response.

Typically questions are grouped into domains. For example, OA is a disease that causes inflammation and pain at one of the major joints (e.g., the hip or knee joint). The Western Ontario and McMaster Universities Arthritis Index (WOMAC) (Bellamy, 1995) is a QoL index for patients with OA. The WOMAC consists of three domains: pain, stiffness, and physical function. The pain domain asks the question "How much pain do you have?" for five situations:

1. Walking on a flat surface
2. Going up or down stairs
3. At night while in bed
4. Sitting or lying
5. Standing upright

There are two questions in the stiffness domain and 17 questions in the physical function domain for a total of 24 questions. Responses to each question involve a five-point Likert scale: "None," "Mild," "Moderate," "Severe," and "Extreme" with a score of 0, 1, 2, 3, and 4 assigned to each of these ratings, respectively. A domain score is calculated by summing the scores from each domain. Thus, the pain domain score ranges from 0 to 20, the stiffness domain score ranges from 0 to 8, and the physical function domain score ranges between 0 and 68, with higher scores indicating a lower quality of life. Thus, for a trial participant, improvement is reflected in a reduction of the domain score. The total WOMAC score and each domain score can be evaluated in clinical trials. Such summary scores are typical for many QoL instruments.

Criticisms of the use of these scores include the fact that they are usually constructed to provide equal weight to each question, but some questions may be more important than others. Interpreting the numbers can also be challenging (i.e., how the numbers relate to function, etc.). Scales often need calibration with other measures for which the clinical impact is known. QoL instruments should be carefully validated to ensure important properties including content validity (the extent to which the items reflect the intended domain of interest) and reliability (the extent to which results are repeatable and consistent under similar conditions).

9.3.2 Issues in Design and Analyses

QoL measures are often used as secondary endpoints in clinical trials given their subjective nature. However, patients often perceive QoL measures to be more important than a trial's primary endpoint. For example, trials evaluating interventions to treat asthma or chronic obstructive pulmonary disease (COPD) typically use a primary endpoint of forced expiratory volume at 1 s (FEV_1). However, patients may be more interested in how well they feel, how difficult breathing is, and how well they can function with respect to daily activities (e.g., whether they can work or exercise). For this reason QoL measures have become increasingly utilized as primary endpoints in trials, for example, QoL measures have been the primary endpoint in recent trials evaluating interventions for erectile dysfunction, overactive bladder, OA, and fibromyalgia. Evans et al. (2007) used pain as a primary endpoint in a trial evaluating an intervention for the treatment of painful peripheral neuropathy.

Special considerations are necessary when designing and analyzing studies with QoL/PRO evaluations since data are primarily subjective. For example, pain is a particularly important QoL/PRO measure as high levels of pain can have dramatic effects on patient function and well-being. Pain is also particularly challenging to measure given its potential to be influenced by placebo effects. The initiative on methods, measurement, and pain assessment in clinical trials (IMMPACT) is a group that strives to develop consensus reviews and recommendations for improving the design, execution, and interpretation of clinical trials of treatments for pain (www.immpact.org). More broadly the Patient Reported Outcomes Harmonization Group conducts efforts to harmonize the use of patient reported outcomes (http://www.eriqa-project.com/pro-harmo/home.html).

When conducting multinational trials, one needs to be aware of translation issues with QoL instruments. Language limitations may result in a question being asked in a slightly different way. Careful consideration of cultural differences is important as well. Culture may affect the acceptability of answering a question with a specific response. Culture can also influence response variation. Robertson et al. (2012) reported the results of the semantic verbal fluency test (a test which asks participant to name as many, for example, boys' names as possible in a timed period) as part of a comprehensive neuropsychological evaluation. Some cultures have only a few boys names whereas others have many. Responses in cultures with few boys names have very little variation. This can make responses to these tests challenging to interpret.

9.3.3 Patient Preferences

A particularly challenging issue when analyzing QoL data is that the preferences or weights that patients and trial participants put on various aspects of well-being and function may vary from person to person. Evaluation of trial outcomes as the weights on important factors vary can be conducted.

The QoL time without symptoms or toxicity (Q-TWiST) method was originally used to evaluate interventions for cancer when survival time was a primary endpoint (Gelber et al., 1989; Glasziou et al., 1990). Q-TWiST is a two-step process. In the first step, overall survival is partitioned into three clinical disease states: time with toxicity due to therapy (TOX), time without toxicity or progression (TWiST), and time after progression (REL). In the second step, interventions are compared with respect to a survival time where the three clinical disease states are weighted by desirability. The time without toxicity or progression is typically given full weight (w = 1) but the toxicity time, and the time after progression are down-weighted.

$$Q\text{-}TWiST = (wtox \times TOX) + (1 \times TWiST) + (wrel \times REL)$$

$$\text{where} \quad 0 \leq wtox \leq 1 \text{ and } 0 \leq wrel \leq 1.$$

Noting that the weights may vary depending on patient preference or clinical opinion, sensitivity analysis is then conducted across all possible choices of weights. For example, a patient that treasures every moment of life may select weights of wtox = wrel = 1 (i.e., no down-weighting) whereas a patient that does not wish to endure any suffering may select weights of wtox = wrel = 0. A threshold utility plot displays contours of similar treatment effects (e.g., adjusted-months of survival gained or lost) as a function of the selected weights. The vertical axis of the plot is the weight of toxicity (wtox) while the horizontal axis is the weight for the time after relapse (wrel). Thus, an individual patient and his or her clinician can select their own weights and select an intervention strategy accordingly. Q-TWiST was applied in the Eastern Cooperative Oncology Group (ECOG) trial 1684 comparing interferon Alpha-2b versus observation for high-risk resected melanoma, where treatment was estimated to generally improve Q-TWiST, but significant differences were seen only in those patients whose subjective values wtox are sufficiently large and sufficiently wrel small (Cole et al., 1996). Figure 9.15 provides an example of such an analysis.

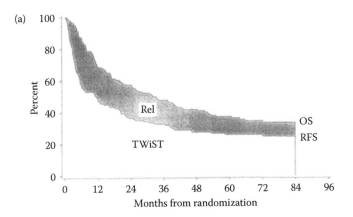

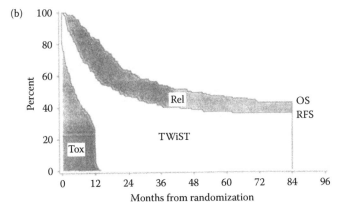

FIGURE 9.15
From (a) to (b): Survival plots for control and active treatment, partitioned into the three components of Q-TWiST (Tox, TWiST, and Rel), and net "utility" gain (in months) due to treatment, conditional on specific values of utility weights for progression/relapse and toxicity. A more general form of these analyses has created the measure quality-adjusted life-years (QALYs) but weighting various disease states of patients.

References

Alemayehu D, Whalen E. 2006. A new paradigm for deriving and analyzing number needed to treat. *J Biopharm Stat* 16(2):181–92.
Altman DG. 1998. Confidence intervals for the number needed to treat. *Br Med J* 317:1309–1312.
Altman DG, Andersen PK. 1999. Calculating the number needed to treat where the outcome is time to an event. *Br Med J* 319:1492–1495.
Antithrombotic Trialists' (ATT) Collaboration. 2009. Aspirin in the primary and secondary prevention of vascular disease: Collaborative meta-analysis of individual participant data from randomised trials. *Lancet* 373:1849–1860.
Bellamy N. 1995. *WOMAC User's Guide*, Suite 303, Colborne Building 3, Victoria Hospital, 375 South Street, London ON N6A 4G5, Canada.
Bender R. 2005. Number needed to treat (NNT). In *Encyclopedia of Biostatistics*, 2nd edn., P. Armitageand T. Colton (eds.). Chichester, UK: Wiley. Vol. 6, pp. 3752–3761.
Boers M, Brooks P, Fries JF, Simon LS, Strand V, Tugwell P. 2010. A first step to assess harm and benefit in clinical trials in one scale. *J Clin Epidemiol* 63:627–632.
Chuang-Stein C. 1994. A new proposal for benefit-less risk analysis in clinical trials. *Controlled Clin Trials* 15:30–43.
Chuang-Stein C, Entsuah R, Pritchett Y. 2008. Measures for conducting comparative benefit:risk assessment. *Drug Inf J* 42:3:223–233.
Chuang-Stein C, Mohberg NR, Sinkula MS. 1991. Three measures for simultaneously evaluating benefits and risks using categorical data from clinical trials. *Stat Med* 10:1349–1359.
Clarke B. February 2010. Getting personal. *Int Clin Trials* (15):16–21.
Clifford DB, Evans SR, Yang Y, Acosta E, Goodkin K, Tashima K, Simpson D, Dorfman D, Ribaudo H, Gulick RM. 2005. Impact of Efavirenz on neuropsychological performance and symptoms in HIV-infected individuals. *Ann Intern Med* 143:714–721.
Cole BF, Gelber RD, Kirkwood JM. 1996. Quality-of-life-adjusted survival analysis of interferon alfa-2b adjuvant treatment of high-risk resected cutaneous melanoma: An Eastern Cooperative Oncology Group study. *J Clin Oncol* 14:2666–2673.
Cook RJ, Sackett DL. 1995. The number needed to treat: A clinically useful measure of treatment effect. *BMJ* 310:452–454.
Duncan BW, Olkin I. 2005. Bias of estimate of number needed to treat. *Stat Med* 24:1837–1848.
Eichler HG, Abadie E, Raine JM, Salmonson T. 2009. Safe drugs and the cost of good intentions. *NEJM* 360:14:1378–1380.
European Medicines Agency (EMEA) Committee for Medicinal Products for Human Use (CHMP). 2008. Reflection paper on benefit-risk assessment methods in the context of the evaluation of marketing authorization application of medicinal products for human use, London, 19 March 2008. Doc. Ref. EMEA/CHMP/15404/2007.
Evans SR. 2008. Benefit:risk evaluation in clinical trials. *Drug Inf J* 42(3):221.
Evans SR. 2014. *Using Endpoint Data to Analyze Patients Rather than Patient Data to Analyze the Endpoints*. Boston, MA: JSM.
Evans SR, Rubin D, Follmann D, Pennello G, Huskins WC, Powers JH, Schoenfeld D et al. 2015. Desirability of outcome ranking (DOOR) and response adjusted for

days of antibiotic risk (RADAR). *Clinical Infectious Diseases* 2015, doi: 10.1093/cid/civ495.

Evans SR, Simpson D, Kitch DW, King A, Clifford DB, Cohen BA, McArthur JC. 2007. A randomized trial evaluating Prosaptide™ for HIV-associated sensory neuropathies: Use of an electronic diary to record neuropathic pain. *PLoS ONE* 2(7):e551. doi:10.1371/journal.pone.0000551.

Federal Coordinating Council for Comparative Effectiveness Research, HHS (FCC for CER). 2009. Report to the President and the Congress. http://www.hhs.gov/recovery/programs/cer/cerannualrpt.pdf.

Fleming T. 2008. Identifying and addressing safety signals in clinical trials. *NEJM* 359(13):1400–1402.

Follman D. 2002. Regression analysis based on pairwise ordering of patients' clinical histories. *Stat Med* 21:3353–3367.

Freedman L, Anderson G, Kipnis V, Prentice R, Wang CY, Rossouw J, Wittes J, DeMets D. 1996. Approaches to monitor the results of long-term disease prevention trials: Examples from the women's health initiative. *Controlled Clin Trials* 17:509–525.

Froud R, Eldridge S, Lall R, Underwood M. 2009. Estimating the number needed to treat from continuous outcomes in randomised controlled trials: Methodological challenges and worked example using data from the UK Back Pain Exercise and Manipulation (BEAM) trial. *BMC Med Res Methodol* 9:35.

Gelber RD, Gelman RS, Goldhirsch A. 1989. A quality-of-life oriented endpoint for comparing treatments. *Biometrics* 45:781–795.

Gilbert GS, Ting N, Zubkoff L. 1991. A statistical comparison of durg safety in safety in controlled clinical trials: The genie score as an objective measure of lab abnormalities. *Drug Inf J* 25:81–96.

Glasziou PP, Simes RJ, Gelber RD. 1990. Quality adjusted survival analysis. *Stat Med* 9:1259–1276.

Grieve A. 2003. The number needed to treat: A useful clinical measure or a case of the Emperor's new clothes? *Pharm Stat* 2:87–102.

Haas M, Schneider M, Vavrek D. 2010. Illustrating rating risk difference and number needed to treat from a randomized controlled trial of spinal manipulation for cervicogenic headache. *Chiropr Osteopat* 18:9.

Hughes DA, Bayoumi AM, Pirmohamed M. 2007. Current assessment of risk-benefit by regulators: Is it time to introduce decision analysis? *Clin Pharmacol Ther* 82:2:123–127.

Hutton JL. 2000. Number needed to treat: Properties and problems. *J R Stat Soc* 163:403–419.

ICH E1. 1994. The extent of population exposure to assess clinical safety for drugs intended for long-term treatment of non-life threatening conditions, International Conference on Harmonization.

ICH E14. 2005. The clinical evaluation of QT/QTc interval prolongation and proarrhythmic potential for non-antiarrrhythmic drugs, International Conference on Harmonization.

ICH E2a. 1994. Clinical safety data management: Definitions and standards for expedited reporting, International Conference on Harmonization.

ICH 2c. 2012. Periodic benefit-risk evaluation report, International Conference on Harmonization.

ICH E2d. 2003. Post-approval safety data management: Definitions and standards for expedited reporting, International Conference on Harmonization.

ICH 2e. 2004. Pharmacovigilance planning, International Conference on Harmonization.

IOM. 2007. *Understanding the Benefits and Risks of Pharmaceuticals: Workshop Summary.* Washington, DC: The National Academies Press.

IOM (Institute of Medicine). 2009. *Initial National Priorities for Comparative Effectiveness Research.* Washington, DC: The National Academies Press.

Jachuck SJ, Brierly H, Jachuck S, Wilcox PM. 1982. The effect of hypotensive drugs on the quality of life. *J R Coll Gen Pract* 32:103–105.

Kent DM, Hayward RA. 2007. Limitations of applying summary results of clinical trials to individual patients: The need for risk stratification. *JAMA* 298:10:1209–1212.

Lagakos SW. 2006. Time-to-event analyses for long-term treatments—The APPROVe Trial. *NEJM* 355(2):113–117.

Lesaffre E, Pledger G. 1999. A note on the number needed to treat. *Controlled Clin Trials* 20:439–447.

Mancini GBJ, Schulzer M. 1999. Reporting risks and benefits of therapy by use of the concepts of unqualified success and unmitigated failure: Application to highly cited trials in cardiovascular medicine. *Circulation* 99:377–383.

Mascalchi M, Belli G, Zappa M, Picozzi G, Falchini M, Della Nave R, Allescia G et al. 2006. Risk-benefit analysis of X-ray exposure associated with lung cancer screening in the Italung-CT trial. *AJR Am J Roentgenol* 187:421–429.

McAlister FA. 2008. The "number needed to treat" turns 20—And continues to be used and misused. *CMAJ* 179(6):549–553.

Minelli C, Abrams KR, Sutton AJ, Cooper NJ. 2004. Benefits and harms associated with hormone replacement therapy: Clinical decision analysis. *BMJ* 328:371.

Norton JD. 2011. A longitudinal model and graphic for benefit-risk analysis, with case study. *Drug Inf J* 45:741–747.

O'Callaghan T. 2010. Are cholesterol drugs a good idea for healthy people? *Time*, March 2010, http://healthland.time.com/2010/03/31/are-cholesterol-drugs-a-good-idea-for-healthy-people.

O'Neil RT. 2007. A perspective on characterizing benefits and risks derived from clinical trials: Can we do more? *Drug Inf J* 42:3:235–245.

Payne JT, Loken. MK. 1975. A survey of the benefits and risks in the practice of radiology. *CRC Crit Rev Clin Radiol Nucl Med* 6:425–439.

Pfeffer MA, Jarcho JA. 2006. The charisma of subgroups and the subgroups of CHARISMA. *NEJM* 354;16:20:1744–1746.

Pritchett Y, Tamura R. 2008. The application of global benefit-risk assessment in clinical trial design and some statistical considerations. *Pharm Stat* 7:170–178.

Robertson et al. 2012. Improved Neuropsychological and Neurological Functioning Across Three Antiretroviral Regimens in Diverse Resource-Limited Settings: AIDS Clinical Trials Group Study A5199, the International Neurological Study. *CID* 55(6):868–876.

Sackett DL, Haynes RB, Guyatt GH, Tugwell P. 1991. *Clinical Epidemiology: A Basic Science for Clinical Medicine*, 2nd edn. Boston, MA: Little Brown.

Savor JL. 2004. Number needed to treat estimates incorporating effects over the entire range of clinical outcomes. *Arch Neurol* 61:1066–1070.

Schneider LS, Tariat PN, Dagerman KS. 2006. Effectiveness of atypical antipsychotic drugs in patients with Alzheimer's disease. *N Engl J Med* 355:1528–1538.

Shaffer ML, Watterberg KL. 2006. Joint distribution approaches to simultaneously quantifying benefit and risk. *BMC Med Res Methodol* 6:48.

Shakespeare TP, Gebski VJ, Veness MJ, Simes J. 2001. Improving interpretation of clinical studies by use of confidence levels, clinical significance curves, and risk-benefit contours. *Lancet* 357:1349–1353.

Siddiqui O. 2009. Statistical methods to analyze adverse events data of randomized clincial trials. *J Biopharm Stat* 19:889–899.

Temple R. 2007. Quantitative decision analysis: A work in progress. *Clin Pharmacol Ther* 82:2:127–130.

Thabane L. 2003. A closer look at the distribution of number needed to treat (NNT): A Bayesian approach. *Biostatistics* 4:3:365–370.

Thall PF, Cook JD, Estey EH. 2006. Adaptive dose selection using efficacy-toxicity trade-offs: Illustrations and practical considerations. *J Biopharm Stat* 16:623–638.

Walter SD. 2001. Number needed to treat (NNT) estimation of a measure of clinical benefit. *Stat Med* 20:3947–3962.

Walter SD, Irwig L. 2001. Estimating the number needed to treat (NNT) index when the data are subject to error. *Stat Med* 20:893–906.

Wittkowski K. 2004. Novel methods for multivariate ordinal data applied to olympic medals, risk profiles, genomic pathways, genetic diplotypes, pattern sensitivity, and array normalization. *Comput Sci Stat* 35:626–646.

10

Publishing Trial Results

Clinical trial results are communicated to the medical community primarily in the form of publications in the medical literature. Appropriate reporting of clinical trial results is crucial for safe and effective use of an intervention, and thus for ensuring optimal patient care. However, appropriate reporting is challenging, and poor reporting practices are very common. Reporting the results of clinical trials requires careful thought. The keys to appropriate reporting of clinical trials are *honesty* and *transparency*.

The results of clinical trials should be reported so that the medical community is aware of the results. However, a recent study by Ross et al. (2011) found that less than half of trials funded by the NIH are published in a peer-reviewed journal indexed by Medline within 30 months of trial completion and nearly a third of trials remained unpublished 51 months after trial completion. Prayle et al. (2011) reviewed trials with drugs that had received at least one FDA approval noting mandatory reporting of such trials by law, and found that 163/738 (22%) had reported results within 1 year of completion of the trial consistent with the law. This was compared with 76/727 (10%) for trials that were not subject to mandatory reporting. As a result of these studies, the U.S. Congress has contacted the FDA (responsible for enforcing the statutory drug trial reporting requirements) and the NIH (responsible for maintaining the clinicaltrials.gov database) regarding improvement of enforcing noncompliance and ways to improve compliance.

Selective reporting of results is a very common concern. One particularly problematic issue is the under-reporting or delayed reporting of negative evidence. If trial results are negative, researchers often elect not to publish them, partly perhaps, because medical journals consider the results unexciting and thus unworthy of publication. However, if several trials are conducted to evaluate the effectiveness of an intervention, and only one trial is positive, and furthermore, is the only trial that is published, then the medical community is left with a distorted view of the evidence of effectiveness of the intervention. Examples of selective/delayed reporting include the following:

- Results from a clinical trial of Vytorin, a popular cholesterol drug, indicated that Vytorin provided no benefit. However, publication of the results was delayed for years while the drug continued to be marketed until a congressional investigation in 2008.
- A clinical trial evaluating Multaq, a drug for the treatment of an irregular heartbeat was halted prematurely due to increased mortality vs.

placebo in 2003. However, trial results were not published until 5 years later.

- Independent meta-analyses of the diabetes drug Avandia found that the drug increased the rate of heart attacks and cardiovascular deaths. However, 35 of 42 studies included in the analyses remained unpublished.
- A clinical trial of Infuse, a bone growth stimulating back surgery product, was not published until 5 years after the trial was prematurely stopped due to unwanted bone growth near the spines of trial participants.

Bourgeois et al. (2010) found that trials funded by industry were less likely to be published within 2 years of study completion and were more likely to report positive outcomes than trials funded by other sources. Negative evidence should be reported with equal vigor in a timely fashion to ensure optimal patient treatment.

TABLE 10.1

Common Poor Reporting Practices

- *Failure to clearly state*:
  - The objectives
  - Justification for the sample size
  - Methods of randomization
  - Assumptions of the trial design or analyses
  - Analysis methods
  - Trial conclusions
  - Limitations of the trial
- *Failure to provide*:
  - Confidence intervals with p-values
  - Estimates of variation with point estimates
  - Multiplicity context (e.g., how many subgroups were evaluated)
  - Measures of relative and absolute risk
  - Summaries of all benefits and harms
  - A reference of comparison to aid in decision-making
  - Estimates of between-arm contrasts, in addition to arm-specific summaries
- Poor interpretation of p-values, for example, nonsignificant p-value implies "no effect" and significant p-values imply clinically important effects
- Use of too many decimal places (e.g., $p < 0.00001$) or stating $p = NS$ (not significant)
- Inconsistency across presented tables (e.g., different sample sizes)
- Failing to present all data on plots or in tables
- Providing voluminous tables when figures are more interpretable
- Failure to use confidence intervals to interpret "negative" results by ruling out effect sizes with reasonable confidence
- Selective reporting of outcomes, for example, reporting the results from positive secondary endpoints but not from negative primary endpoints
- Drawing conclusions that are not supported by the data (e.g., inferring causality from association; failure to account for important sources of bias; interpreting nonrandomized trials as randomized; interpreting small trials as large ones)

However, merely reporting trials is not enough. Many trials are reported with spin, often overstating the trial results or drawing conclusions that are not supported by the data. Boutron et al. (2010) reviewed trials reported in 2006 and found that in trials with nonsignificant primary outcomes, the reporting and interpretation of findings were frequently inconsistent with the trial results. Trial results must be reported in an honest and responsible manner. To address this concern, some journals are now request or even require submission of the data along with the manuscript. For example, the *British Medical Journal* (BMJ) announced that beginning in January 2013 it will no longer publish the results of a clinical trial unless the researchers agree to provide trial data upon request, hoping to prod industry to make data available so that independent researchers can vet the data. In 2013, a group of cardiovascular journals issued a statement urging investigators to use language consistent with the integrity of the study that was conducted (e.g., observational study vs. randomized controlled trial).

Some of the more common issues are listed in Table 10.1 (many from Altman, 1998; Altman and Bland, 1998; Berwick et al., 1981; Evans, 2010; Glantz, 1980; Horton and Switzer, 2005; Weiss and Samet, 1980; West and Ficalora 2007; Windish et al., 2007; Wulff et al., 1986), while suggestions to consider for improving the reporting are provided in Table 10.2; Bailar and Mosteller (1988) provided guidelines for statistical reporting in medical publications.

TABLE 10.2

Suggestions for Better Reporting

- Strive to describe and illuminate rather than to convince
- Provide estimates of variation along with estimates of the effect
- Provide confidence intervals with p-values or point estimates of the effect
- Provide estimates of absolute AND relative risk as they are interpreted within the context of each other (e.g., 2 in 10 events vs. 1 in 10 events has a RR = 2; 2 in 1000 events vs. 1 in 1000 events also has RR = 2; but the first is more important than the second)
- Provide summaries for each important outcome (benefits AND harms)
- Provide information for a reference comparison (i.e., risk without the intervention or risk for an alternative intervention); provide comparison data for risks of events that are well-understood such as risks for transportational, recreational, or occupational hazards
- Provide accounting for all randomized participants
- Describe the duration of follow-up (consider a Kaplan–Meier plot) and the reasons for loss-to-follow-up discussing the potential bias from this loss
- Include a table of key definitions
- Provide transparency of the multiplicity context (e.g., how many tests were conducted or how many subgroups were examined)
- Summarize heterogeneity of effects across subgroups, if appropriate
- Consider using creative figures instead of tabular summaries as they are often easier to digest and display trends and signals than voluminous listings and tables (e.g., when displaying trends over time, use longitudinal plots rather than tables of summary statistics; forest plots for displaying confidence interval estimates for various subgroups; and flow diagrams to display participant flow through the trial)
- Discuss assumptions of the design and analyses and evaluation of their validity
- Discuss the limitations of the trial
- Know your audience as this helps determine when less vs. more details are needed

10.1 Guidelines for Reporting Clinical Trial Results

Although each medical journal may have its own reporting manuscript style or template, expert groups have developed guidelines for the reporting of different types of trials. These include the consolidated standards of reporting trials (CONSORT) statement for reporting randomized trials, transparent reporting of evaluations with nonrandomized designs (TREND) for reporting nonrandomized trials, and the standards for reporting of diagnostic accuracy (STARD) for reporting diagnostic trials.

10.1.1 The CONSORT Statement

The CONSORT statement (www.consort-statement.org) is an evidence-based, minimum set of recommendations for reporting randomized clinical trials. It outlines a standard approach for authors to prepare reports of trial results, facilitating complete and transparent reporting, and aiding critical appraisal and interpretation. It is considered an evolving document and currently consists of two parts: (1) a 22-item checklist (Table 10.3) and (2) a flow diagram (Figure 10.1). The checklist items focus on reporting how the trial was designed, analyzed, and interpreted while the flow diagram displays the progress of all participants through the trial (Moher et al., 2010).

The main CONSORT statement is based on a standard two-arm parallel group superiority trial design. However, several extensions to the CONSORT statement have been developed to address other trial designs (e.g., cluster-randomized trials, noninferiority (NI) and equivalence trials, and pragmatic trials), specific interventions (e.g., acupuncture, herbal medicine, and non-pharmacologic treatment), or data such as harms.

10.1.2 Reporting of Harms Data

One particularly important extension to the CONSORT statement describes the reporting of harms data (Ioannidis et al., 2004). There is often an asymmetry with respect to the attention to efficacy vs. harms data with the latter only being analyzed using simple counts and percentages. Toxicities are often not reported in the abstract of publications, leaving the details of the safety evaluations to the main body of the manuscript.

Common poor reporting practices regarding harms data are summarized in Table 10.4 but include using generic or vague statements such as, "the drug was generally well tolerated," reporting only the adverse events observed at a certain frequency or rate threshold (e.g., 3%) regardless of seriousness or severity, reporting only the adverse events that reach a p-value threshold in the comparison of the randomized arms (e.g., $p < 0.05$), reporting measures of central tendency (e.g., means or medians) for continuous variables without any information on extreme values, and failing to provide

TABLE 10.3

CONSORT Checklist

Section/Topic	Item No	Checklist Item	Reported on Page No
Title and abstract			
	1a	Identification as a randomized trial in the title	___ ___
	1b	Structured summary of trial design, methods, results, and conclusions (for specific guidance see CONSORT for abstracts)	___ ___
Introduction			
Background and objectives	2a	Scientific background and explanation of the rationale	___ ___
	2b	Specific objectives or hypotheses	___ ___
Methods			
Trial design	3a	Description of trial design (such as parallel, factorial) including allocation ratio	___ ___
	3b	Important changes to methods after trial commencement (such as eligibility criteria), with reasons	___ ___
Participants	4a	Eligibility criteria for participants	___ ___
	4b	Settings and locations where the data were collected	___ ___
Interventions	5	The interventions for each group with sufficient details to allow replication, including how and when they were actually administered	___ ___
Outcomes	6a	Completely defined prespecified primary and secondary outcome measures, including how and when they were assessed	___ ___
	6b	Any changes to trial outcomes after the trial commenced, with reasons	___ ___
Sample size	7a	How sample size was determined	___ ___
	7b	When applicable, explanation of any interim analyses and stopping guidelines	___ ___
Randomization:			
Sequence generation	8a	Method used to generate the random allocation sequence	___ ___
	8b	Type of randomization; details of any restriction (such as blocking and block size)	___ ___

(Continued)

TABLE 10.3 (*Continued*)

CONSORT Checklist

Section/Topic	Item No	Checklist Item	Reported on Page No
Allocation concealment mechanism	9	Mechanism used to implement the random allocation sequence (such as sequentially numbered containers), describing any steps taken to conceal the sequence until interventions were assigned	
Implementation	10	Who generated the random allocation sequence, who enrolled participants, and who assigned participants to interventions	
Blinding	11a	If done, who was blinded after assignment to interventions (e.g., participants, care providers, those assessing outcomes), and how	
	11b	If relevant, description of the similarity of interventions	
Statistical methods	12a	Statistical methods used to compare groups for primary and secondary outcomes	
	12b	Methods for additional analyses, such as subgroup analyses and adjusted analyses	
Results			
Participant flow (a diagram is strongly recommended)	13a	For each group, the numbers of participants who were randomly assigned, received intended treatment and were analyzed for the primary outcome	
	13b	For each group, losses and exclusions after randomization, together with reasons	
Recruitment	14a	Dates defining the periods of recruitment and follow-up	
	14b	Why the trial ended or was stopped	
Baseline data	15	A table showing baseline demographic and clinical characteristics of each group	
Numbers analyzed	16	For each group, number of participants (denominator) included in each analysis and whether the analysis was by original assigned groups	
Outcomes and estimation	17a	For each primary and secondary outcome, results for each group and the estimated effect size and its precision (such as 95% confidence interval)	
	17b	For binary outcomes, presentation of both absolute and relative effect sizes is recommended	

(*Continued*)

TABLE 10.3 (*Continued*)

CONSORT Checklist

Section/Topic	Item No	Checklist Item	Reported on Page No
Ancillary analyses	18	Results of any other analyses performed, including subgroup analyses and adjusted analyses, distinguishing prespecified from exploratory	_____
Harms	19	All important harms or unintended effects in each group (for specific guidance see CONSORT for harms)	_____
Discussion			
Limitations	20	Trial limitations, addressing sources of potential bias, imprecision, and, if relevant, multiplicity of analyses	_____
Generalizability	21	Generalizability (external validity, applicability) of the trial findings	_____
Interpretation	22	Interpretation consistent with results, balancing benefits and harms, and considering other relevant evidence	_____
Other Information			
Registration	23	Registration number and name of trial registry	_____
Protocol	24	Where the full trial protocol can be accessed if available	_____
Funding	25	Sources of funding and other support (such as supply of drugs), role of funders	_____

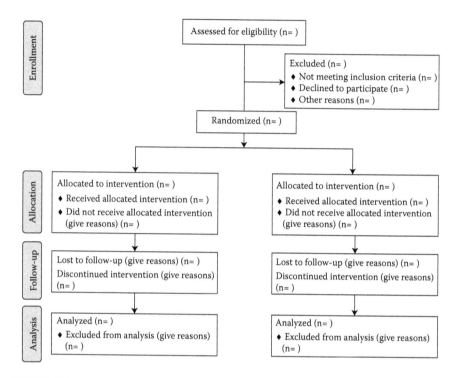

FIGURE 10.1
CONSORT flow diagram.

TABLE 10.4

Common Poor Reporting Practices Regarding Harms Data

- Using generic or vague statements, such as "the drug was generally well tolerated" or "the comparator drug was relatively poorly tolerated"
- Failing to provide separate data for each study arm
- Providing summed numbers for all adverse events for each study arm, without separate data for each type of adverse event
- Providing summed numbers for a specific type of adverse event, regardless of severity or seriousness
- Reporting only the adverse events observed at a certain frequency or rate threshold (e.g., >3% or >10% of participants)
- Reporting only the adverse events that reach a P-value threshold in the comparison of the randomized arms (e.g., $p < 0.05$)
- Reporting measures of central tendency (e.g., means or medians) for continuous variables without any information on extreme values
- Improperly handling or disregarding the relative timing of the events, when timing is an important determinant of the adverse event in question
- Not distinguishing between patients with one adverse event and participants with multiple adverse events
- Providing statements about whether data were statistically significant without giving the exact counts of events
- Not providing data on harms for all randomly assigned participants

TABLE 10.5

CONSORT Recommendations for Reporting Harms in Clinical Trials

- If the study collected data on harms and benefits, the title or abstract should so state
- If the trial addresses both harms and benefits, the introduction should so state
- List addressed adverse events with definitions for each (with attention, when relevant, to grading, expected vs. unexpected events, reference to standardized and validated definitions, and a description of new definitions)
- Clarify how harms-related information was collected (mode of data collection, timing, attribution methods, intensity of ascertainment, and harms-related monitoring and stopping rules, if pertinent)
- Describe plans for presenting and analyzing information on harms (including coding, handling of recurrent events, specification of timing issues, handling of continuous measures, and any statistical analyses)
- Describe for each arm the participant withdrawals that are due to harms and the experience with the allocated treatment
- Provide the denominators for analyses on harms
- Present the absolute risk of each adverse event (specifying type, grade, and seriousness per arm), and present appropriate metrics for recurrent events, continuous variables and scale variables, whenever pertinent
- Describe any subgroup analyses and exploratory analyses for harms
- Provide a balanced discussion of benefits and harms with emphasis on study limitations, generalizability, and other sources of information on harms

data on harms for all randomly assigned participants in accordance with the ITT principle.

The authors outline 10 recommendations (Table 10.5) for better reporting of harms in clinical trials. Highlights include describing for each arm the participant withdrawals that are due to harms and the experience with the allocated treatment; providing the denominators for analyses on harms; presenting the absolute risk of each adverse event (specifying type, grade, and seriousness per arm), and presenting appropriate metrics for recurrent events, continuous and scale variables, whenever pertinent; and providing a balanced discussion of benefits and harms with emphasis on study limitations, generalizability, and other sources of information on harms. Lagakos (2006) has suggested use of Kaplan–Meier curves for estimating cumulative incidence of harms when risk changes over time.

10.1.3 The TREND Statement

The TREND statement (www.cdc.gov/trendstatement/) is a suggested set of reporting standards for nonrandomized evaluations of interventions (Des Jarlais et al., 2004). It also consists of a 22 item checklist.

10.1.4 The STARD Statement

The STARD statement (www.STARD-statement.org) was developed to improve the quality of the reporting of diagnostic studies. It consists of a checklist of 25 items (Table 10.6) and a flow diagram (Figure 10.2) (Bossuyt et al., 2003).

TABLE 10.6

STARD Checklist

Section and Topic	Item No.		On Page No.
TITLE/ABSTRACT/KEYWORDS	1	Identify the article as a study of diagnostic accuracy (recommend MeSH heading "sensitivity and specificity")	
INTRODUCTION	2	State the research questions or study aims, such as estimating diagnostic accuracy or comparing accuracy between tests or across participant groups	
METHODS			
Participants	3	The study population: The inclusion and exclusion criteria, setting and locations where data were collected	
	4	Participant recruitment: Was recruitment based on presenting symptoms, results from previous tests, or the fact that the participants had received the index tests or the reference standard?	
	5	Participant sampling: Was the study population a consecutive series of participants defined by the selection criteria in item 3 and 4? If not, specify how participants were further selected	
	6	Data collection: Was data collection planned before the index test and reference standard were performed (prospective study) or after (retrospective study)?	
Test methods	7	The reference standard and its rationale	
	8	Technical specifications of material and methods involved including how and when measurements were taken, and/or cite references for index tests and reference standard	
	9	Definition of and rationale for the units, cut-offs and/or categories of the results of the index tests and the reference standard	
	10	The number, training and expertise of the persons executing and reading the index tests and the reference standard	
	11	Whether or not the readers of the index tests and reference standard were blind (masked) to the results of the other test and describe any other clinical information available to the readers	

(Continued)

TABLE 10.6 (*Continued*)

STARD Checklist

Section and Topic	Item No.		On Page No.
Statistical methods	12	Methods for calculating or comparing measures of diagnostic accuracy, and the statistical methods used to quantify uncertainty (e.g., 95% confidence intervals)	
	13	Methods for calculating test reproducibility, if done	
RESULTS			
Participants	14	When study was performed, including beginning and end dates of the recruitment	
	15	Clinical and demographic characteristics of the study population (at least information on age, gender, spectrum of presenting symptoms)	
	16	The number of participants satisfying the criteria for inclusion who did or did not undergo the index tests and / or the reference standard; describe why participants failed to undergo either test (a flow diagram is strongly recommended)	
Test results	17	Time interval between the index tests and the reference standard and any treatment administered in between	
	18	Distribution of severity of disease (define criteria) in those with the target condition; other diagnoses in participants without the target condition	
	19	A cross tabulation of the results of the index tests (including indeterminate and missing results) by the results of the reference standard; for continuous results, the distribution of the test results by the results of the reference standard	
	20	Any adverse events from performing the index tests or the reference standard	
Estimates	21	Estimates of diagnostic accuracy and measures of statistical uncertainty (e.g., 95% confidence intervals)	
	22	How indeterminate results, missing data and outliers of the index tests were handled	
	23	Estimates of variability of diagnostic accuracy between subgroups of participants, readers or centers, if done	
	24	Estimates of test reproducibility, if done	
DISCUSSION	25	Discuss the clinical applicability of the study findings	

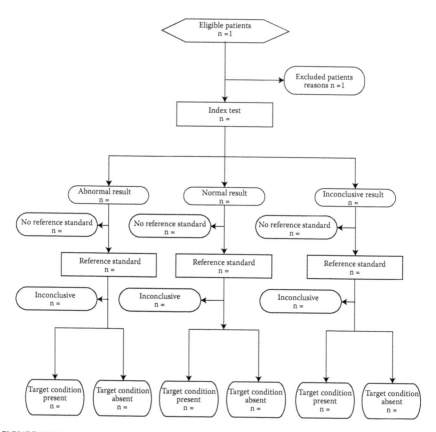

FIGURE 10.2
STARD flow diagram.

10.2 Reporting the Results of Subgroup Analyses

The second international study of infarct survival (the ISIS-2 trial) was a randomized placebo-controlled factorial trial evaluating streptokinase and aspirin after myocardial infarction on vascular mortality (ISIS-2 Collaborative Study Group, 1988). Both interventions demonstrated improvements, for example, aspirin reduced vascular mortality by 23%, p < 0.0001. However, the authors reported that aspirin increased mortality in two subgroups, Geminis and Libras. The authors reported these results to make a point: the results of subgroup analyses should be reported and interpreted with great care.

The reporting of subgroup analyses in the medical literature has generally been poor (Wang et al., 2007) and results of subgroup analyses are often overinterpreted (Assman et al., 2000). Subgroup analyses often have low power to detect effects within subgroups (thus increasing type II error) but are also subject to multiplicity concerns (thus also increasing type I error).

This has triggered *The New England Journal of Medicine* to develop guidelines (Table 10.7) for the reporting of subgroup analyses. They provide suggestions for the abstract, methods, results, and discussion sections of manuscripts. In the abstract, they suggest presenting subgroup results only if the subgroup analyses were based on a primary study outcome, if they were prespecified, and if they were interpreted in light of the totality of prespecified subgroup analyses undertaken. In the methods section, they suggest the following:

1. Indicating the number of prespecified subgroup analyses that were performed and the number of prespecified subgroup analyses that are reported (distinguishing a specific subgroup analysis of special interest from the multiple subgroup analyses typically done to assess the consistency of a treatment effect on various patient characteristics), and for each reported analysis, indicating the endpoint that was assessed and the statistical method that was used to assess the heterogeneity of treatment differences.

TABLE 10.7

Guidelines for the Reporting of Subgroup Analyses in the NEJM

Abstract
- Present subgroup results only if they were based on a primary study outcome if they were prespecified, and if they were interpreted in light of the totality of prespecified subgroup analyses undertaken

Methods
- Indicate the number of prespecified subgroup analyses that were performed and the number of prespecified subgroup analyses that are reported. Distinguish a specific subgroup analysis of special interest from the multiple subgroup analyses typically done to assess the consistency of a treatment effect on various patient characteristics. For each reported analysis, indicate the endpoint that was assessed and the statistical method that was used to assess the heterogeneity of treatment differences
- Indicate the number of post hoc subgroup analyses that were performed and the number of post hoc subgroup analyses that are reported. For each reported analysis, indicate the endpoint that was assessed and the statistical method used to assess the heterogeneity of treatment differences. Detailed descriptions may require a supplementary appendix
- Indicate the potential effect on type I errors (false positives) due to multiple subgroup analyses and how this effect is addressed. If formal adjustments for multiplicity were used, describe them; if no formal adjustment was made, indicate the magnitude of the problem

Results
- When possible, base analyses of the heterogeneity of treatment effects on tests for interaction, and present them along with effect estimates (including confidence intervals) within each level of each baseline covariate analyzed. A forest plot is an effective method for presenting this information

Discussion
- Avoid overinterpretation of subgroup differences. Be appropriately cautious in appraising their credibility, acknowledge the limitations, and provide supporting or contradictory data from other studies, if any

2. Indicating the number of post hoc subgroup analyses that were performed and the number of post hoc subgroup analyses that are reported and for each reported analysis, indicating the endpoint that was assessed and the statistical method used to assess the heterogeneity of treatment differences.

3. Indicating the potential effect on type I errors (false positives) due to multiple subgroup analyses and how this effect is addressed (if formal adjustments for multiplicity were used, describe them; if no formal adjustment was made, indicate the magnitude of the problem).

In the results section, they suggest basing analyses of the heterogeneity of treatment effects on tests for interaction, and presenting them along with effect estimates (including confidence intervals) within each level of each baseline covariate analyzed. They note that a forest plot is an effective method for presenting this information. In the discussion section, they suggest avoiding overinterpretation of subgroup differences and being cautious in appraising their credibility, acknowledge the limitations, and provide supporting or contradictory data from other studies.

In general, when subgroup analyses are conducted then they should be reported regardless of significance (i.e., avoidance of selective reporting), the number of subgroup analyses conducted should be transparent so that results can be interpreted within the appropriate multiplicity context, and which subgroup analyses were prespecified vs. post hoc should also be made clear. Subgroup analyses should generally be considered exploratory analyses rather than confirmatory and labeled as such in publications.

10.3 Reporting Benefits and Risks

The comprehensive presentation of both benefits and risks as well as how they interact is crucial. Researchers should ask, "How can data summaries be presented to help patients, clinicians, regulators, and other consumers of the information so that informed decisions can be made?" Summaries should be interpretable for nonresearchers as well as seasoned researchers.

The Pharmaceutical Research and Manufacturers of America (PhRMA) organized the Benefit Risk Action Team (BRAT) to develop a structured and transparent framework for the benefit:risk assessment (CIRS-BRAT Framework) (Coplan et al., 2011). The BRAT approach culminates in an informative benefit:risk summary table that consists of the important beneficial and harmful outcomes, estimated risks for the intervention and control and forest plots for the risk difference and relative risk (see two examples in Figure 10.3 comparing treatments for migraine headaches and two statins

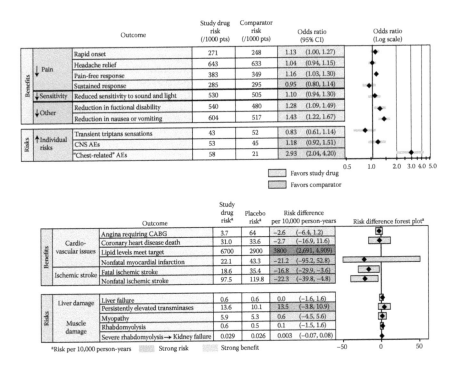

		Outcome	Study drug risk (/1000 pts)	Comparator risk (/1000 pts)	Odds ratio (95% CI)		Odds ratio (Log scale)
Benefits	Pain	Rapid onset	271	248	1.13	(1.00, 1.27)	
		Headache relief	643	633	1.04	(0.94, 1.15)	
		Pain-free response	383	349	1.16	(1.03, 1.30)	
		Sustained response	285	295	0.95	(0.80, 1.14)	
	Sensitivity	Reduced sensitivity to sound and light	530	505	1.10	(0.94, 1.30)	
	Other	Reduction in fuctional disability	540	480	1.28	(1.09, 1.49)	
		Reduction in nausea or vomiting	604	517	1.43	(1.22, 1.67)	
Risks	Individual risks	Transient triptans sensations	43	52	0.83	(0.61, 1.14)	
		CNS AEs	53	45	1.18	(0.92, 1.51)	
		"Chest-related" AEs	58	21	2.93	(2.04, 4.20)	

0.5 1.0 2.0 3.0 4.0 5.0

Favors study drug
Favors comparator

		Outcome	Study drug risk[a]	Placebo risk[a]	Risk difference per 10,000 person-years		Risk difference forest plot[a]
Benefits	Cardio-vascular issues	Angina requiring CABG	3.7	64	-2.6	(-6.4, 1.2)	
		Coronary heart disease death	31.0	33.6	-2.7	(-16.9, 11.6)	
		Lipid levels meet target	6700	2900	3800	(2,691, 4,909)	
		Nonfatal myocardial infarction	22.1	43.3	-21.2	(-95.2, 52.8)	
	Ischemic stroke	Fatal ischemic stroke	18.6	35.4	-16.8	(-29.9, -3.6)	
		Nonfatal ischemic stroke	97.5	119.8	-22.3	(-39.8, -4.8)	
Risks	Liver damage	Liver failure	0.6	0.6	0.0	(-1.6, 1.6)	
		Persistently elevated transaminases	13.6	10.1	13.5	(-3.8, 10.9)	
	Muscle damage	Myopathy	5.9	5.3	0.6	(-4.5, 5.6)	
		Rhabdomyolysis	0.6	0.5	0.1	(-1.5, 1.6)	
		Severe rhabdomyolysis → Kidney failure	0.029	0.026	0.003	(-0.07, 0.08)	

[a] Risk per 10,000 person-years   Strong risk   Strong benefit

−50   0   50

FIGURE 10.3
Examples of PhRMA benefit:risk summary tables.

for prevention of cardiovascular disease). The table provides a comprehensive but concise summary of estimated benefits and harms.

It is helpful to provide information for a reference comparison (e.g., risk without the intervention or risk for an alternative intervention) when summarizing benefits and risks. Consider the example of a summary of the benefits and harms of Tamoxifen for the prevention of breast cancer in Table 10.8. It is a very informative table since it provides (1) a reference comparison of benefits and harms with and without Tamoxifen allowing women and their doctors to weigh the benefits and risks of both options (i.e., taking Tamoxifen or not), (2) information regarding absolute and relative risks, and (3) summaries for benefits and harms stratified by outcome severity to aid with intuitive weighting of different outcomes.

It is also helpful to display how the benefits and harms interact. Did the benefits and harms occur in the same participants or within different participants? Chuang-Stein et al. (1991) proposed the *global benefit:risk score*, a multinomial outcome based on efficacy and safety data to evaluate participant-specific outcomes. For example, the following ordered categories may be defined (in order of desirability): efficacy without serious side effects, efficacy with serious side effects, no efficacy with a lack of serious side effects, no efficacy with serious side effects, toxicity leading to dropout. An algorithm

TABLE 10.8

Five-Year Risk of Invasive Breast Cancer with and without Tamoxifen for 40-Year-Old White Women with UTERI

Severity	Event	Expected Cases in 10,000 Untreated Women	Cases Prevented (+) or Caused (−) in 10,000 Treated Women
Life-threatening			
	Invasive breast cancer	200	+97
	Hip fracture	2	+1
	Endometrial cancer	10	−16
	Stroke	22	−13
	Pulmonary embolism	7	−15
	NET		**+54 (prevented)**
Severe			
	In situ breast cancer	106	+53
	Deep vein thrombosis	24	−15
	NET		**+38 (prevented)**

Source: Adapted from Gail, M.H. et al., 1999, *JNCI*, 91:1829–1846.

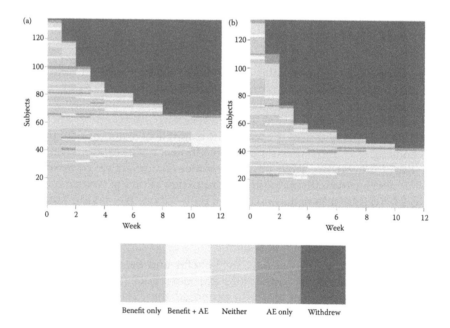

FIGURE 10.4

Example of longitudinal benefit:risk graphic. Individual response profiles for patients randomized to hydromorphone (a) or placebo (b). N = 134 in each arm. (Adapted from Norton, J.D., 2011, *Drug Inf J*, 45:741–747.)

could be created to classify each patient objectively, or a blinded adjudication committee could be used. Norton (2011) extended this idea to account for longitudinal changes within-patient (i.e., each participant is classified at each time interval of the trial). He then developed a graphic that displays within-patient changes longitudinally, the temporal relationship and association between benefits and harms, as well as the aggregate intervention effects. An example from a randomized placebo-controlled trial of hydromorphone for the treatment of pain is displayed in Figure 10.4.

10.4 Reporting NI Trials

The reporting of NI trials has been suboptimal in the medical literature. Greene et al. (AIM, 2000) reviewed 88 studies claiming NI trials but noted that 67% of these studies inappropriately claimed NI based upon nonsignificant superiority tests. Furthermore, only 23% of the studies prespecified an NI margin. Piaggio et al. (JAMA, 2006) published an extension of the CONSORT statement to outline appropriate reporting of NI trials. Details can be found at www.consort-statement.org.

10.5 Reporting Adaptive Designs

When used appropriately, adaptive designs can be efficient and informative. However, when used inappropriately, they can threaten trial integrity via loss of control of statistical error rates and the induction of operational bias. Adaptive designs thus require careful reporting so that rational and methods for adaptation are transparent and that the evaluation of these concerns can be made. Guidelines for the reporting of adaptive designs are provided in Table 10.9.

10.6 Reporting Bayesian Designs

The reporting of a clinical trial should thus distinguish between the reporting of the trial results (a summary of the data itself) and guided interpretation provided by a Bayesian analyses. Bayesian analyses of a clinical trial thus use data from the trial to transform prior beliefs about the effect of an intervention prior to the trial to beliefs after the trial is conducted. Hence, prior beliefs have an explicit role in interpreting the potential value of an

TABLE 10.9

Guidelines for the Reporting of Adaptive Designs

Clearly describe:
- Present subgroup results only if they were based on a primary study outcome if they were prespecified, and if they were interpreted in light of the totality of prespecified subgroup analyses undertaken
- The adaptation
- Whether the adaptation was planned or unplanned
- The rationale for the adaptation
- When the adaptation was made
- The data upon which adaptation is based and whether the data were unblinded
- The planned process for the adaptation including who made the decision regarding adaptation
- Deviations from the planned process
- Consistency of results before vs. after the adaptation

Discuss:
- Potential biases induced by the adaptation
- Adequacy of firewalls to protect against operational bias
- The effects on error control and multiplicity context

---

intervention. Thus, Bayesian analyses of clinical trials should be reported carefully. Sensitivity of the interpretation to the choice of the prior is paramount. This can be achieved by employing several well-defined priors based on other trial results and expert opinion. If the qualitative interpretation is very sensitive to the selected prior, then interpretation can be challenging. One analysis should use a noninformative prior. Use of a noninformative prior implies that the posttrial beliefs are effectively determined by the data and thus is the most objective approach. However, this may not be realistic (i.e., investigators often have views, based on experiences on the range of effects that would be considered possible). Graphically displaying the posttrial belief of a beneficial intervention effect as the prior probability varies from no effect can illustrate the strength of the evidence for a beneficial intervention effect.

Bayesian analyses should include a presentation of the entire posterior distribution (not a single point estimate) as it represents a summary of the beliefs/evidence about the parameter of interest. It is informative to provide the probability that the intervention effect is within various meaningful regions (e.g., region of NI). Bayesian or "credible intervals" for an intervention effect can be constructed. Many believe that the credible interval is more interpretable to nonstatistical researchers. A 95% credible interval is interpreted as "the probability that the parameter lies in the interval is 95%" (note that this is distinct from the confidence interval that would have 95% probability of covering the true intervention effect). Bayes factor, a measure of how much evidence in favor of alternative hypothesis has changed after observing the data (not an absolute measure of support for a hypothesis) may also be presented.

References

Altman DG. 1998. Statistical reviewing for medical journals. *Stat Med* 17:2661–2674.

Altman DG, Bland JM. 1998. Improving doctors' understanding of statistics. *J R Stat Soc A* 154:223–267.

Assman SF, Pocock SJ, Enos LE, Kasten LE. 2000. Subgroup analysis and other (mis) uses of baselin data in clinical trials. *Lancet* 355:1064–1069.

Bailar JC, Mosteller F. 1988. Guidelines for statistical reporting in articles for medical journals. *Ann Intern Med* 108:266–273.

Berwick D, Fineber HV, Weinstein MC. 1981. When doctors meet numbers. *Am J Med* 71:991–998.

Bossuyt PM, Reitsma JB, Bruns DE Gatsonis PP, Glasziou PP, Irwig LM, Lijmer JG, Moher D, Rennie D, De Vet HC. 2003. Standards for reporting of diagnostic accuracy. Toward complete and accurate reporting of studies of diagnostic accuracy: The STARD initiative. Standards for reporting of diagnostic accuracy. *Ann Intern Med* 138:40–44.

Bourgeois FT, Murthy S, Mandl KD. 2010. Outcome reporting among drug trials registered in clinicaltrials.gov. *Ann Intern Med* 153:158–166.

Boutron I, Dutton S, Ravaud P, Altman DG. 2010. Reporting and interpretation of randomized controlled trials with statistically nonsignificant results for primary outcomes. *JAMA* 303(20):2058–2064.

Chuang-Stein C, Mohberg NR, Sinkula MS. 1991. Three measures for simultaneously evaluating benefits and risks using categorical data from clinical trials. *Stat Med* 10:1349–1359.

CIRS-BRAT Framework. CIRS: Centre for Innovation in Regulator Science. www.cirs-brat.org

Coplan PM, Noel RA, Levitan BS, Feguson J, Mussen F. 2011. Development of a framework for enhancing the transparency, reproducibility, and communication of the benefit-risk balance of medicines. *Clin Pharmacol Ther* 89(2): 312–315.

Des Jarlais DC, Lyles C, Crepaz N. 2004. Improving the reporting quality of nonrandomized evaluations of behavioral and public health interventions: The TREND statement. *Am J Public Health* 94(3):361–366.

Evans SR. 2010. Common statistical concerns in clinical trials. *J Exp Stroke Transl Med* 3(1):1–7.

Gail MH, Costantino JP, Bryant J, Croyle R, Freedman L, Helzlsouer K, Vogel V. 1999. Weighing the risks and benefits of Tamoxifen treatment for preventing breast cancer. *JNCI* 91:1829–1846.

Glantz SA. 1980. How to detect, correct and prevent errors in the medical literature. *Biostatistics* 61:1–7.

Greene WL, Concato J, Feinstein AR. 2000. Claims of equivalence in medical research: Are they supported by the evidence? *Ann Intern Med* 132:715–722.

Horton NJ, Switzer SS. 2005. Statistical methods in the journal (letter). *N Engl J Med* 353:1977–1979.

Ioannidis JPA, Evans SJW, Gotzsche PC, O'Neill RT, Altman DG, Schulz K, Moher D. 2004. Better reporting of harms in randomized trials: An extension of the CONSORT statement. *Ann Intern Med* 141:781–788.

ISIS-2 Collaborative Study Group. 1988. Randomized trial of intravenous strptoki-
    nase, oral aspirin, both, or neither among 17,187 cases of suspected acute myo-
    cardial infarction: ISIS-2. *Lancet* 2(8607):349–360.
Lagakos SW. 2006. Time-to-event analyses for long-term treatments—The APPROVe
    Trial. *NEJM* 355(2):113–117.
Moher D, Hopewell S, Schulz KF, Montori V, Gøtzsche PC, Devereaux PJ, Elbourne
    D, Egger M, Altman DG. 2010. CONSORT 2010 Explanation and Elaboration:
    Updated guidelines for reporting parallel group randomised trials. *BMJ*
    340:c869. doi: 10.1136/bmj.c869.
Norton JD. 2011. A longitudinal model and graphic for benefit:risk analysis with case
    study. *Drug Inf J* 45:741–747.
Piaggio G, Elbourne DR, Altman DG, Pocock SJ, Evans SJW. 2006. Reporting of non-
    inferiority and equivalence randomized trials: An extension of the CONSORT
    statement. *JAMA* 295:1152–1160.
Prayle AP, Hurley MN, Smyth AR. 2011. Compliance with mandatory reporting of
    clinical trial results on ClinicalTrials.gov: Cross sectional study. *BMJ* 344:d7373.
    doi: 10.1136/bmj.d7373.
Ross JS, Tse T, Zarin DA, Xu H, Zhou L, Krumholz HM. 2011. Publication of NIH
    funded trials registered in ClinicalTrials.gov: Cross sectional analysis. *BMJ*
    344:d7292. doi: 10.1136/bmj.d7292.
Wang R, Lagakos SW, Ware JH, Hunter DJ, Drazen JM. 2007. Statistics in medicine—
    reporting of subgroup analyses in clinical trials. *N Engl J Med* 357(21):2189–2194.
Weiss ST, Samet JM. 1980. An assessment of physician knowledge of epidemiology
    and biostatistics. *J Med Educ* 55:692–697.
West CP, Ficalora RD. 2007. Clinician attitudes towards biostatistics. *Mayo Clin Proc*
    82:939–943.
Windish DM, Huot SJ, Free ML. 2007. Medicine residents' understanding of the bio-
    statistics and results in he medical literature. *JAMA* 298:1010–1022.
Wulff HR, Andersen B, Brandenhoff P, Guttler F. 1986. What do doctors know about
    statistics? *Stat Med* 6:3–10.

Appendix: Excerpts from the Lipitor® Drug Label

A.1 Lipitor (Atorvastatin Calcium) Tablets

A.1.1 Description

LIPITOR (atorvastatin calcium) is a synthetic lipid-lowering agent. Atorvastatin is an inhibitor of 3-hydroxy-3-methylglutaryl-coenzyme A (HMG-CoA) reductase.

A.1.2 Clinical Pharmacology

A.1.2.1 Pharmacodynamics

Atorvastatin as well as some of its metabolites are pharmacologically active in humans. The liver is the primary site of action and the principal site of cholesterol synthesis and low-density lipoprotein (LDL) clearance. Drug dosage rather than systemic drug concentration correlates better with low-density lipoprotein cholesterol (LDL-C) reduction. Individualization of drug dosage should be based on therapeutic response.

A.1.2.2 Pharmacokinetics and Drug Metabolism

Absorption: Atorvastatin is rapidly absorbed after oral administration; maximum plasma concentrations occur within 1–2 h. Extent of absorption increases in proportion to atorvastatin dose. The absolute bioavailability of atorvastatin (parent drug) is approximately 14% and the systemic availability of HMG-CoA reductase inhibitory activity is approximately 30%. The low systemic availability is attributed to presystemic clearance in gastrointestinal mucosa and/or hepatic first-pass metabolism. Although food decreases the rate and extent of drug absorption by approximately 25% and 9%, respectively, as assessed by maximum concentration (Cmax) and area under the curve (AUC), LDL-C reduction is similar whether atorvastatin is given with or without food. Plasma atorvastatin concentrations are lower (approximately 30% for Cmax and AUC) following evening drug administration compared with morning. However, LDL-C reduction is the same regardless of the time of day of drug administration.

Distribution: Mean volume of distribution of atorvastatin is approximately 381 L. Atorvastatin is ≥98% bound to plasma proteins. A blood/plasma ratio of approximately 0.25 indicates poor drug penetration into red blood cells. Based on observations in rats, atorvastatin is likely to be secreted in human milk.

Metabolism: Atorvastatin is extensively metabolized to ortho- and parahydroxylated derivatives and various beta-oxidation products. *in vitro* inhibition of HMG-CoA reductase by ortho- and parahydroxylated metabolites is equivalent to that of atorvastatin. Approximately 70% of circulating inhibitory activity for HMG-CoA reductase is attributed to active metabolites. *in vitro* studies suggest the importance of atorvastatin metabolism by cytochrome P450 3A4, consistent with increased plasma concentrations of atorvastatin in humans following coadministration with erythromycin, a known inhibitor of this isozyme. In animals, the ortho-hydroxy metabolite undergoes further glucuronidation.

Excretion: Atorvastatin and its metabolites are eliminated primarily in bile following hepatic and/or extra-hepatic metabolism; however, the drug does not appear to undergo enterohepatic recirculation. Mean plasma elimination half-life of atorvastatin in humans is approximately 14 h, but the half-life of inhibitory activity for HMG-CoA reductase is 20–30 h due to the contribution of active metabolites. Less than 2% of a dose of atorvastatin is recovered in urine following oral administration.

Special populations: This section discusses clinical pharmacology regarding the following special populations—geriatric, pediatric, renal insufficiency, hemodialysis, and hepatic insufficiency populations.

A.3 Clinical Studies

A.3.1 Prevention of Cardiovascular Disease

In the Anglo-Scandinavian cardiac outcomes trial (ASCOT), the effect of LIPITOR (atorvastatin calcium) on fatal and nonfatal coronary heart disease was assessed in 10,305 hypertensive patients 40–80 years of age (mean of 63 years), without a previous myocardial infarction and with total cholesterol (TC) levels ≤251 mg/dL (6.5 mmol/L). Additionally all patients had at least three of the following cardiovascular risk factors: male gender (81.1%), age >55 years (84.5%), smoking (33.2%), diabetes (24.3%), history of CHD in a first degree relative (26%), TC:HDL > 6 (14.3%), peripheral vascular disease (5.1%), left ventricular hypertrophy (14.4%), prior cerebrovascular event

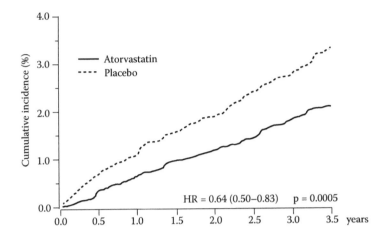

FIGURE A.1
Effect of LIPITOR 10 mg/day on cumulative incidence of nonfatal myocardial infarction or coronary heart disease death (in ASCOT-LLA).

(9.8%), specific electrocardiogram (ECG) abnormality (14.3%), proteinuria/ albuminuria (62.4%). In this double-blind, placebo-controlled study patients were treated with anti-hypertensive therapy (goal BP < 140/90 mm Hg for nondiabetic patients, <130/80 mm Hg for diabetic patients) and allocated to either LIPITOR 10 mg daily (n = 5168) or placebo (n = 5137), using a covariate adaptive method which took into account the distribution of nine baseline characteristics of patients already enrolled and minimized the imbalance of those characteristics across the groups. Patients were followed for a median duration of 3.3 years.

LIPITOR significantly reduced the rate of coronary events (either fatal coronary heart disease [46 events in the placebo group vs. 40 events in the LIPITOR group] or nonfatal multiple imputation [MI] [108 events in the placebo group vs. 60 events in the LIPITOR group]) with a relative risk reduction of 36% ([based on incidences of 1.9% for LIPITOR vs. 3.0% for placebo], p = 0.0005 [see Figure A.1]). The risk reduction was consistent regardless of age, smoking status, obesity, or presence of renal dysfunction. The effect of LIPITOR was seen regardless of baseline LDL levels. Due to the small number of events, results for women were inconclusive.

LIPITOR also significantly decreased the relative risk for revascularization procedures by 42%. Although the reduction of fatal and nonfatal strokes did not reach a predefined significance level (p = 0.01), a favorable trend was observed with a 26% relative risk reduction (incidences of 1.7% for LIPITOR and 2.3% for placebo). There was no significant difference between the treatment groups for death due to cardiovascular causes (p = 0.51) or noncardiovascular causes (p = 0.17).

TABLE A.1

Dose–Response in Patients with Primary Hypercholesterolemia (Adjusted Mean% Change from Baseline)[a]

Dose	N	TC	LDL-C	Apo B	TG	HDL-C	Non-HDL-C/HDL-C
Placebo	21	4	4	3	10	−3	7
10	22	−29	−39	−32	−19	6	−34
20	20	−33	−43	−35	−26	9	−41
40	21	−37	−50	−42	−29	6	−45
80	23	−45	−60	−50	−37	5	−53

[a] Results are pooled from two dose–response studies.

A.3.2 Hypercholesterolemia (Heterozygous Familial and Nonfamilial) and Mixed Dyslipidemia (Fredrickson Types IIa and IIb)

LIPITOR reduces total-C, LDL-C, very-low-density lipoprotein cholesterol (VLDL-C), apolipoprotein (apo B), and triglycerides (TG), and increases high-density lipoprotein cholesterol (HDL-C) in patients with hypercholesterolemia and mixed dyslipidemia. Therapeutic response is seen within 2 weeks, and maximum response is usually achieved within 4 weeks and maintained during chronic therapy. LIPITOR is effective in a wide variety of patient populations with hypercholesterolemia, with and without hypertriglyceridemia, in men and women, and in the elderly. Experience in pediatric patients has been limited to patients with homozygous familial hypercholesterolemia (FH). In two multicenter, placebo-controlled, dose–response studies in patients with hypercholesterolemia, LIPITOR given as a single dose over 6 weeks significantly reduced total-C, LDL-C, apo B, and TG (pooled results are provided in Table A.1).

In patients with *Fredrickson* types IIa and IIb hyperlipoproteinemia pooled from 24 controlled trials, the median (25th and 75th percentile) percent changes from baseline in HDL-C for atorvastatin 10, 20, 40, and 80 mg were 6.4 (−1.4, 14), 8.7 (0, 17), 7.8 (0, 16), and 5.1 (−2.7, 15), respectively. Additionally, analysis of the pooled data demonstrated consistent and significant decreases in total-C, LDL-C, TG, total-C/HDL-C, and LDLC/HDL-C.

A.3.3 Hypertriglyceridemia (Fredrickson Type IV)

The response to LIPITOR in 64 patients with isolated hypertriglyceridemia treated across several clinical trials is shown in Table A.2. For the atorvastatin-treated patients, median (min, max) baseline TG level was 565 (267–1502).

Clinical data supporting the following indications are also discussed here—*Dysbetalipoproteinemia* (*Fredrickson* type III), *homozygous familial hypercholesterolemia*, and *heterozygous familial hypercholesterolemia* in *pediatric patients*.

TABLE A.2

Combined Patients with Isolated Elevated TG: Median (min, max) Percent Changes from Baseline

	Placebo (N = 12)	Atorvastatin 10 mg (N = 37)	Atorvastatin 20 mg (N = 13)	Atorvastatin 80 mg (N = 14)
Triglycerides	−12.4 (−36.6, 82.7)	−41.0 (−76.2, 49.4)	−38.7 (−62.7, 29.5)	−51.8 (−82.8, 41.3)
Total-C	−2.3 (−15.5, 24.4)	−28.2 (−44.9, −6.8)	−34.9 (−49.6, −15.2)	−44.4 (−63.5, −3.8)
LDL-C	3.6 (−31.3, 31.6)	−26.5 (−57.7, 9.8)	−30.4 (−53.9, 0.3)	−40.5 (−60.6, −13.8)
HDL-C	3.8 (−18.6, 13.4)	13.8 (−9.7, 61.5)	11.0 (−3.2, 25.2)	7.5 (−10.8, 37.2)
VLDL-C	−1.0 (−31.9, 53.2)	−48.8 (−85.8, 57.3)	−44.6 (−62.2, −10.8)	−62.0 (−88.2, 37.6)
Non-HDL-C	−2.8 (−17.6, 30.0)	−33.0 (−52.1, −13.3)	−42.7 (−53.7, −17.4)	−51.5 (−72.9, −4.3)

A.4 Indications and Usage

A.4.1 Prevention of Cardiovascular Disease

In adult patients without clinically evident coronary heart disease, but with multiple risk factors for coronary heart disease such as age, smoking, hypertension, low HDL-C, or a family history of early coronary heart disease, LIPITOR is indicated to

- Reduce the risk of myocardial infarction
- Reduce the risk of stroke
- Reduce the risk for revascularization procedures and angina

In patients with type 2 diabetes, and without clinically evident coronary heart disease, but with multiple risk factors for coronary heart disease such as retinopathy, albuminuria, smoking, or hypertension, LIPITOR is indicated to

- Reduce the risk of myocardial infarction
- Reduce the risk of stroke

In patients with clinically evident coronary heart disease, LIPITOR is indicated to

- Reduce the risk of nonfatal myocardial infarction
- Reduce the risk of fatal and nonfatal stroke
- Reduce the risk for revascularization procedures
- Reduce the risk of hospitalization for congestive heart failure (CHF)
- Reduce the risk of angina

A.4.2 Hypercholesterolemia

Contraindications: Pregnancy and Lactation

Warnings: Liver dysfunction, skeletal muscle

Precautions: Drug interactions—Drug interactions between Lipitor and the following drugs are studied—*Antacid, Antipyrine, Colestipol, Cimetidine, Digoxin, Erythromycin, Oral Contraceptives,* and *Warfarin Endocrine function, central nervous system (CNS) toxicity, Carcinogenesis, Mutagenesis, Impairment of fertility, Pregnancy, Pregnancy category X, Nursing mothers, Pediatric use, Geriatric use, Use in Patients with Recent Stroke* or *transient ischemic attack (TIA).*

Adverse reactions: LIPITOR is generally well-tolerated. Adverse reactions have usually been mild and transient. In controlled clinical studies of 2502 patients, <2% of patients were discontinued due to adverse experiences attributable to atorvastatin. The most frequent

TABLE A.3

Adverse Events in Placebo-Controlled Studies (% of Patients)

Body System/ Adverse Event	Placebo N = 270	Atorvastatin 10 mg N = 863	Atorvastatin 20 mg N = 36	Atorvastatin 40 mg N = 79	Atorvastatin 80 mg N = 94
Body as a Whole					
Infection	10.0	10.3	2.8	10.1	7.4
Headache	7.0	5.4	16.7	2.5	6.4
Accidental injury	3.7	4.2	0.0	1.3	3.2
Flu syndrome	1.9	2.2	0.0	2.5	3.2
Abdominal pain	0.7	2.8	0.0	3.8	2.1
Back pain	3.0	2.8	0.0	3.8	1.1
Allergic reaction	2.6	0.9	2.8	1.3	0.0
Asthenia	1.9	2.2	0.0	3.8	0.0
Digestive System					
Constipation	1.8	2.1	0.0	2.5	1.1
Diarrhea	1.5	2.7	0.0	3.8	5.3
Dyspepsia	4.1	2.3	2.8	1.3	2.1
Flatulence	3.3	2.1	2.8	1.3	1.1
Respiratory System					
Sinusitis	2.6	2.8	0.0	2.5	6.4
Pharyngitis	1.5	2.5	0.0	1.3	2.1
Skin and Appendages					
Rash	0.7	3.9	2.8	3.8	1.1
Musculoskeletal System					
Arthralgia	1.5	2.0	0.0	5.1	0.0
Myalgia	1.1	3.2	5.6	1.3	0.0

adverse events thought to be related to atorvastatin were constipation, flatulence, dyspepsia, and abdominal pain.

Clinical adverse experiences: Adverse experiences reported in ≥2% of patients in placebo-controlled clinical studies of atorvastatin, regardless of causality assessment, are shown in Table A.3.

Dosage and administration:

Heterozygous familial hypercholesterolemia in pediatric patients (10–17 years of age)

Homozygous familial hypercholesterolemia

Concomitant therapy

Dosage in patients with renal insufficiency

Index

A

AAAS, *see* American Association for Advancement of Science (AAAS)

ABE, *see* Average bioequivalence (ABE)

ABECB, *see* Acute bacterial exacerbation of chronic bronchitis (ABECB)

ABOM, *see* Acute bacterial otitis media (ABOM)

ABS, *see* Acute bacterial sinusitis (ABS)

Absorption, 122, 321–322

Absorption, distribution, metabolism and excretion (ADME), 122

AC, *see* Active control (AC)

Academic training, 60

AC approach, *see* Adjudication committee approach (AC approach)

Accelerated approval, 34–35

Acceptable type I error rate, 83–84

Acceptable type II error rate, 84

ACM, *see* Advisory Committee Meeting (ACM)

ACTG, *see* AIDS Clinical Trials Group (ACTG)

Active control (AC), 103–104, 132

Acute bacterial exacerbation of chronic bronchitis (ABECB), 135

Acute bacterial otitis media (ABOM), 135

Acute bacterial sinusitis (ABS), 135

Acute myelogenous leukemia (AML), 164

Acute studies, 28

Adaptive designs, 85–86, 151
    adaptation, 153
    changing endpoints, 156–160
    implementation, 152–153
    sample size recalculation, 160–161
    two-stage designs, 153–155

Adaptive dose–response designs, 95

Adaptive randomization, 95–96

Adherence, 239
    monitoring, 240

    nonadherence, 241
    pill count methods, 240

Adjudication committee approach (AC approach), 289

ADME, *see* Absorption, distribution, metabolism and excretion (ADME)

Adolescent trials network (ATN), 183

ADR reports, *see* Adverse drug reaction reports (ADR reports)

Adult longitudinally linked randomized treatment trials (ALLRT trials), 73

Adverse drug reaction reports (ADR reports), 52

Adverse event (AE), 262, 263
    analysis issues, 266–270
    coding, 264–265
    by dose and time, 269
    by dose and time of dose, 268
    SAE, 264
    severity, 263
    signs, 263
    spontaneous *vs.* active collection, 265–266
    targeted AEs, 266

Advisory Committee Meeting (ACM), 48, 277
    examples, 50
    FDA, 48, 49
    preparation, 51

AE, *see* Adverse event (AE)

AIDS Clinical Trials Group (ACTG), 8, 72, 226, 286

Albuterol, 103

ALLRT trials, *see* Adult longitudinally linked randomized treatment trials (ALLRT trials)

ALS, *see* Amyotrophic lateral sclerosis (ALS)

Alzheimer's disease, 286

American Association for Advancement of Science (AAAS), 66

Index

Office for Human Research Protections
(OHRP), 19
OHARA, *see* Oral HIV AIDS Research
Alliance (OHARA)
OHRP, *see* Office for Human Research
Protections (OHRP)
OMERACT, *see* Outcome measures in
rheumatology (OMERACT)
One dimension approaches
comparative measures, 279–282
within-intervention measures,
278–279
Open-hidden design, 98
Open-label, 99
Oral candidiasis (OC), 136
Oral HIV AIDS Research Alliance
(OHARA), 73, 136
Ordinal endpoints, 76–77
Orphan drug, 35
Orphan Drug Act, 35
OS, *see* Overall survival (OS)
Osteoarthritis (OA), 132, 277
Outcome measures in rheumatology
(OMERACT), 287
Overall survival (OS), 202

P

p-values, 116
confidence intervals *vs.*, 215–217
poor interpretation, 215–216
PAC, *see* Postapproval commitment (PAC)
Parallel group designs, 105–106, 107
Patient-Reported Outcomes (PROs),
77, 291
Patient preferences, 294–295
PBE, *see* Population bioequivalence (PBE)
PDR, *see* Population difference ratio
(PDR)
PDUFA, *see* Prescription Drug User Fee
Act (PDUFA)
Pediatric investigational plan (PIP), 171
Periodic safety update reports
(PSURs), 273
Per protocol (PP), 139, 206
analyses, 224
ITT *vs.*, 139, 206–212
powering for, 86
Person–time, 267

PET, *see* Positron emission tomography
(PET)
PFS, *see* Progression-free survival (PFS)
Pharmaceutical and Medical Devices
Agency (PMDA), 39
Pharmaceutical Research and
Manufacturers of America
(PhRMA), 314, 315
Pharmacoeconomic studies, 37
Pharmacokinetics and
pharmacodynamics (PK/PD),
26, 122–123
Pharmacology, 27–28
Phase I clinical trials, 8, 9, 30–31, 121
bioavailability, 123–125
MTD estimation, 125–126
PK/PD designs, 122–123
Phase II clinical trials, 10, 31–33, 126
adaptive designs, 151–161
biomarker designs, 148–151
diagnostic device trials, 164–169
dose-finding study designs, 127–132
dynamic treatment regimes, 161–164
factorial designs, 144–148
futility designs, 143–144
proof of concept study, 126–127
Phase III clinical trials, 10, 31–33, 126
adaptive designs, 151–161
biomarker designs, 148–151
dose-finding study designs, 127–132
dynamic treatment regimes, 161–164
factorial designs, 144–148
futility designs, 143–144
proof of concept study, 126–127
Phase IV clinical trial, 10, 169
for label-extension purposes, 170
PIP, 171
PhRMA, *see* Pharmaceutical Research
and Manufacturers of America
(PhRMA)
PHS, *see* Physicians Health Study (PHS)
Physicians Health Study (PHS), 145
PI, *see* Principal investigator (PI)
"Pill count" methods, 240
PIP, *see* Pediatric investigational plan
(PIP)
PIPs, *see* Predicted interval plots (PIPs)
PIs, *see* Predicted intervals (PIs)
Pittsburgh sleep quality index (PSQI), 292